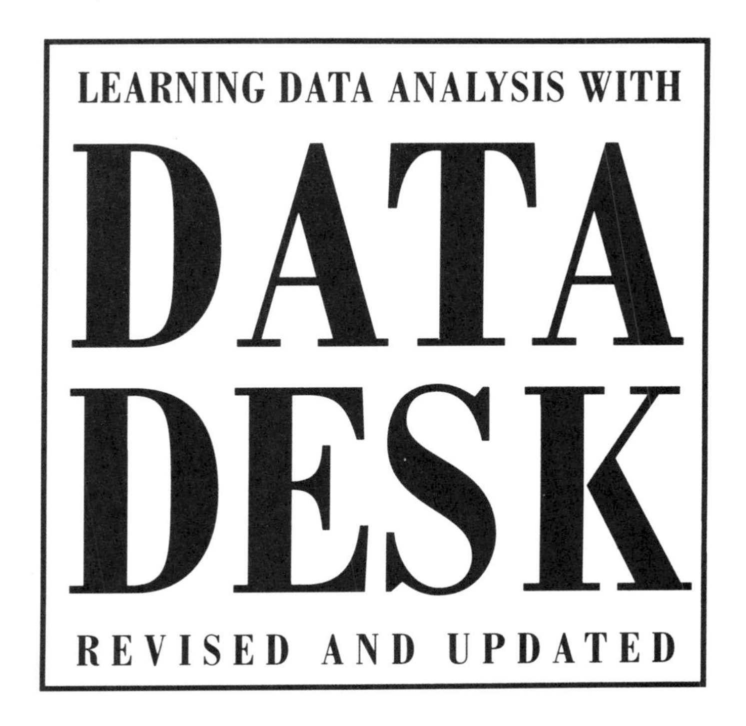

PAUL F. VELLEMAN
CORNELL UNIVERSITY

W. H. FREEMAN AND COMPANY
NEW YORK

Library of Congress Cataloging-in-Publication Data

Velleman, Paul F., 1949-
Learning data analysis with Data Desk / by Paul F. Velleman.
p. cm.
Includes index.
ISBN 0-7167-2201-1
1. Data desk. 2. Mathematical statistics—Data Processing.
I. Title
QA276.4.V445 1993 92-42310
519.5'0285'56369—dc20 CIP

Printed in the United States of America.
3 4 5 6 7 8 9 0 VB 9 9 8 7 6 5 4 3

Contents

Preface x

1. Introduction

- 1.1 What Statistics Do I Need to Know to Use Data Desk 3
- 1.2 What Computing Background Do I Need 4
- 1.3 A Brief History of Statistics Software 4
- 1.4 Data Analysis 5
- 1.5 The Macintosh Revolution 6
- 1.6 Data Desk's Graphics and Statistics 6
- Appendix 1A Macintosh Terms and Operations 8

2. Basic Concepts

- 2.1 The Data Desk Desktop 13
- 2.2 Menus and Submenus 14
- 2.3 Datafiles 15
- 2.4 Windows 15
- 2.5 Icons 17
- 2.6 Variables 17
- 2.7 Folders 19
- 2.8 The Results Folder 20
- 2.9 Relations 20
- 2.10 Selecting 21
- 2.11 Result Windows, Plots, and Hot Result™ 21
- 2.12 Defaults and Preferences 22
- 2.13 Leaving Data Desk 23

3. Data

- 3.1 Basic Concepts 27
- 3.2 Relations, Variables, Values and Data Structure 28
- 3.3 Missing Data 29
- 3.4 Infinities 30
- 3.5 Outliers, Blunders, and Rogues 30
- 3.6 Numerals and Numeric Values 30
- Appendix 3A Kinds of Data 32
- Appendix 3B Matrices, Tables, and Relations 34

4. Data Desk Concepts

- 4.1 Selecting Variables 37
- 4.2 HyperViews 38
- 4.3 Updating Windows 38
- 4.4 Derived Variables 39

4.5 The Clipboard 40
4.6 ScratchPads 40
4.7 Jot Notes 41
4.8 Information Records 42
4.9 Sliders 42
4.10 Dependencies 42
4.11 Datafile Details 43
4.12 Memory Requirements 44
4.13 Multiple Window Modifier 45
Appendix 4A Hints and Shortcuts 46
Appendix 4B Hot, Warm and Cold Objects 47
Appendix 4C Data Desk Limits 49

5. Entering and Editing Data

5.1 Entering Data for One Variable 53
5.2 Editing a Variable 54
5.3 Extended and Discontinuous Selection 55
5.4 Editing Several Variables 56
5.5 The Editing Sequence 56
5.6 Moving Around 57
5.7 Cutting, Copying, and Pasting 58
5.8 Undo 58
5.9 Updating Relations 58
5.10 Scrolling 59
5.11 Managing Windows 59
5.12 Details of Editing 59
5.13 Shifting Cases 60
5.14 More on Linking 60
5.15 Finding and Replacing Cases 61
5.16 Store and Revert 62
5.17 Printing Variables 63
5.18 Example 64
5.19 Example: Editing Data 66
Appendix 5A Configure Editing 69

6. Importing and Exporting

6.1 Data Tables 73
6.2 Copying Variables to a Data Table 74
6.3 Pasting Variables from a Data Table to the Data Desk Desktop 74
6.4 Alternative Delimiters 76
6.5 Writing a Data Table to a Text File 76
6.6 Reading Text Files 76
6.7 Importing from Several Files 77
6.8 Appending Cases in Data Desk 77
6.9 Copying Data Desk Results 78

6.10 Printing from Data Desk 79
6.11 Layout Windows 80
6.12 Saving Screen Images 80

7. Simple Summaries

7.1 Measures of Center 83
7.2 An Example 84
7.3 Measures of Spread 84
7.4 Order Statistics 85
7.5 General Summaries 86
7.6 Moments 87
7.7 HyperViews 87
7.8 Summaries As Variables 88
7.9 Summaries As Variables by Group 88
Appendix 7A The Biweight, a Robust Center 90
Appendix 7B Coefficient of Skewness and Kurtosis 91

8. Displaying Data

8.1 Data Analysis Displays 95
8.2 Plotting Conventions 96
8.3 Histograms 99
8.4 Recentering and Rescaling Histograms 100
8.5 Dotplots 102
8.6 Dotplots of Separate Variables 103
8.7 Boxplots 104
8.8 Bar Charts 106
8.9 Pie Charts 107
8.10 Scatterplots 109
8.11 Lineplots 110
8.12 Normal Probability Plots 112
Appendix 8A Boxplot Definitions 113
Appendix 8B Probability Plots 114

9. Working with Displays

9.1 Chapter Organization 117
9.2 Organization of Features and Commands in Data Desk 118
9.3 Basic Plot Actions 119
9.4 Select 120
9.5 Link 121
9.6 Brush and Slice 122
9.7 Identify 123
9.8 Move 123
9.9 Isolate 124
9.10 Resize 126
9.11 Substitute 126

9.12 Symbols 128
9.13 Colors 128
9.14 Axes 130
9.15 Scale 130
9.16 Visibility 133
9.17 Lines 133
9.18 Lines and Color 135
9.19 Working with Displays 135
Appendix 9A Plot Tool Shortcuts 137

10. Brushing and Slicing

10.1 Brushing and Slicing 141
10.2 Principles of Brushing and Slicing 141

11. Derived Variables

11.1 The Transform Submenus 145
11.2 Typing Derived Variable Expressions 146
11.3 Updating 148
11.4 Expression Types 148
11.5 Expression Conventions 149
11.6 Dependencies 149
11.7 Working with Open Derived Variables 150
11.8 Calculating in ScratchPads 150
11.9 Boolean Expressions 151
11.10 IF/THEN/ELSE 152
11.11 Subscripting 153
11.12 Identifying Variables by Name 154
11.13 Dynamic Parameters 155
11.14 Re-expressing Data to Improve Analyses 157
11.15 An Example 159
11.16 Rules for Re-expression 161
11.17 Efficiency 162
11.18 Indicator Variables and Logical Expressions 162
11.19 Working with Several Relations 163
11.20 Subtle Points 163
11.21 Common Errors and How to Avoid Them 164
Appendix 11A Derived Variable Expressions 166
Appendix 11B Casewise Functions 169
Appendix 11C Collapsing Functions 175

12. Manipulating Variables

12.1 Sorting 179
12.2 Ranking 181
12.3 Generating Patterned Variables 181
12.4 Appending and Splitting Variables 183
12.5 Transpose 184
12.6 Selectors 185

12.7 Performing Analyses Group by Group 186
12.8 Samples 188
12.9 Duplicating Icons 189
12.10 Data Tables 189
12.11 Copying and Printing Results 190
Appendix 12A Missing Values 191

13. Integrated Analyses

13.1 Context-based Expertise 197
13.2 Data Analysis Expertise 197
13.3 Global and Context-sensitive HyperViews 198
13.4 Updating Window 199
13.5 Following the Links 199
13.6 Case Identities 200
13.7 Consistency 201
13.8 Recording the Analysis History 201
13.9 Avoiding Window Overload 201

14. Tables

14.1 Frequency Counts and Percentages 205
14.2 Two Factors 206
14.3 Contingency Tables 207
14.4 Table Contents 207
14.5 Independence and Chi Square 210
14.6 HyperViews in Tables 211
14.7 Copying and Printing Tables 212
Appendix 14A Equations 213
Appendix 14B Multi-Way Tables 214

15. Random Numbers and Simulation

15.1 Randomness 217
15.2 Creating Random Samples 218
15.3 Distributions 219
15.4 Generating Random Samples 220
15.5 Bernoulli Trials 221
15.6 Binomial Distribution 221
15.7 Poisson Distribution 221
15.8 Uniform Distribution 222
15.9 Normal Distribution 222
15.10 The Law of Large Numbers 223
15.11 Sampling Distribution 224
15.12 Central Limit Theorem 226
Appendix 15A Details and Formulas 228

16. Simple Inference

16.1 Confidence Intervals 231
16.2 Confidence Intervals for μ When σ is Known 232

16.3 Confidence Intervals for μ When σ is Unknown 233
16.4 Where Is the Randomness? 234
16.5 Multiple Intervals and the Bonferroni Adjustment 234
16.6 Testing Hypotheses 235
16.7 Hypothesis Test for μ When σ is Known 235
16.8 Hypothesis Test for μ When σ is Unknown 236
16.9 Multiple Hypothesis Tests and the Bonferroni Adjustment 237
16.10 Chi-Square Test of Individual Variances: Inference for Spread 237
Appendix 16A Example: Simulating Confidence Interval Performance 239
Appendix 16B Example: Simulating Hypothesis Test Performance 241

17. Comparing Two Samples

17.1 Displaying Differences 245
17.2 Comparing Two Means When the Variances Are Assumed Equal 245
17.3 Example 246
17.4 Formulas 247
17.5 Confidence Intervals for Pooled Variance 247
17.6 Comparing Two Means When the Variances Are Not Assumed Equal 247
17.7 Paired Data 248
17.8 Example 248
17.9 Comparing Multiple Means When the Variances are Assumed Equal 249

18. One-Way ANOVA

18.1 Example 253
18.2 Comparing Several Groups Graphically 253
18.3 One-way Analysis of Variance 253
18.4 The ANOVA Table 255

19. Multi-Way ANOVA

19.1 Notation 259
19.2 Interaction in ANOVA 260
19.3 One Observation per Cell 261
19.4 Working with ANOVA Tables 262
19.5 Notes on Computing ANOVA 263
19.6 ANOVA Options 264

20. Simple Regression

20.1 Coefficients in Regression 269
20.2 Least Squares Regression 269
20.3 Predicted Values and Residuals 270

20.4 Performing a Regression: An Example 270
20.5 Inference for Regression Coefficients 271
20.6 The ANOVA Table for Regression 273
20.7 R^2 and Adjusted R^2 273
20.8 Examining Residuals 274
20.9 Checking Assumptions 275

21. Correlation

21.1 Pearson Product-Moment Correlation 279
21.2 Linear Association 280
21.3 Correlation and Regression 280
21.4 Correlation and Standard Scores 281
21.5 Correlation Tables 281
21.6 Spearman Rank Correlation (Rho) 282
21.7 Kendall's Tau 283
21.8 Covariance 284
21.9 Missing Values and Correlation 284
21.10 Extracting Values from Correlation Tables 284

22. Multiple Regression

22.1 Multiple Regression 287
22.2 Example 287
22.3 Interpreting the Regression Table 289
22.4 Predicted Values and Residuals 289
22.5 Missing Values in Regression 290
22.6 Regression Options 290
22.7 Working with Regression Summary Tables 291
22.8 Interactive Regression Model 293
22.9 Stepwise Regression 295

23. Regression Diagnostics

23.1 Checking Basic Assumptions 300
23.2 Partial Regression Plots 301
23.3 Leverage 303
23.4 Studentized Residuals 306
23.5 Distance Measures 308
23.6 Regression Options 311
23.7 A Note on Identifying Cases 312
23.8 Collinearity 313

Appendix: Special Keys 315

Index 323

Exercises E1

Preface

WELCOME TO DATA DESK. *Learning Data Analysis with Data Desk* and the program that goes with it embody a new way to apply statistics to data. The program, the Student Version of Data Desk®, works graphically so there is no new language to learn. Together, the book and program help you to visualize abstract concepts and to see them applied to real data. Whatever your background and whatever path you are following to learn more about statistics, you will find that it is easier when you can see, touch, and experiment with new concepts and methods. That is what Data Desk and this book are designed to do.

The Student Version of Data Desk is a special version of the popular Data Desk statistics program available from Data Description, Inc of Ithaca, NY. The Student Version is designed to provide the functionality needed by students learning statistics with all of the ease and convenience of Data Desk but without multivariate statistics and complex data manipulation that might distract from learning. When you want to move up from coursework to analyzing research data, you will probably want to upgrade to the full Version. Contact Data Description for upgrade details.

The Student Version of Data Desk included with this book is based on the 4.0 version of Data Desk. It can open and work with any Data Desk 4.0 datafile that has no more than 1000 cases and 15 variables. The Student Version can use only Data Desk 4.0 datafiles. It will not read or write files in other formats or datafiles created for the 3.0 or earlier versions of the full Data Desk program.

The Student Version includes two disks. The first holds the Data Desk program. The second disk contains datafiles with sample data sets. These data sets are used in the examples and exercises in the book. You should make a copy of the datafiles to work with. **Never work with the original copy of a datafile.**

Data Desk works on any Macintosh computer with at least one megabyte of memory, an 800K disk drive, and another disk drive (either a hard disk or a second floppy disk). If your Macintosh has a special arithmetic chip, often called a *floating point unit* , Data Desk will automatically use the chip for substantially increased speed. If your Macintosh has a color screen, Data Desk will display data using color.

Whatever system you use, your first action should be to make a copy of the disks and put the original disks in a safe place. **Never work with the original disks. Always use a copy.** This will ensure that when the disk you use for your work fails, becomes damaged, or is lost, you will still have the original and will be able to make a fresh copy. (You read that correctly. I said *when* the disk becomes damaged, not *if* it does. Disks wear out. Disks carried around with your books wear out even faster. Please protect your investment and work with a copy, not with the original disk.)

To the Teacher

This book has been designed to integrate smoothly into almost any basic statistics course. The examples, discussions, and exercises supplement a wide range of texts — from the most traditional to the most modern. Each topic is reviewed and illustrated, providing an additional description that can help students to grasp difficult concepts. The chapters dealing with specific methods can be covered in almost any reasonable order.

Data Desk can serve two key functions in a statistics course. First, it provides a laboratory that gives students practical, hands-on experience with randomness far beyond what they could gain otherwise without years of analyzing data. Many of the examples and exercises in this book rely on random numbers generated with known properties to illustrate concepts of statistics and probability.

Second, Data Desk is a powerful calculator that enables students to apply even complex methods to real data and then try out alternatives. The hope, of course, is that the student will see past the computer and understand more about how statistics helps to describe the world. I have found in my teaching that seeing a practical application of statistical methods helps students to understand them. But *applying the methods for themselves* helps students to really learn them.

You may find it helpful to have a copy of the full Data Desk 4.0 program so that you can import data from other sources to create new exercises for your students. Data Description can supply this version of the program at special education prices.

Some schools have chosen to establish computer laboratory facilities for students to use and to equip them with software. The Student Version of Data Desk is intended for individual student use and *cannot be* licensed for multiple use in a laboratory or library. To equip a laboratory facility or software lending library with Data Desk you must obtain the appropriate licenses from Data Description.

To the Student

I have been teaching statistics for over 20 years, and I know that you may be facing this course with some trepidation. Statistics can be confusing and intimidating at first. If you have not used a computer before, you may be concerned about the additional burden of learning that new skill. While I cannot dispel all of your concerns, I can assure you from experience with literally thousands of students that Data Desk will actually make it easier for you to understand the concepts of statistics.

If you make the effort early on to become comfortable with the computer and with Data Desk, you will even find that it can be fun to experiment with the concepts and methods of statistics. I realize that in saying this I am risking my credibility with some of you. The thought of a coherent sentence using both the words "statistics" and "fun" strains credulity. Nevertheless, statistics can help us to learn about the world around us, and the computer makes that connection effortless. I can imagine no mysteries more interesting than those of how the world around us works, where the patterns are, and what they look like.

The discussions and examples in this book form a bridge between the basic statistics concepts and methods taught in class and the practical, powerful tool in your computer. The Data Desk program is designed to be easy to learn in stages. Once you learn basic skills (if you know how to use a Macintosh, you have most of these already), you can learn the rest bit by bit as you need it. Along the way, you will find that Data Desk is consistent — the same actions accomplish the same results in a variety of circumstances — so it is easy to guess what to do even if you don't know for sure.

One remark worth making at the start is to remind you that you cannot hurt the computer or the program by any ordinary use, and you can only alter a datafile when you explicitly choose to do so. Ordinary commands, anything you can type at the keyboard, and virtually anything you can do with the mouse can do no real harm. You should feel free to experiment and explore; it is the best way to become comfortable with the computer and with Data Desk.

Synopsis

Chapter 1 presents the background material that will help you to get started with Data Desk.

Chapter 2 defines Data Desk's underlying concepts and terms. It is the most important chapter to read before starting to use the program.

Chapter 3 discusses data and its many forms. It defines Data Desk terms and relates them to others that you may already know from other programs.

Chapter 4 discusses more powerful features of Data Desk.

Chapter 5 shows how to enter data into Data Desk by typing it. Chances are you will not need this chapter at all. The datasets provided on your disk are all you need to reproduce the examples in the book or to do any of the exercises. Your professor may provide additional datasets keyed to another text or to class lectures.

Chapter 6 discusses how to move data in and out of Data Desk using the Import, Export, Copy and Paste commands. It also shows how to print and copy out results.

Chapter 7 discusses Data Desk's summary statistics, such as means and standard deviations.

Chapters 8, 9, and 10 document Data Desk's graphics.

- Chapter 8 defines each kind of plot. If you just want to make a particular kind of plot, you can go directly from entering your data to the appropriate section of Chapter 8.
- Chapter 9 discusses the plot tools that you will use to modify and work with plots.
- Chapter 10 discusses plot brushing, slicing and rotation, illustrates its use, and relates it to traditional statistics methods.

Chapters 11 and 12 provide background that is used in some of the exercises for later chapters. Even if you are skipping around in the book, it is a

cises for later chapters. Even if you are skipping around in the book, it is a good idea to read them before going on.

- Chapter 11 defines Data Desk's transformation and calculation features using derived variables, and documents the **Transform** submenus.
- Chapter 12 discusses data manipulations such as sorting, ranking, appending, splitting, and substituting.

Chapter 13 presents HyperViews and the related ability to update plots and analyses to reflect changes in the data. These are extremely powerful features and can be used to compare before and after views. HyperViews are easy enough to use without reading about them first, but you may want to read Chapter 13 to learn about their more powerful features after you have some experience working with Data Desk.

Chapter 14 discusses frequency breakdowns and contingency tables.

Chapter 15 deals with random numbers and simulation. Both of these concepts are used in later chapters to illustrate how methods work, so it is best to read this chapter before moving on.

Chapter 16 discusses simple inference, and provides many examples.

Chapter 17 extends inference to the comparison of two samples.

Chapter 18 moves on to comparing three or more groups with one-way analysis of variance (ANOVA).

Chapter 19 takes the next step to two-way and multi-way ANOVA.

Chapter 20 introduces regression analysis.

Chapter 21 discusses correlation.

Chapter 22 brings us to multiple regression, presenting a way to understand multiple regression that uses the computer to make difficult multivariate concepts easier to grasp.

Chapter 23 provides definitions and examples of modern regression diagnostic methods.

The Appendix following Chapter 23 provides some tips to help you use Data Desk more efficiently.

The final section of the book contains exercises for many of the chapters. In the past, courses that used statistics packages simply asked students to "hand in your output". But with Data Desk, the graphics and analyses happen on the screen, so you must think a bit about what to print and what to hand in. There is usually little need to print more than an occasional plot. The exercise sheets are designed to guide you through examples. They prompt you to observe and record the key aspects of each example so that you can work efficiently.

How Data Desk Came To Be

Data Desk was conceived when the Macintosh was first announced. The design grew out of the author's 15 years of professional experience teaching statistics; writing, using, and evaluating statistics software; and consulting for computer package users. The initial implementation was a program to support teaching, which has been used in classes at Cornell University and elsewhere since 1985.

Data Desk Professional 1.0, released in 1986, extended the student program to a form useful for research and business. Data Desk 2.0 and 3.0 followed in 1988 and 1990 respectively. In late 1992 we announced Data Desk 4.0, the program on which this Student Version is based. We built upon our experience with both students and professionals to enhance and refine the design, and to expand the capabilities of the program substantially without sacrificing the natural interface that made Data Desk so easy to learn and use.

We are very proud of our new Student Version. This program is more capable, faster, and easier to use than the original. The book and exercises have benefited from our experience and from the comments and suggestions of others who have used the program.

We will continue to develop and extend Data Desk. We welcome your suggestions, comments, and criticisms. Some of the best features currently in Data Desk were originally suggestions or requests from users.

CHAPTER 1

Introduction

1.1	What Statistics Do I Need to Know to Use Data Desk	*3*
1.2	What Computing Background Do I Need	*4*
1.3	A Brief History of Statistics Software	*4*
1.4	Data Analysis	*5*
1.5	The Macintosh Revolution	*6*
1.6	Data Desk's Graphics and Statistics	*6*

APPENDIX

1A	Macintosh Terms and Operations	*8*

DATA DESK EMBODIES a new way to apply statistics to data. It works graphically so there is no new language to learn. Together, the books documenting the program, and program itself help you to visualize abstract concepts and to apply them to real data. Whatever your background and whatever kinds of data you work with, you will find it easier to discover and understand the patterns in your data when you can see and touch the data in a variety of ways. That is what Data Desk is designed for.

This manual is your guide to Data Desk and to both new and traditional methods of looking at data. While it is always helpful to understand some statistics theory, it is far more important that you know something about your data — for example, what sort of patterns would be interesting or meaningful to you. Data Desk's natural interface, HyperView suggestions, and the tutorial examples in this book can ease you over any gaps in your statistics knowledge, but *you* must supply the understanding of your data.

Data Desk differs from traditional statistics packages because it is designed for the entire process of data analysis rather than for the computing of specific statistics. Data analysis is the process of discovering, describing, and confirming structure or patterns in data. The process itself is often a way to learn about data; one step of an analysis can reveal things about the data that suggest what to do next. Data Desk is designed to make this process convenient, comfortable, and informative. It provides powerful tools, it permits any operation at any point in the analysis, and it is transparent; after a little practice you "see through" the program and feel that you are working directly with the data rather than giving instructions to a computer program to tell *it* to work with the data.

For example, Data Desk depicts identifiable parts of the data and products of analyses with graphic *icons*. You touch the icons with the mouse to open, select, move, alter, or otherwise work with them. Data Desk also makes abstract concepts such as a "cloud of datapoints" into a physical reality that you can touch (with the mouse) and manipulate.

In short, Data Desk offers a unique approach to understanding data.

Although Data Desk is easy to learn, we ask you to read the first two chapters of this book before diving in. This Introduction describes the philosophy behind Data Desk. Chapter 2 defines Data Desk's objects and conventions.

1.1 *What Statistics Do I Need to Know to Use Data Desk*

Data Desk is designed to be easy to learn. Every important command is available in the menus at the top of the screen. All of the data you might work with is represented on the screen with icons that you select by pointing to them with the mouse cursor and clicking the mouse button.

Most important of all, Data Desk is consistent. Once you learn how to do something in Data Desk, it is usually safe to assume that the same methods will work for all similar operations. Thus, for example, the methods you use to edit data will work to edit the names on icons and to fill in items in dialogs. The way in which you select icons by pointing and clicking (and the way in which you add to the set of selected icons by holding the Shift key and clicking) are the same as the way in which you select points in a plot or bars in a histogram.

Even the basic rules for using Data Desk generalize. In Data Desk you first select the items you want to work with (icons on the desktop, points in a plot, text words and characters, etc.) and then choose a command from a menu. As you learn each new part of Data Desk, you will find the next part still easier to learn because it works just like what you already know.

As you become comfortable with Data Desk, we hope that you will experiment with it. Data analysis can be an adventure with each step revealing new wonders and suggesting new paths to explore. Some of these paths may lead you to learn about new statistics or graphics. This book was written with such explorations in mind. It discusses statistics concepts, methods, and terminology as well as instructions for using Data Desk, so you can review statistics that you may have forgotten, learn new methods as you find a use for them, or find new ways to put together old ideas.

1.2 *What Computing Background Do I Need?*

This book assumes that you are familiar with the basic principles of Macintosh operation. Specifically, you should know how to:

- Point, select, and drag with the mouse (using clicks, shift-clicks, and drags)
- Use the icons on the Macintosh Finder desktop
- Use windows, including moving, resizing, scrolling, and closing windows
- Pull down menus from the menu bar and choose commands

The Appendix of this chapter gives basic definitions.

If you are already familiar with the Macintosh you will be able to use Data Desk's many features simply by "doing what comes naturally". Even if you plan to learn Data Desk by diving in, you should read the initial chapters of this book to learn about the objects and basic concepts on the Data Desk Desktop.

1.3 *A Brief History of Statistics Software*

Data Desk owes much to early statistics packages that pioneered new and better ways to analyze data. Many commonly used statistics were known by the 1930's, but had to wait for the invention of computers to become practical. Statisticians were eager users of early computers, and programs to compute statistics were among the first general-purpose programs. The Macintosh on your desk is more powerful than any of the large, expensive computers of the 1950's that were first used for statistics.

By the early 1960's, statistics programs that could share the same data were gathered together into libraries and documented in large reference volumes. Users specified commands by punching numeric codes in particular columns of a computer card. The cards were read by the computer and the results of the analysis printed some time later.

In the late 1960's and early 1970's mnemonic control languages were introduced. These provided verbal commands like *Scatterplot* and *Regression* and referred to variables by name rather than by numeric code. However, the packages still relied on "batch" computing in which the user wrote a program and the package executed the entire program all at once. Users still might have to wait for hours to see their results or error reports.

Interactive computing, in which you type a command at a computer terminal and the package executes it immediately, became available in the late 1970's. Interactive computing made it practical to analyze data step-by-step, letting things learned during the earlier steps of the analysis guide the later steps. For the first time it was possible to learn about data from the *experience* of analyzing it rather than simply from reading a finished analysis. Most statistics packages today work interactively, although many of them were originally designed for entire-program-at-a-time batch computing.

1.4 Data Analysis

In the 1970's some statisticians turned their attention from computing isolated calculations to the entire process of analyzing data. They questioned the traditional statistical analysis paradigm in which one first forms a hypothesis, then collects data, and finally tests the hypothesis. Instead, they advocated the open-ended exploration of data with few preconceived assumptions. The father of such "exploratory data analysis" was John W. Tukey of Princeton University and AT&T Bell Labs.

Tukey suggested that we examine our data as a detective would examine the scene of a crime — not with a hypothesis ("I'll bet the butler did it."), but with an open mind and as few assumptions as possible. By letting the data speak to us we hope to learn the truths hidden beneath the random fluctuations, errors, and general confusion seen in real data.

This way of analyzing data is quite different from the traditional paradigm, and traditional statistics packages support it poorly. In his path breaking book, *Exploratory Data Analysis,* Tukey advocates building each step of an analysis upon the knowledge gained from earlier steps. Some traditional statistics packages have added some of the plots Tukey developed to their libraries of procedures, but without the free-form interaction required by modern data analysis they cannot support the full data analysis process.

1.5 The Macintosh Revolution

Every computer program creates an environment in which you work or play. Computer game designers were among the first to design program environments to resemble physical or imaginary environments. Most mainframe programs create crude environments in which you only communicate with the computer by typing and in which your choices of action at any time are quite limited.

The 1980's were the decade of the personal computer. At first, this meant that mainframe packages were ported to personal computers. Vendors trumpeted "all the power of a mainframe statistics package on your personal computer". But the Macintosh took personal computers beyond imitating mainframes. The features of the Macintosh fundamental to this revolution are:

- High-quality graphics
- A mouse, with which you can touch objects on the screen to communicate with the program
- Windows that provide local work environments
- Consistent rules for how programs should look and work so that you need not learn new methods for each program
- Mode-free program design so you won't be surprised by what a program does
- Natural ways to use several programs together including support for transferring data and graphics from one program to another.

Data Desk takes full advantage of these features. It encourages you to communicate with graphs and pictures, creates realistic images of otherwise abstract concepts, actively supports your analyses with suggested steps, and lets you do almost anything at any time. Variables, results, plots, and other parts of a data analysis are depicted as icons on a desktop. They can be moved, reordered, opened, and selected for use. You can even drag the icon of a variable into a plot or analysis window to add it to that plot or analysis.

As you use Data Desk, you learn new things about your data — things you never suspected and never would have thought to test for in a traditional statistical analysis. Thus, Data Desk supports not only the initial, exploratory phase of data analysis, but the entire analysis process.

1.6 Data Desk's Graphics and Statistics

Data Desk does the things you expect of a statistics package. It computes means and standard deviations, it makes scatterplots and histograms, it performs regression analysis, and so on. However, Data Desk does many of these things in new ways and does many other things not found in traditional statistics packages.

For example, a Data Desk plot is not merely a picture of your data, but is also a way to talk to your data. With the mouse you can touch datapoints in a plot to ask questions about them or to restrict another analysis to a subset of the data. Data Desk allows you to *rescale* or *reposition* datapoints at the touch of the mouse to help find interesting patterns, relationships and anomalies. Data Desk's windows *link* together so that several aspects of the same individuals can be seen simultaneously in adjacent windows. You can add new variables to analyses by dragging the variables' icons any plot or table. Data Desk offers to *update* displays and analyses when underlying variables have been modified. And Data Desk will *suggest* analysis paths to take given your progress so far.

Statistics has never had a reputation for being fun. But data analysis is detective work. Analyzing data can be like playing a computer game except that the goal is to puzzle out the world rather than to match wits with a game designer — the beauty of it all is that there is always another good puzzle after you solve the current one.

Appendix 1A
Macintosh Terms and Operations

DESKTOP
The Macintosh and Data Desk use a *Desktop metaphor*. That is, the screen depicts objects as if they were on a large desk. Objects can overlap other objects. Displays can be opened for viewing or put away. Whenever we refer to "the desktop" in this book we mean the Data Desk Desktop. If necessary, we may refer explicitly to the "Finder desktop".

ICON

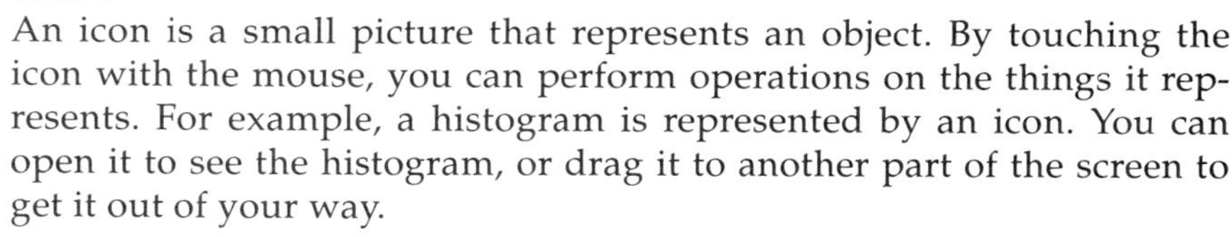

An icon is a small picture that represents an object. By touching the icon with the mouse, you can perform operations on the things it represents. For example, a histogram is represented by an icon. You can open it to see the histogram, or drag it to another part of the screen to get it out of your way.

TRASH

To discard an icon (and its contents) drag it into the Trash. The trash icon is in the lower right corner of the Data Desk Desktop. It looks slightly different from the Finder's trash icon, and it works the same way.

As with other icons, you can open the trash icon. Inside you will find icons that have been dragged into the Trash but not yet finally discarded.

WINDOW
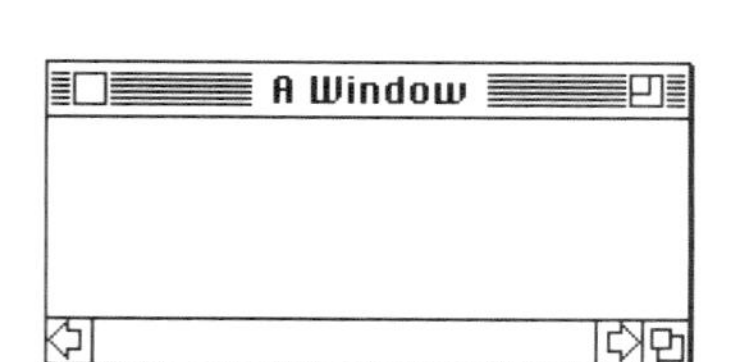

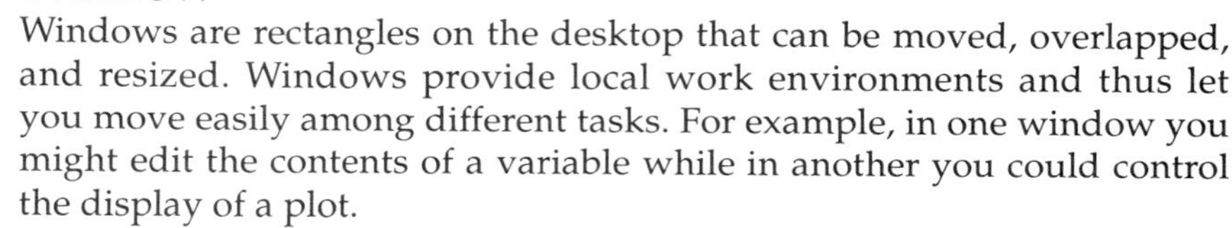

Windows are rectangles on the desktop that can be moved, overlapped, and resized. Windows provide local work environments and thus let you move easily among different tasks. For example, in one window you might edit the contents of a variable while in another you could control the display of a plot.

CLICK
To click an object or at a particular point on the screen, move the mouse until the cursor (usually a little arrow or an I-beam) is at the point you want, press the mouse button and release it.

SELECT
You ordinarily select one or more icons before choosing a command. The command then operates on the selected icons. To select an icon, move the cursor until the tip of the arrow touches the middle of the icon, and click. The icon will highlight, turning black with white markings.

SHIFT-CLICK
Ordinarily, clicking *de*-selects anything that has already been selected. Shift-clicks on text or data in open variables extend selections inclusively, selecting all cases between the insertion point and the shift-clicked case. Shift-clicking icons selects them specifically as *x*-variables.

Option-click

An option-click in an icon window selects icons as *y*-variables for a plot or analysis. Like a Shift-click, it extends the selection, leaving already selected icons selected.

Command-click

A command-click in a variable window extends a selection *discontinuously*, adding the clicked case to the selection without selecting the intervening cases. When a command-click is appropriate, pressing the ⌘ key changes the cursor to a ⌘.

Drag

To drag, click, hold the mouse button down, and move the mouse. If you have clicked an icon, you will drag it with you. Dragging in text selects a range of text. Dragging in a variable editing window selects cases. Dragging in an icon window draws a rectangle around some icons — they are selected when you stop dragging. To stop a drag, release the mouse button. Dragging a variable's icon into plot or summary table adds that variable to the plot or summary table.

Menu

Menus are lists of commands. Menu titles appear in the *menu bar* across the top of the screen. To choose a command from a menu, click the menu title and drag down the menu until the command you want is highlighted. Then release the button. Many commands have *command-key equivalents* that provide a way to issue the command from the keyboard. The menu item will have the ⌘ symbol and a letter next to it. Holding down the ⌘ key and pressing the letter is the same as choosing the command from the menu.

Submenu

Any command with a ▶ to its right in the menu list (where the command key might otherwise be) is the top of a submenu. If you pause over the command a submenu will drop to the right or left of your mouse position. Drag across to the submenu to continue the selection just as you would for a menu. Submenu commands are always grouped according to their type and function.

Palette

A palette is a special window that stays in front of any other window on the desktop. Typically, palettes hold options or tools (click on a tool to "pick it up"). For example, the Plot Tools palette holds 12 tools for manipulating plots. The Plot Colors Palette offers 64 colors; select points and click a color to set the selected points to that color.

Double-click

One way to open an icon is to select it and then choose the **Open Icons** command from the **Data** menu. Another way is to point to the icon and press the mouse button twice quickly. This is called a double-click.

CHAPTER 2

Basic Concepts

2.1	The Data Desk Desktop	*13*
2.2	Menus and Submenus	*14*
2.3	Datafiles	*15*
2.4	Windows	*15*
2.5	Icons	*17*
2.6	Variables	*17*
2.7	Folders	*19*
2.8	The Results Folder	*20*
2.9	Relations	*20*
2.10	Selecting	*21*
2.11	Result Windows, Plots, and Hot Result™	*21*
2.12	Defaults and Preferences	*22*
2.13	Leaving Data Desk	*23*

THIS CHAPTER DEFINES the basic objects and actions that make Data Desk work. If you are an experienced Macintosh user, Data Desk operations should be familiar and natural to you. If this is your first Macintosh experience, what you learn here will help you to learn other Macintosh programs. Even if you have used the Macintosh before, and find that you can use basic Data Desk functions without reading about them, it is a good idea to read through this chapter to learn about concepts, commands, and icons that are special to Data Desk.

2.1 *The Data Desk Desktop*

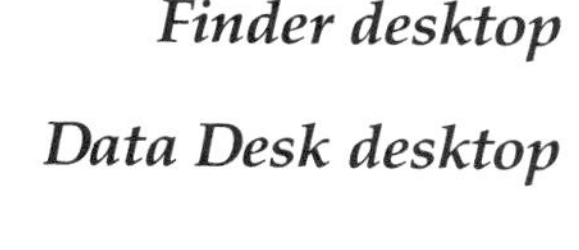

The desktop that the Macintosh displays when you first turn it on is managed by a program called the *Finder*. The *Data Desk desktop* looks and acts very much like the Finder desktop, except that the menu bar is different, there are different icons on the Data Desk desktop (including a different trash icon), and there is an identifying signature in the lower left corner. Most of the ways you select and drag icons, position and resize windows, select menu items, and edit text are the same on Data Desk as on the Finder's desktop. (Appendix 4B notes a few ways in which the Data Desk desktop differs from the Finder desktop.)

Basic operations on the desktop include:

- Selecting icons by touching them with the mouse and clicking the mouse button
- Extending the selection with a "shift-click"
- Selecting several adjacent icons by dragging across them
- Dragging one or more selected icons to another place on the desktop
- Selecting menu commands.

If you are unfamiliar with any of these concepts, you should review the introductory material in your Macintosh manual.

The Data Desk desktop can be reduced to cover only part of the screen. Click the small white rectangle in the lower right corner of the desktop (where a size box would be if it were an ordinary window) and drag it up and to the left. The top of the desktop automatically drops down from under the menu bar so you can reposition the desktop as well. If you are using System 7 or MultiFinder, shrinking the Data Desk desktop reveals the Finder desktop beneath. The datafile icon at the upper right and Data Desk trash icon at the lower right always stay on the Data Desk desktop, moving with the Desktop as you resize or reposition it. To resize the Data Desk desktop back to full screen, click its title bar to highlight it, then click the zoom box in the upper right corner.

> TIP
>
> When working in MultiFinder or System 7, the Data Desk Desktop can be resized and repositioned. Click its zoom box to resize to full screen.

2.2 *Menus and Submenus*

Data Desk menus are organized according to function as follows:

🍎 File Edit Data Special Modify Manip Calc Plot

🍎

The 🍎 menu holds desk accessories. It also has the **Help** command, which enters Data Desk's on-line help system and the **About Data Desk** command, which provides information about the datafile and disk space, and identifies the program and version number.

File

File commands deal with the Macintosh file system. They include opening, closing, saving, and deleting datafiles, printing data and results, importing and exporting datafiles, and quitting the program.

Edit

Edit holds commands for editing, searching, replacing data, and setting preferences. Some Edit commands edit plots as well.

Data

Data commands create, open, close, duplicate icons on the Data Desk desktop. The Data menu is roughly equivalent to the Finder's File menu for manipulating Data Desk icons, but commands in the Data menu do not directly affect your datafiles on the disk.

Special

Special functions include commands for controlling selector and group buttons, the result log, emptying the Trash, and working with windows. The Windows submenu includes a complete list of all open windows to help you find a window that may have become hidden behind other windows on the Data Desk desktop.

Modify

Modify commands manipulate displays. They are grouped into submenus according to the aspect of the display that they affect.

Manip

Manip functions include sorting, ranking, splitting, generating and appending variables. In general, manipulation commands work with variables and produce new, modified variables as results.

Calc

Calc commands compute statistics. They generally work with variables and produce statistics calculations, result variables, and supporting plots.

Plot

Plot commands work with variables and produce displays.

submenus

Data Desk *submenus* group related commands together. Submenus accommodate more functions than could ever fit in standard menus and organize them so they are easy to find. A submenu is indicated by a ▶ symbol in the right margin of the menu. Position the mouse over that menu item and hold down the mouse button. A new menu automatically drops either to the right or to the left of the main menu. To enter this submenu, slide the mouse to the side (still holding down the button),

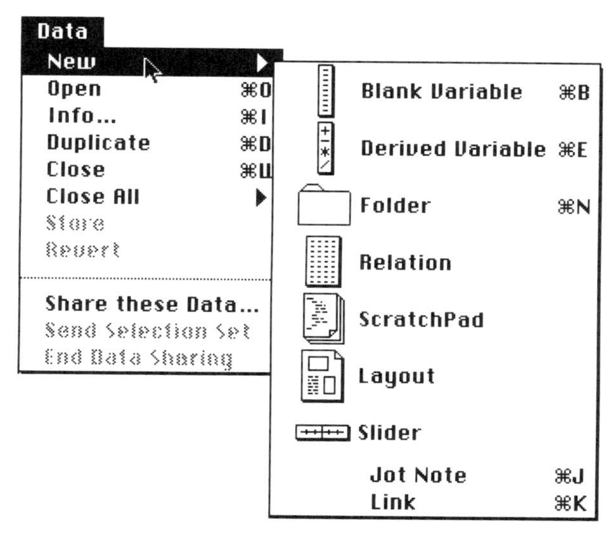

Figure 2-1. Submenus drop automatically from menu items with a ▶ on the right. The ***New*** *submenu lists all the icons you can create.*

and make your menu selection as usual. You need not slide exactly horizontally, but instead can move diagonally, directly to the subcommand you want.

Some submenus in the Modify menu are tear-off palettes. You may use them as ordinary submenus, dragging to the right from the menu and selecting an item. Alternatively, you can drag to the right and down off the bottom of the palette to tear it off the menu bar and place it on the desktop. When torn off, palettes remain on the desktop and are not covered by other windows, so you can select items simply by clicking on them.

Data Desk provides menus attached to places other than the main menu bar. When you see a box with a drop shadow, it indicates that a menu will pop up when you click the box. Other pop-up menus are attached to parts of Data Desk tables and plots, and are called HyperView® menus.

2.3 *Datafiles*

Datafiles store data between work sessions on Data Desk. You can start Data Desk by opening a Data Desk datafile from the Finder desktop. Datafiles contain icons that represent objects such as variables, displays, and tables.

When you quit Data Desk, you can save the entire state of your analysis —including all data and results— in a datafile on the Finder desktop. The icons of variables and parts of analyses belong to Data Desk and cannot appear by themselves on the Finder desktop. The icon that appears on the Finder desktop is the datafile icon. Datafiles can be copied to other disks, duplicated, renamed, or discarded on the Finder desktop.

To open a Data Desk datafile from within Data Desk, choose **Open...** from Data Desk's **File** menu.

When you export variables to a text file, a similar file icon appears on the Finder desktop. If you start Data Desk by opening or double-clicking the icon of a Data Desk text file, Data Desk copies the data from the text file, and closes the text file.

The **New Datafile** command in the **File** menu closes the current datafile and creates a new, empty datafile named "Untitled".

2.4 *Windows*

Data Desk icons open into windows that reveal their contents. Windows can be moved around, overlapped, and resized with the mouse. Some windows contain icons, some contain the text of a variable's values, some contain output from a statistics procedure, and some contain plots. This book sometimes uses the terms "icon window," "variable editing window," "output window," and "plot window," respectively, for these four types of windows.

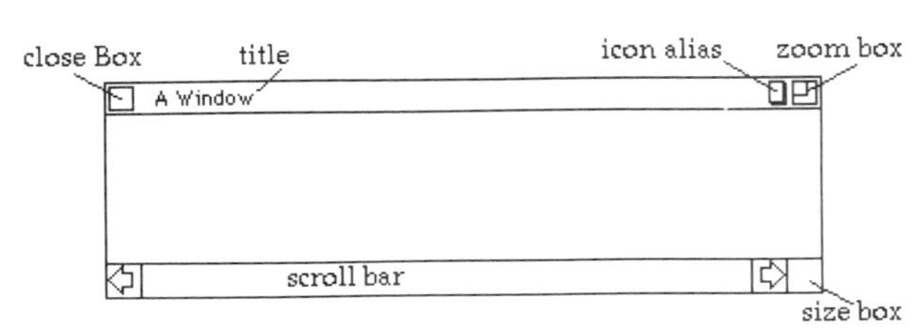

Figure 2-2. *The parts of a window.*

Each window has a *title bar* across the top. The upper left corner of the title bar has a small *close box*; clicking the box closes the window. Most windows also have a *size box* in the lower right corner. To resize the window, drag this box until the gray outline of the window is the right size. A *zoom box* in the upper right corner of many windows provides a quick way to expand the window to its maximum size and then shrink it back to its original size.

Data Desk windows look slightly different from standard Macintosh windows. Their title bars highlight by inverting to white writing on a black background. The small rectangle on the right of the title bar is a miniature icon that behaves like the window's icon. You can select the window's icon by clicking on the *icon alias*. Double-clicking on the icon alias locates the window's icon, making it visible on the desktop and selecting it. A window may also have a small triangle on the left of the title bar (▷). This is the window's global HyperView. Section 4.2 discusses HyperViews.

Windows that hold icons only resize horizontally. Resizing some windows, such as a window displaying a scatterplot, alters the way their contents are displayed. Resizing other windows only alters the amount of their contents that is visible. The *scroll bar*, located across the bottom or on the right border of the window, provides a way to slide the window across its contents. A gray scroll bar indicates that some of the contents of the window are beyond the bounds of the window's rectangle and invites you to scroll the window to see them. In addition, dragging a selection rectangle off the right or left edge of an icon window or off the bottom or top of an editing window scrolls the window.

active window

When two windows overlap, one falls in front of the other. The window in front has its title bar highlighted (black with white writing). The front window is called the *active window*, because it is the window that reflects your actions. To bring a window to the front, click any part of it or double-click the (gray) icon from which it opened.

To move a window to another part of the desktop, drag it by its title bar. A window can be dragged so that part of it extends off the screen. By dragging and resizing, you can position windows next to each other on the desktop in any arrangement you wish. Windows remember their position on the desktop when they are closed, and return to that position when reopened, if possible.

TIP
Most windows remember their position on the desktop when they are closed and return to that position when reopened. Hold Option while opening an icon to "forget" its old position.

The **Windows** submenu in the **Special** menu lists all open windows, front to back. When you select the name of a window in this menu, the window is moved to the front. This is a handy way to retrieve a window buried under other windows and therefore not available for clicking. The {Special ▶ Arrange} **Tile Windows** command repositions and resizes variable, plot, and output windows so that they fit neatly on the screen. When many windows are open, the tiled windows may be quite small. You can enlarge any window with its zoom box if you want to examine it more thoroughly, and zoom it back when you are done. The **Stack**

TIP

The {Special ▸ Arrange} **Send Window Behind** command makes the frontmost window the bottom window without the need to click on another window.

Windows command makes each window large and fans them into a stack that shows an identifiable section of each window with just enough room for clicking.

There is no limit to the number of windows Data Desk can have open at once, but it is a good idea to keep your Data Desk desktop neat by closing unneeded windows.

With MultiFinder or System 7, Data Desk windows can be anywhere on the Macintosh screen. Even though they may be moved away from the Data Desk desktop, they remain Data Desk windows. Clicking on any Data Desk window makes Data Desk the active application.

2.5 *Icons*

Each of the principal objects that Data Desk works with or produces is represented on the desktop by an icon. You can tell what kind of object you have by what its icon looks like.

To do anything with an icon you must first select it by moving the tip of the cursor's arrow on top of the icon and clicking. Selected icons highlight. To move an icon, click on it and drag. You can move several icons by selecting them all and then dragging any one of them. The icons need not be in the same window. You may place icons on the desktop, but it is usually more convenient to leave them in the windows that ordinarily hold them. As we shall see below, it can make a difference which window an icon is in.

Icons provide a convenient place to put away data, plots, and analyses so that they do not clutter the screen but are still readily available. Whenever you close a window it will close into an icon. Whenever you open an icon, it will open into a window. To open an icon, select it and choose **Open** from the **Data** menu, or double-click on it.

TIP

You can work with a gray (open) icon as you would with any other icon.

When an icon is open to show its window, the icon appears gray. Gray icons are still active; you can move them, discard them, and use them as you would any other icon.

To rename an icon, click on its current name and edit it as you would any text on the Macintosh. Pressing the Tab key selects the name of the next icon to the right in the same window.

2.6 *Variables*

variable

A *variable* contains data. A typical variable might have numbers recording measurements or observations about some individuals, organized as a column of values. While variables often hold numbers, they can also hold text or a mixture of numbers and text. To see the contents of a variable, open its icon. The variable opens into a window displaying its contents, and the icon of the variable turns gray to indicate that the variable is open. You can enter new data or alter the data in the window with the standard Macintosh editing features. Chapter 5 gives details and shows examples of how to enter and edit data in variables.

A Variable

The icon for a variable looks like a column of values. Ordinarily you can leave variables closed and work with the icons. After all, statistics is about the relationships among the variables not about calculations on the numbers, so you rarely need to see the numbers themselves. By leaving the variables closed, you can keep your screen much less cluttered.

variable names

Each variable has a name. You may use almost any name you can type, including names with punctuation marks, spaces, and numbers. For example, "Wages + Tips", "123", and "Σ{random values}" are all legal variable names. Variable names can have up to 33 characters, but it is a good idea to choose short, evocative names. If a variable name is too long to fit neatly under its icon on the desktop, the name is abbreviated. Click the abbreviated name to see the full name.

To change the name of a variable (or of any icon) click on the name to select it and type the new name. Press the Tab key to advance to the next icon and rename it as well. To create a new variable choose {Data ▶ New} **Blank Variable**. Chapter 5 discusses data entry and editing in detail.

Most Data Desk operations use one or more variables to plot or compute something. You specify the variables by clicking on their icons to select them. For example, to make the histogram of a variable, click its icon and choose **Histograms** from the **Plot** menu. When you select a variable, its icon highlights black and it is branded with a "Y".

HOW-TO

To select multiple icons:
- Click and shift-click each in turn.

or
- Drag a selection rectangle around them.

To select a second and third variable, hold down the Shift key and click their icons in turn. They will highlight with "X" brands. Alternatively, if variables are adjacent in a window, you can point to one side of them, hold the mouse button down, and *drag* the mouse across the variables. An outline box will follow the mouse. When you release the mouse button, all icons covered by the box are selected. (It is important to start dragging while the tip of the mouse arrow points to the side of a variable. Otherwise, if the arrow touches a variable icon, you will select that variable and drag it with you.) You may select icons from several different windows.

When several variables are selected, some may highlight differently from the others. Some variables have solid black highlighting with "Y", and others are highlighted grey with diagonal stripes and "X" brand.

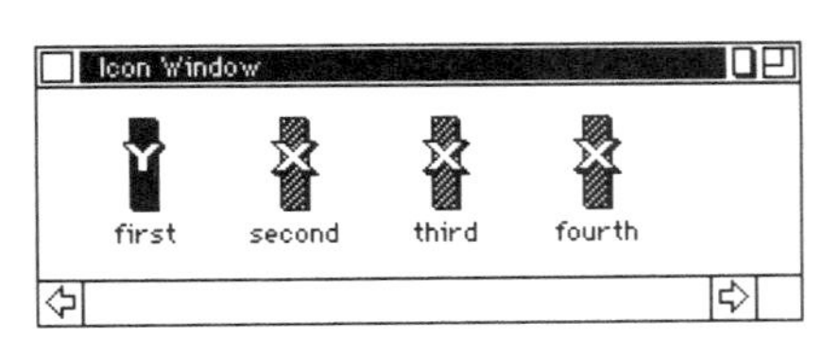

Figure 2-3. *y- and x- highlighted icons.*

Y-highlighted variables play a special role for some commands in Data Desk. For example, they are the dependent or predicted variables in a regression and the y-axis in a scatterplot. Typically, the first variable you select is a y-variable and subsequently selected variables are x-variables. You can explicitly select a y-variable at any time by holding down the Option key while selecting the variable. The mouse cursor changes to **Y** to indicate y-selection. Similarly, holding the Shift key while selecting variables changes the cursor to **X** to indicate x-selection. Both of these cursors select with the point of the arrow rather than with the middle of the cursor.

To change an already selected icon from y-highlighting to x-highlighting, Shift-click the icon. To change to y-highlighting, Option-click. To deselect an icon hold down both the Shift and Option keys and click it.

REMEMBER

Option-click to select y.
Shift-click to select x.
Option-shift-click to de-select.

When you *drag* across several variables, they are selected in left-to-right order. The first variable will be the one on the left, even if you drag from right to left.

To discard a variable, drag its icon to the Trash. The variable can be retrieved by opening the trash icon and dragging it back out. The **Empty Trash** command in the **Special** menu finally discards variables placed in

the Trash. The Data Desk trash icon looks slightly different from the Finder's trash icon. If you resize the Data Desk desktop on MultiFinder or System 7, you can see both trash icons. You can only discard a Data Desk icon in the Data Desk Trash. If the Trash doesn't accept an icon, check that you have dragged the icon to Data Desk's trash icon and not the Finder's trash icon.

2.7 Folders

When using a small number of variables, you may want to arrange them in a single icon window and select them as needed. For more complex analyses or larger collections of data, it is better to organize variables into groups so that you can deal with them easily.

Several icons may belong together because they describe the same individuals or circumstances, because they contain related quantities, because you plan to use them together in an analysis, or because you want to group them together to clean up the desktop. In Data Desk, icons can be grouped into *folders* for any of these reasons.

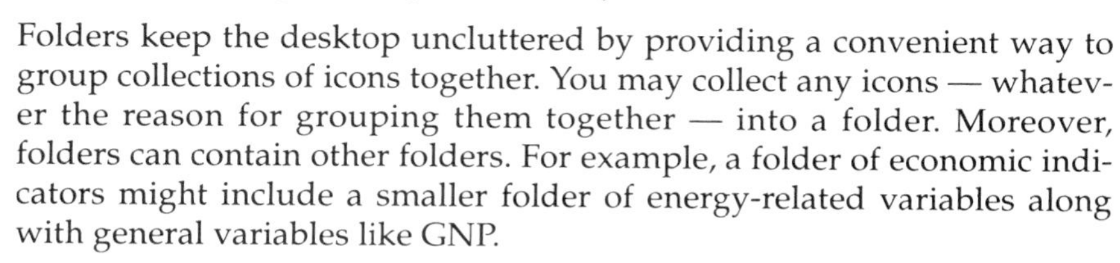

Folders keep the desktop uncluttered by providing a convenient way to group collections of icons together. You may collect any icons — whatever the reason for grouping them together — into a folder. Moreover, folders can contain other folders. For example, a folder of economic indicators might include a smaller folder of energy-related variables along with general variables like GNP.

Folder windows resemble Relation windows, which are discussed in Section 2.9.

Feel free to make many folders to keep the desktop clear and organized. It is easier to find an icon by opening an aptly named folder than by scrolling a window back and forth. An icon buried several layers down in nested folders can be retrieved with the **Locate** commands in the **Special** menu. In general, each new analysis you perform on a dataset deserves its own folder and an appropriate, descriptive name (for example, *Regression on sales*). Each of these folders, in turn, may have several sub-folders to hold plots and output icons separately, for example. Data Desk automatically groups the results of many analyses together in their own folders.

Data Desk folders differ from the folders on the Finder desktop in one major way. Data Desk's folders keep icons in a strict left-to-right order. It is always clear which item is the first (i.e. leftmost), which is the second, and so on. This order can be important to the statistics and display operations in Data Desk. So a second reason for using folders is to keep variables in a particular order.

Folders also provide a convenient way to manipulate groups of variables. They can simplify advanced analyses by providing a way to group variables together. When you select a folder's icon, Data Desk selects all of the icons it contains in left-to-right order.

TIP

Using folders helps keep the Data Desk Desktop uncluttered.
Retrieve a buried folder icon with the **Locate** commands in the **Special** menu.

To create a new folder choose the {Data ▶ New} **Folder** command and provide a name. The command-key ⌘N creates a new folder as well. The new folder is added to the frontmost icon window on the right.

An open folder shows a short, wide window that holds a single row of icons.

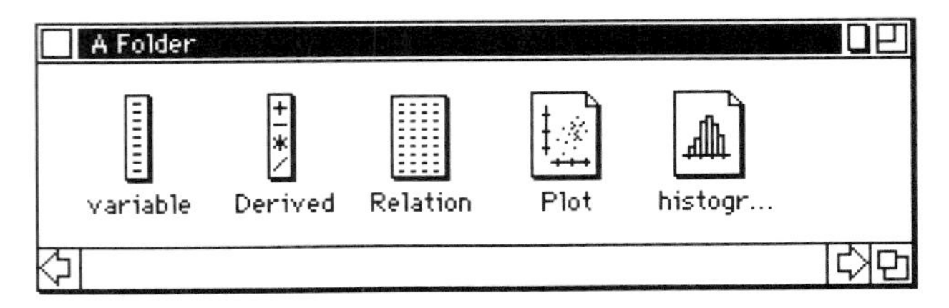

Figure 2-4. A Folder window holds icons.

> REMEMBER
>
> Selecting a folder automatically selects all of the icons it contains in left-to-right order.

If there are too many icons to fit in the window, its scroll bar across the bottom can move them left and right. If you drag a selection rectangle off the side of a folder, the icons automatically scroll away from you and continue the selection. You can drag the icons to new positions in the window, on the desktop, or to other folders.

Icons dragged into the icon of a folder are appended to the right of the icons in the folder. The folder icon highlights when the dragged icon is over it. Release the mouse button to drop the dragged icons into the folder. You can drop icons into a closed folder icon or into the gray icon of an open folder.

> REMEMBER
>
> Icons dragged into the icon of a folder are appended to the right of the icons in the folder. This is true even if the folder window is open.

To discard a folder, drag its icon to the Trash. Every icon in the folder will be discarded as well; any icons that are open close their windows immediately. Of course, a discarded folder can be retrieved until the Trash is emptied. When the gray icon of an open folder is dragged into the Trash, the folder's window closes and follows it into the Trash.

2.8 The Results Folder

Most Data Desk operations create new icons. Every plot, table, and variable has its own icon. Data Desk places these new icons in a folder that has been designated as the Results Folder.

Ordinarily, Data Desk will create a Results folder and place it on the desktop. If you wish, you may make a new Folder and designate it as the Results folder. Click the folder icon (or the icon alias in the folder's window title bar) and choose **Results Log▶ Set** from the **Special** menu. Data Desk results windows open automatically so you can keep the Results log closed to keep the desktop cleaner.

2.9 Relations

Most datasets are rectangular. There are variables (usually represented as columns) and cases (usually represented as rows). Each case has a value recorded for each variable. The recorded value may be a value defined as "missing" rather than a number or a category name. Because each case has a value for each variable and each variable has a value at each case, the array of data can be shown as a rectangular table of values.

Data analyses typically relate two or more variables to each other. This is only possible when the variables hold data for the same cases in the

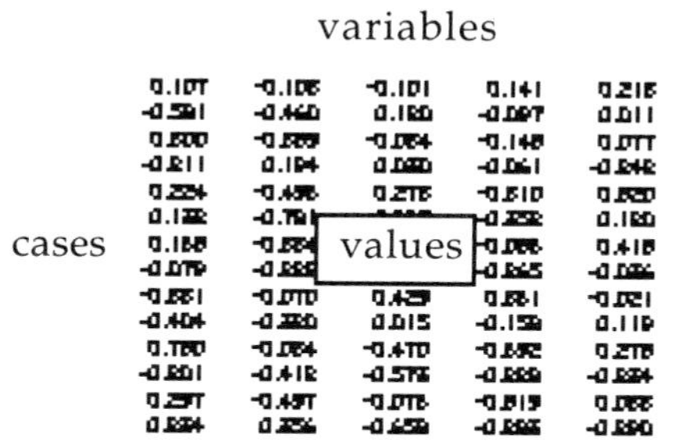

variables

cases

values

Figure 2-5. A relation typically has variables in columns by cases in rows.

same order. If a variable recording median education in each of the 50 states was arranged in alphabetical order, it would make no sense to plot it against a variable holding median income in each state that was ordered from West to East. Or against a variable that recorded income by region rather than by state.

This rectangular structure is known in database theory as a *relation*, and Data Desk adopts this terminology. Formally, each row in a relation must be unique. Data Desk accomplishes this by assigning a unique case number to each row in order from top to bottom.

If your dataset is a standard rectangular data table, calling it a relation changes nothing. (Like the *Bourgeois Gentillehomme*, who was pleased to learn that he spoke in prose, you can be assured that you already use relations.) However, if your data includes variables recorded for several relations, you will find that Data Desk's relational data management abilities let you structure, enter, and work with your data in more natural ways.

For most datasets, Data Desk uses relations to make your life easier automatically. For example, if your data form a simple relation, Data Desk automatically keeps cases aligned in your variables. Thus, if you cut a case out of one variable, Data Desk offers to delete that case from all variables in the relation to preserve your ability to analyze the variables together.

Most analyses that deal with more than one variable make sense only when the variables are in the same relation. You cannot combine variables from two different relations in the same plot or calculation, but Data Desk provides ways to refer from one relation to another so that the resulting variables are properly matched.

2.10 *Selecting*

TIP

⌘-**A** is a convenient shortcut for **Select All**.

The concept of selecting is fundamental to Data Desk operations. Icons, cases, and parts of plots can all be selected. Data Desk commands and plot tools operate on these selected objects. To select something ordinarily you touch it with the mouse and click. If you hold down the Shift key while clicking, selected objects accumulate; if not, the new selection replaces any previous selection. When selecting cases, hold the Shift key to select the range of cases between the current insertion point and the click. Hold the Option key to add the clicked case to the selection discontinuously.

The {Edit} **Select All** command selects all selectable objects in the frontmost window. That is, it selects all cases in variable editing windows, all icons in icon windows, and all datapoints in plot windows. The ⌘-A combination is a keyboard shortcut for Select All.

2.11 *Result Windows, Plots, and HotResults™*

Result Window

Most Data Desk commands produce some sort of output — usually one or more icons that open to reveal tables of numbers or plots. Some commands produce variables, which can then be used in other commands. Others produce *result windows*. A result window is a window that contains the results of a command. The icon of a result window looks like a document.

The icon opens into a window containing words and numbers. Although result windows contain text, they cannot be edited like variables. However, you can copy the text with the **Copy Window** command, which is active in the **Edit** menu when the result window is active, and transfer it to a ScratchPad for editing. Tables of results have columns separated by tab characters, so you can reposition them in a word processor or paste them into a spreadsheet program.

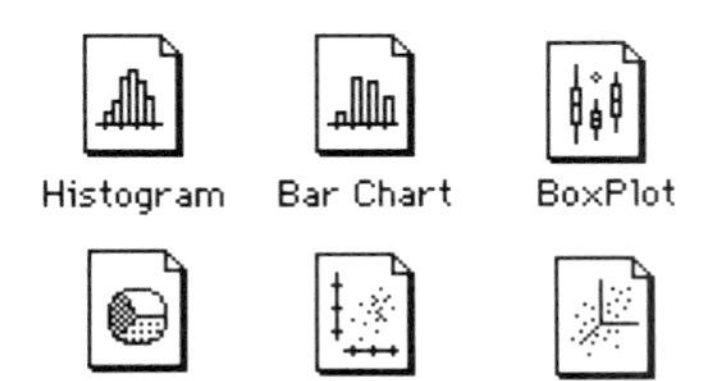

Data Desk makes many kinds of plots. Each one has an icon with the standard shape of a result window but a picture that varies according to the particular kind of plot. Chapter 8 describes the various kinds of plots available in Data Desk.

Data Desk plots are interactive tools for the data analyst. Each type of plot provides ways to select and identify datapoints, and to work with or modify the plot. Chapters 9 and 10 discuss these methods.

Most Data Desk commands that produce variables holding computed results, create a special kind of variable called a *HotResult*™. HotResults can be used exactly like variables except that their values depend upon the analysis that produced them. If you change any of the variables used in that analysis, or the analysis itself, its HotResults update to reflect the new state of the data or analysis. However, you cannot open a HotResult icon and edit its values.

HotResults behave very much like derived variables, which are discussed in Section 4.4. Their ability to remain consistent with the data when you alter or correct it is an example of Data Desk's ability to maintain consistency throughout analyses and plots.

2.12 *Defaults and Preferences*

Data Desk comes with standard defaults for all its Manipulation, Calculation, and Plot options. You set your own defaults by selecting the desired options from each Options dialog and clicking on the Set Defaults box. These changes are saved with the program and are active for analyses generated in the current and any future datafile.

For example, the initial default settings for Summary Statistics Options are Mean, Standard Deviation and the number of Numeric cases (See Chapter 7 for more information on Summary Statistics). To add additional settings, like Median and Variance, to this list, choose {Calc ▶ Calculation Options} **Select Summary Statistics....** Click on the boxes corresponding to Median, Variance and Set Defaults, then click on OK. Any future **Summaries...** command will include the new settings. Of course you can always add new settings to a specific analysis without adding them to your set of defaults by choosing the new settings from the Options dialog and not clicking on the Set Defaults box.

Data Desk has a **Preferences...** dialog, which is accessed from the **Edit** menu. It has three sections. The first section affects your editing actions and is explained in Chapter 5.

Data Desk saves intermediate results in a "Scratch file" on your disk. Ordinarily, you need not think about this file at all. Sometimes, however, Data Desk may try to put the scratch file on a shared disk. The second section of the **Preferences...** dialog allows you to change the location of the scratch file, an intermediate working file created for each datafile until it is saved. Data Desk automatically finds the most efficient location for this file. If that location causes any problem, you can change the location of the scratch file. Click the **Change** button and select the new location for the file. To put the scratch file back in its usual location, click the **Automatic** button.

The third section of the **Preferences...** dialog lets you control the precision and presentation of numbers in any derived variable, generated variable or scratchpad. The standard setting is for Fixed notation with 8 significant digits. Change the notation to Scientific or Engineering by clicking on the **Notation** pop-up menu and choosing the desired format. Change the number of digits to be displayed by typing a new value in the **Significant Digits** box. When using scientific or engineering notation, Data Desk provides for the rounding of data to eliminate spurious values close to zero. You control this feature by entering the number of digits to round to in the **Round to Nearest** box. For example, the number 0.1234 displayed with scientific notation, 3 significant digits and rounded to 0.0001 is displayed as 1.23e-2. The number 0.0001234 with the same parameters is displayed 1.00e-4. Changing the rounding parameter to 0.000001 displays the number as 1.23e-4.

2.13 *Leaving Data Desk*

To leave Data Desk, choose **Quit** from the **File** menu. Quit closes all open windows, saves changes in the main datafile (if you approve) and returns to the Finder. If the datafile is named "Untitled" Data Desk asks for a new datafile name.

CHAPTER 3

Data

3.1 Basic Concepts *27*
3.2 Relations, Variables, Values and Data Structure *28*
3.3 Missing Data *29*
3.4 Infinities *30*
3.5 Outliers, Blunders, and Rogues *30*
3.6 Numerals and Numeric Values *30*

APPENDICES
3A Kinds of Data *32*
3B Matrices, Tables, and Relations *34*

BOOKS ABOUT STATISTICS, graphics, and data analysis almost never discuss or define data. While this book is not the place for a long discourse, an understanding of basic terms and principles can help anyone analyze data better.

3.1 *Basic Concepts*

Most people think of data as numbers and category names, but if that was all we knew about our data we would be unable to analyze it in a sensible way. Data must be *about* something for any analysis to be meaningful. Analysts who concentrate on the numbers but lose sight of what the data are about can easily go astray, producing impressive but worthless graphs and tables, and missing important patterns or exceptions. Indeed, we can only understand what kinds of patterns are important if we know why we are looking at the data.

case

Data are values measuring or reporting information about each individual in a group along with details of how the values go together and what they report. The values can be numbers or names, or even a mixture of the two. The individuals are called *cases* in Data Desk.

variable

Data values that record the same thing about each case are gathered into a *variable*. In formal statistics, a variable is a particular kind of mathematical entity. In data analysis, a variable is both the data values and the underlying phenomenon they record. Typically, a variable is represented as a column of values, with a row for each case.

relation

In Data Desk, variables holding values about the same cases in the same order are gathered into a *relation*. Because the rows of each variable in a relation hold data about the same cases, the columns representing the variables are often placed side-by-side to form a rectangular table. The individual cases reported on by these variables can usually be described generally. For example, they may be survey respondents, cars, countries, or months. Because each variable in a relation describes the same individuals, they characterize the relation, so it is often wise to name the relation after them.

Data values are not simply numbers or categories. They come from some source and record information about some phenomenon. To analyze the data intelligently, we need to know more about the phenomenon and how it has been recorded. In colloquial terms we need to know the reporter's familiar who, what, where, when, and why.

In this case, the *who* names the observer, recorder, or source of the information. This should be part of the information you report along with your analysis, and can be placed in the information record of a relation or folder.

The what, when, and where define the matter recorded, the time, and the location. You might need one, two, or all three to describe particular data values. Variables are often named with their *what*. Measurements of sales would be in a variable named *Sales*. Sometimes, the *when* or *where* should be part of the variable name, especially when the same matter is recorded for different times or places in related variables. Sales for stores in each of two years (when) might be in variables named *Sales87* and *Sales88*. Sales for stores in each of four regions (where) might be in variables named *SalesNE, SalesSE, SalesNW*, and *SalesSW*.

It is also important to have a sense of *why* the data were recorded. The *why* information may help to guide your analysis toward particular displays or methods, or alert you to possible biases. For example, data on smoking and health attributed to a tobacco industry researcher in the 1950's should be interpreted differently than data on the same subject reported by a respected health research institute in the 1980's. It is also important to know whether you are using data for a purpose other than the original intent of the observer.

3.2 *Relations, Variables, Cases, Values and Data Structure*

Data Desk is a general tool that can work with many kinds of data from a wide variety of sources; as such, it uses a good deal of general terminology. Many fields and some kinds of programs use specialized terminology. We catalogue here some alternative terms:

data table

rectangular dataset

- RELATIONS are usually represented in statistics programs as rectangular tables of data in which each row represents a case and each column represents a variable. They may be referred to operationally as *data tables* or even (by programs that do not imagine more than a single relation) as *datasets*. Formally, a relation is the set of cases, the set of variables, and the values recorded for each case on each variable. A spreadsheet is a typical representation of a single relation. Some writers in statistics use the term *rectangular dataset* to mean the same thing.

  Most statistics and graphics operations only make sense for values that are related — that is, for values measured on the same individuals in the same order. Data Desk often requires that variables analyzed together be in the same relation.

field
column
variable

- VARIABLES are usually represented as columns of values. A database program would think of a variable as a *field* in a database. In a spreadsheet, a variable would commonly be a *column*, although some variables might be rows. The term *variable* is quite standard in statistics and is used throughout this book. It suggests that the values gathered together represent some underlying phenomenon, and that patterns we find in the data values for different variables may tell us something about how these variables are related to each other.

respondent
subject
observation
period
record
row

- CASES are the individuals to which the data values refer. A case may be called by another name according to circumstances. Thus, one speaks of a survey's *respondent*, of a psychology experiment's *subject*, of a study's *observation*, or of a *period* in time-sequenced data. Each of these would commonly be a single case in a relation. A database program would think of a case as a *record* with values for each field. When variables are represented as columns of values, cases are represented as *rows* across adjacent variables. This is how they appear when you open variables from the same relation in Data Desk.

sample population

In statistics, the group of cases is often a *sample* drawn from a larger *population*. Data Desk makes no particular assumptions about the sample, except that the inferential statistics it computes assume that the sample is representative of the population and usually that it is drawn at random.

unit

- VALUES are the elements that fill the variable-by-case structure. Each variable in a relation has a value (which may be "missing") for each case, and each case has a value for each variable. Data values are usually recorded in some measurement *unit*, such as dollars, inches, years, and so on. While the units can be converted to other units (as when we convert feet to inches), there may be a standard unit used conventionally to record a particular phenomenon. However, even when a particular unit is standard, good data analysis may require that we transform the data, as when we work with the logarithm of the original values or with the difference or quotient of two values, which may alter the unit.

3.3 *Missing Data*

missing values

Sometimes we cannot obtain a data value for each case. Values are lost, experiments are ruined, reports are not filed, subjects are too subjective, respondents don't respond, and so on. You can indicate a missing data value in Data Desk with a blank or with any nonnumeric character. Any data value that is not a valid number is treated as a missing value by any operation that requires numbers. Thus, you can label missing values with information about the cause of the omission.

For example, Data Desk considers all of the following to be missing values in a numeric variable: "missing", "refused to answer", "not at home", "equipment malfunctioned". All behave as missing values in any numeric computation, but show up in the variable's editing window just as they were typed.

Sometimes it is useful to mark a case as missing temporarily to remove it from an analysis. One easy way to do this is to place a nonnumeric character in front of the case value. A "*" makes a good marker that is easy to find and remove when the numeric value is needed again. The data value "*3.4" is treated as missing, but preserves its original numeric value for later reference.

When Data Desk opens a variable that resulted from an internal computation, it displays missing values with the "•" symbol. You can type this symbol as Option-8. Data consisting of category or group names is considered missing only if the case is empty or consists of a •.

Not a Number NaN

Data Desk represents missing values internally with a construct called a NaN, which is short for "Not A Number". NaN's also result from calculations that involve missing values, or from calculations that yield a nonnumeric result (such as the square root of -1 or the log of 0).

3.4 Infinities

infinity

There are two exceptions to the rule that text is treated as missing values in computations: the text "Infinity", and "–Infinity" are interpreted as infinity and negative infinity, respectively. You may want to enter data with infinite values, but you should take care to use only those techniques that can accommodate infinities (for example, by ranking the data first to compute nonparametric statistics). Computations that cannot accommodate infinities treat them as missing values. Any text that begins with the letters "inf" or with a "+" or "-" followed by "inf" (regardless of capitalization) is interpreted as infinity. Thus the words "inference" or "infrastructure" would also be interpreted as infinity.

3.5 Outliers, Blunders, and Rogues

Real data are dirty. Errors creep in at almost every step of data collection, recording, and organization. Even correct data describe a world that contains individuals so extraordinary that it may be impossible to understand the ordinary course of events without treating them specially.[1]

outlier

Any case that is far from the body of the data is an *outlier*. Some outliers are clearly extreme. Others are more subtle, being extraordinary in some combination of variables but not extreme in any one variable. The medical patient who is 6′3″ and weighs 110 pounds is an example. Neither his height nor his weight alone is extraordinary, but the combination of the two show him to be extraordinarily thin — possibly enough so to warrant excluding him from a medical experiment.

blunder
rogue

Extraordinary cases are sometimes divided into *blunders* and *rogues.* Blunders are clerical or measurement errors in the data that are not informative about the real world. They should be identified and either corrected or omitted.

Rogues are correctly measured and recorded, but are just inherently unusual. Even though a rogue case is correct, you may do well to omit it from your main analysis and treat it specially. A data analysis that describes most of the data well and deals specially with a few exceptions is almost always more useful than an analysis that provides a mediocre fit to all the data but makes no exceptions. In no event should rogues be summarily discarded. A rogue can be worth more to your understanding of the data than all of the ordinary cases because it makes a particular aspect of the data clear.

3.6 Numerals and Numeric Values

Many Data Desk operations calculate with numeric data values. Some, like bar charts, Tables, and Analysis of Variance, use data values to define categories. When a variable is used to define categories, its values are interpreted according to their *text* rather than their *numeric* value so that the numbers are seen as sequences of *numerals.*

numeral

[1]The venerable statistical consultant Cuthbert Daniel has referred to such values as "non-missing" values — that is, values that really should have been missing but aren't.

Some variables can be interpreted reasonably as either numeric or categorical. For example, the Cylinders variable in the Cars data counts the number of cylinders (so that the *average number of cylinders per car* is a reasonable statistic), but might just as well be used to partition cars into four groups: 4-cylinder, 5-cylinder (the Audi 5000), 6-cylinder, and 8-cylinder cars for a pie chart.

When numbers are read as numerals, numbers that are written differently but would evaluate to the same numeric value are *not* equal. For example, the data values "1", "1.0", and "+1.0" all have the same numeric value, but specify categories with different labels in procedures that require categories.

Unlike conventional statistics programs, Data Desk does not require that the values in a variable be all numbers or all text. Nor is it necessary to specify in advance whether numbers are to be interpreted as categories or as numeric values. Data Desk considers your use of the data and determines from that how to interpret the values.

APPENDIX 3A

Kinds of Data

Data come in several different kinds. Traditionally, analysts classified data as Nominal (category names), Ordinal (ordered, but not measured), Interval (suitable for addition and subtraction operations), and Ratio (suitable for multiplication and division operations). While these categories were useful, they concentrate on what we can do with the data values rather than on the nature of those values.

Mosteller and Tukey (*Data Analysis and Regression*, New York, Addison Wesley, 1977) suggest seven categories that more closely represent the nature of the data values:

amounts

AMOUNTS

Amounts are the most common data kind. They cannot be negative. Amounts include collections of things (for example, amounts of money), measurements ("amount" of height), durations (amounts of time), and so on.

counts

COUNTS

Whole numbers enumerating things. They cannot be negative.

counted fractions

COUNTED FRACTIONS

Counted Fractions are ratios with a fixed base, as in "there are 25 workers in this plant and 5 of them are women so 5/25 of the workers are women." The most common counted fractions are *percents,* which are counted fractions with a base of 100.

names

NAMES

Categories taken in no particular order.

ranks

RANKS

Integers reporting the order (but not the value) of cases.

grades

GRADES

Categories with a natural order, such as Freshman, Sophomore, Junior, Senior. Much survey and testing data come as grades because respondents are offered 5 or 7 choices in order from low to high with no indication of spacing or units.

balances

BALANCES

Balances can be positive or negative and may be unbounded in both directions. They are the most general form of data, but may be the least common. Often balances come about as a difference between two amounts (for example, a balance can be used to represent "velocity northward" when a negative value indicates southward travel.) Alternatively, the amount of northward velocity (bounded by zero) and the amount of southward velocity could be reported, with the original value being their difference. The logarithm of an amount that can be arbitrarily close to zero can be a balance.

These kinds of data have properties that are worth noting because they help to determine the kinds of plots and analyses appropriate for the data. For example, variables that are Nominal (Names and Grades) are often best displayed with pie charts and bar charts.

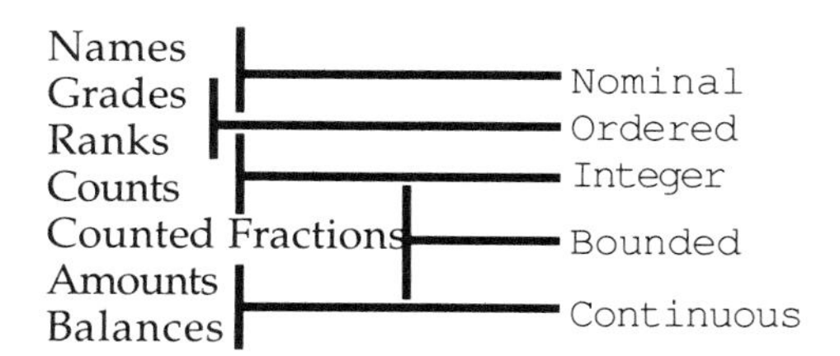

Mosteller and Tukey recommend transformations that can make data analyses simpler based upon the kind of the data:

- Amounts are often better analyzed after taking logs.
- Counts usually benefit from a square root or logarithm transformation.
- Percents almost always are easier to analyze as $\log(p/(1 - p))$, where p is the percent expressed as a fraction between 0 and 1. (Use $\log(p/(100 - p))$ for percents expressed as values between 0 and 100.)
- The ith rank out of n is often best transformed by $\log((i - 1/3)/(n - i + 2/3))$.
- Balances usually should not be transformed.

APPENDIX 3B
Matrices, Tables and Relations

Data analysis requires simple and convenient data manipulation and editing, and powerful numerical computations. A data analysis environment must provide ways to organize, label, manipulate, examine, edit, and compute with data.

matrix

Statisticians often use the mathematical structure called a *matrix* to specify statistics computations. The most common way to do this is to regard each variable as a column of numbers and to join columns together side-by-side in some order to make a matrix. Matrices are powerful tools for specifying and performing statistics calculations, but they are merely mathematical conveniences and are neither fundamental statistics entities nor appropriate data management structures.

table

Because most data tables are rectangular, data analysts often array their data in *tables*. The most common form for these tables is as columns of numbers and text placed side-by-side. Data tables thus look much like data matrices. However, tables are simply ways of arranging data for display and (sometimes) editing. Variable editing windows form a table of data designed for examining and editing data. Because they can be rearranged and resized on the screen they make a more flexible table than the traditional row by column form, but otherwise they can be thought of as a table.

relations

Data Desk keeps variables in icons. The icons look like columns of values and open into windows that show columns of values. *Relations* keep variable icons side-by-side in left-to-right order, and guarantee that when cases are removed, inserted or moved in some variables, the other variables in the relation remain consistent.

folders

Data Desk also provides *folders* to help you organize your icons on the desktop. Folders can hold any Data Desk icons including variables, relations, and other folders. Variables can be grouped into folders to provide a hierarchy in which a single folder icon refers to several related variables.

Data Desk thus assigns the various functions required for data analysis to different views of the data. Variable icons are easy to arrange, organize, and rename. Folders provide a simple structure to make large datafiles more manageable. Variable editing windows open side-by-side to form a data table. Relations offer a clear structure for data and make sophisticated relationships among several relations possible. While Data Desk computations use matrices internally, matrices are not available as objects on the desktop.

CHAPTER 4

Data Desk Concepts

4.1	Selecting Variables	*37*
4.2	HyperViews	*38*
4.3	Updating Windows	*38*
4.4	Derived Variables	*39*
4.5	The Clipboard	*40*
4.6	ScratchPads	*40*
4.7	Jot Notes	*41*
4.8	Information Records	*42*
4.9	Sliders	*42*
4.10	Dependencies	*42*
4.11	Datafile Details	*43*
4.12	Memory Requirements	*44*
4.13	Multiple Window Modifier	*45*

APPENDICES

4A	Hints and Shortcuts	*46*
4B	Hot, Warm and Cold Objects	*47*
4C	Data Desk Limits	*49*

YOU CAN LEARN Data Desk bit-by-bit, starting with simple operations and gradually adding more and more powerful skills. Chapter 2 discusses the basic concepts that every Data Desk user must understand. This chapter goes further, introducing powerful concepts and features that you will want to learn, but which can be skipped at first. Most of the features discussed in this chapter make Data Desk easier and more convenient to use.

4.1 *Selecting Variables*

Ordinarily you need not be concerned about how variables are selected in Data Desk. The first variable you select is the y-variable for scatterplots and regressions, and the second variable is the x-variable.

You can also select y-variables by holding the Option key while clicking, and x-variable by holding the Shift key while clicking. Variables selected with a selection rectangle and those selected by selecting their folder's icon are selected in left-to-right order.

Many windows offer HyperViews to select the variables used in the window. For example, the HyperViews attached to the axis labels of a scatterplot offer to select the variable plotted on that axis. You can hold the Shift or Option key down to select as x or y in this way as well.

When several y-variables and several x-variables have been selected, Data Desk attempts to work with appropriate combinations of variables as follows:

- All one-variable commands ignore selection type and operate on each variable in turn: y-variables first, followed by x-variables.
- Two-variable commands (such as Scatterplots and two-variable transformations) attempt to pair variables:

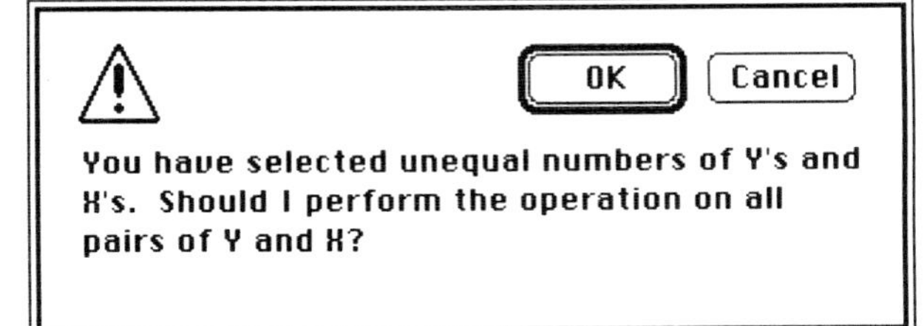

Figure 4-1. *Data Desk offers to work with all pairs of y and x variables when the number of y and x variables does not match.*

If one y-variable and several x-variables are selected, then the y-variable is paired with each x-variable in turn and the command operates on each pair of variables.

If several y-variables and one x-variable are selected, then the x-variable is paired with each y-variable in turn and the command operates on each pair of variables.

If an equal number of y-variables and x-variables is selected, then the first y-variable is paired with the first x-variable, the second y-variable is paired with the second x-variable, etc., and the command operates on each pair of variables.

If an unequal number of y-variables and x-variables (but not exactly one of either kind) is selected, then Data Desk warns you of the mismatch and offers to apply the command to every possible x-y pair.

4.2 *HyperViews*

REMEMBER

The ☝ cursor indicates that the mouse is over a HyperView. Press the mouse button to pop-up the menu.

Data Desk is an open environment. At any time you can do almost anything. Data analysis requires such freedom because any plot or calculation may suggest new things to do. While there is no end to the possible connections and paths, some paths are more common than others. For example, it is common to consider both the scatterplot of two variables and their correlation, so if you are looking at one of these, you might want to see the other.

HyperView® menus suggest such next steps in the data analysis. Each Data Desk output window offers direct paths to additional plots or analyses. Parts of the plot or output text are designated as *HyperView buttons.* When the mouse is over a HyperView button, it changes to a button hand, ☝. Pressing the mouse button at that location pops-up a menu suggesting related plots or analyses. For example, when you press the mouse button with the ☝ over an axis label in a scatterplot, the HyperView menu that pops up offers to locate the icon, make a histogram, or make a Normal probability plot of the variable plotted on that axis. If you press the mouse button with the ☝ over a correlation coefficient in the correlation analysis, the HyperView menu suggests a scatterplot of the underlying variables.

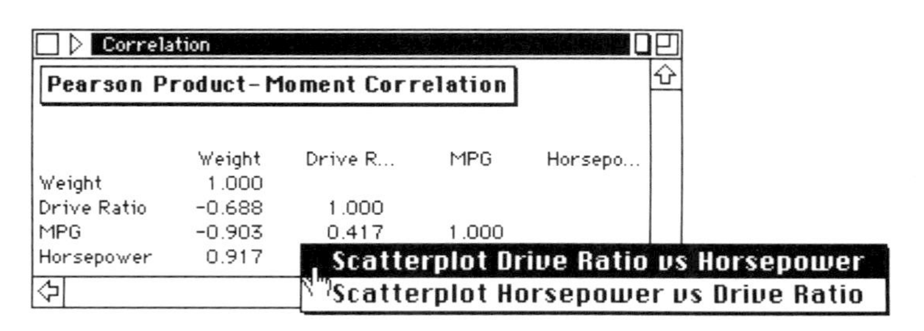

Figure 4-2. *HyperViews suggest additional steps related to the analysis or plot you are viewing.*

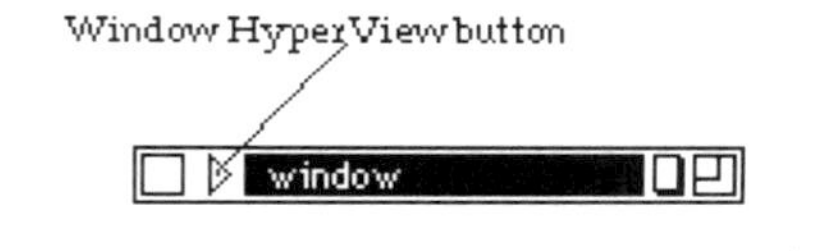

In addition to HyperView buttons on a plot or table, most windows have a window HyperView button that looks like the submenu arrow: ▷ located at thetop of the window, left of the title bar, next to the close box. The window's HyperView menu suggests additional steps that go with the entire window.

HyperView suggestions are a gentle form of built-in statistics expertise in Data Desk. They provide suggestions and advice but they never constrain your options or force unrequested analyses or plots on you. HyperViews and related links among Data Desk operations are discussed in Chapter 13.

While most HyperView commands offer convenient access to standard Data Desk commands, some provide ways to modify or work with the window that are so specific to that window that the commands would have no natural place in the main menu bar. It is usually a good idea to look at the window's HyperView menu of any plot or analysis of interest.

4.3 *Updating Windows*

REMEMBER

A ! indicates that the window is out-of-date.

If you alter any of the underlying variables used by a plot or analysis, Data Desk indicates that the window is outdated by changing the global HyperView arrow to an exclamation mark: !. When a window is outdated, all HyperView menus offer to update the window in place or to redo the analysis or plot in a second window so you can compare outdated and current versions.

The technique of altering an aspect of the data, of an equation, or of an analysis and then updating or recomputing a window is very effective for considering "what if..." scenarios. You can compare two plots or tables to see what would happen if some data values were omitted or altered or what would happen if you analyzed the logarithm of a variable rather than the original values, or what would happen if you included another variable in an analysis.

object-oriented

Using existing windows as examples for new analyses is commonly called *object oriented* interaction because you work with objects that you have created and ask Data Desk for a new instance of the object "just like this one only different in the following way". You will find that this is a very easy and natural way to work because it closely resembles the way you are likely to think about your analysis.

Ordinarily, Data Desk windows only update *when you tell them to*. They always offer to make a parallel window for comparison rather than overwriting the original window. The ❢ marker is the sign that a window is outdated. However, most output windows can also be set to update automatically whenever one of the variables they use changes. Select the **Automatic Update** command in the window's HyperView of a window to have that window recompute automatically. The global HyperView marker, ▷, turns gray to indicate the window's different status.

Automatic updating is particularly useful when combined with sliders to create new dynamic plots. Section 4.8 introduces sliders. The **Manual Update** command in the window's HyperView returns the window to its usual state.

You can also change a window in place by dragging variables into it. When the icon of a variable is dragged over an axis name in most plots, the axis name highlights. Dropping the variable on the axis name causes it to replace the variable originally plotted on that axis and redraw the plot.

Many output table windows can also accept dragged variables. Dragging variables into regression summary tables or correlation tables adds those variables to the analysis.

4.4 *Derived Variables*

Data Desk variables hold data values for plots and analyses. Ordinarily, variables hold values as numbers or text. Derived variables hold instead the *expression* from which data values can be computed. For example, a derived variable might hold the expression *log(income)*. When used as a variable, the derived variable evaluates the logarithm of the values in the variable named *income* so they can be used in the display or calculation.

The derived variable icon resembles a variable icon, but has arithmetic symbols on it to suggest the algebraic expression. Derived variables can be used anytime as a substitute for ordinary variables.

The most common derived variables are simple transformations of other variables or simple arithmetic combinations. The easiest way to generate these is by selecting the variables to be transformed or combined and choosing a function from the {Manip ▶ Transform} submenus.

Derived variables are especially powerful because changing a value in any of the variables used in its expression changes the value of the derived variable. Thus the value of the derived variable holding *log(income)* is always the logarithm of the income values even if you change some income values.

Derived variables are very much like the HotResults created by many Data Desk analyses, except that you can always edit a derived variable expression. You cannot edit a HotResult.

Chapter 11 discusses derived variables at length and lists the functions you can use in derived variable expressions. You will find that derived variables extend Data Desk's capabilities to let you compute many things that might not be built in as menu commands.

4.5 *The Clipboard*

The Clipboard is a standard Macintosh tool. Text, cases, variables, plots and tables are placed on the Clipboard by the **Copy** and **Cut** commands in the **Edit** menu. The {Edit} **Copy** command changes according to the nature of the frontmost window to reflect the kind of object you can copy. For example, it will read **Copy** for text in a single window, **Copy Cases** for cases selected from one or more variable editing windows in a Relation, **Copy Variables** for icon windows, and **Copy Window** when the frontmost window is a plot or output table. The Copy command places the copy on the Clipboard.

The {Edit} **Show Clipboard** command displays the Clipboard window. Resize the window to see more of the Clipboard contents. Ordinarily the Clipboard window displays the contents of the Clipboard automatically. When the Clipboard contains a table of data values copied or cut from several variables, the values may not appear in the Clipboard window. Nevertheless, they are available for **Paste Cases** or **Paste Variables** commands and will be on the Clipboard if you leave Data Desk and enter another Macintosh program or invoke a desk accessory. Chapter 6 discusses the actions of the Clipboard in detail.

ADVANCED CONCEPTS

The discussion in the remainder of this chapter is more advanced. If you are a beginning Macintosh user, you can skip over this material for now. If you are an experienced Macintosh user, you may find some of these comments helpful in getting started with Data Desk.

4.6 *ScratchPads*

ScratchPads are simple editing windows that close into icons. The {Data} **New ScratchPad** command creates a ScratchPad and places it in the Results folder. You can type any message, or paste any text, into a ScratchPad, and edit the text with all the standard Macintosh editing commands (including **Undo**).

ScratchPads are convenient for a number of uses. You may want to keep a ScratchPad on the desktop to hold information about a datafile, or keep one in each relation to describe the data in that relation. ScratchPads are ideal for noting what you did during an analysis and reminding yourself of what you intend to do. They typically hold notes about the data or analysis at large. Jot notes (Section 4.4) provide convenient places for notes about particular windows.

ScratchPads provide a temporary editing environment. For example, the {Edit} **Copy Window** command copies any output table as text. If you paste the text into a ScratchPad, you then can select and copy numbers from the table easily. Similarly, you can import the entire contents of a text file into a scratchpad. The text file might contain data or could, for example, be a text description of a datafile.

ScratchPads also offer a calculator capability closely related to the calculation abilities of derived variables. (See Chapter 11.) Type any expression that would be legal in a derived variable, select it, and type ⌘ = or choose **Evaluate** from the scratchpad's global HyperView menu. (Section 4.6 discusses HyperViews.) The results appear in the ScratchPad just below the expression. Chapter 11 provides more details.

4.7 Jot Notes

Data Desk provides several ways to attach comments or notes to a window. One simple way is to attach a *jot note* to the window. Jot notes provide a simple text editing environment in which you can type any message.

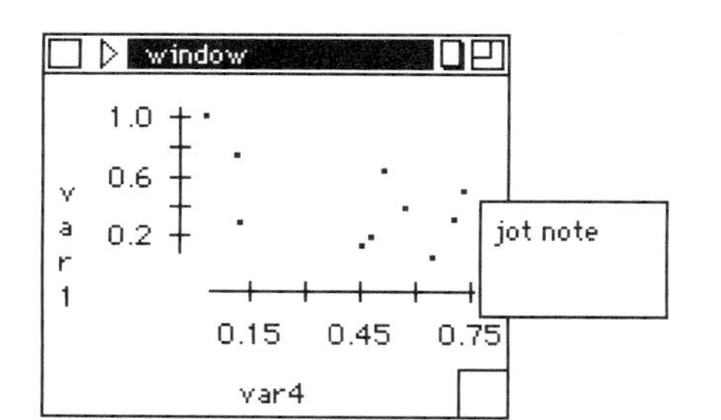

Figure 4-3. *Jot notes let you post comments on any window.*

To create a jot note select {Data ▸ New Icon} **Jot Note**; the new jot note appears on the desktop. Pick it up with the mouse by holding down the Option key as you click anywhere on it, and drag it into place. When you release the mouse, the jot note is tacked to the window directly under the cursor. (Thus, it is best to pick up a jot note near an edge.) You can pick it up again in the same way and reposition it. The cursor shows a tack hammer to indicate the operation. Jot notes open and close along with the window to which they are tacked.

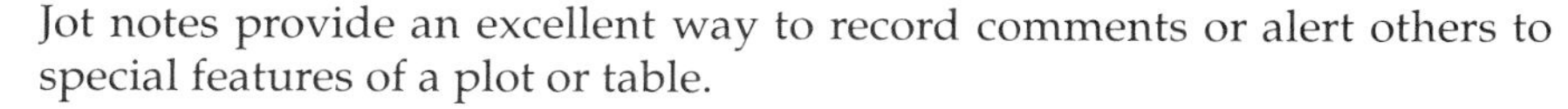

Jot notes provide an excellent way to record comments or alert others to special features of a plot or table.

To discard a jot note, Option-drag it into the Trash. Jot notes dropped in the Trash are immediately discarded; they do not wait for the next Empty Trash command.

If you want to keep a jot note out of the way, resize it by dragging the lower right corner of the note (where a size box would be on an ordinary window) up and left. When you click in the corner, a size box outline appears. You can shrink a jot note to a small square and open it again later to read it.

4.8 Information Records

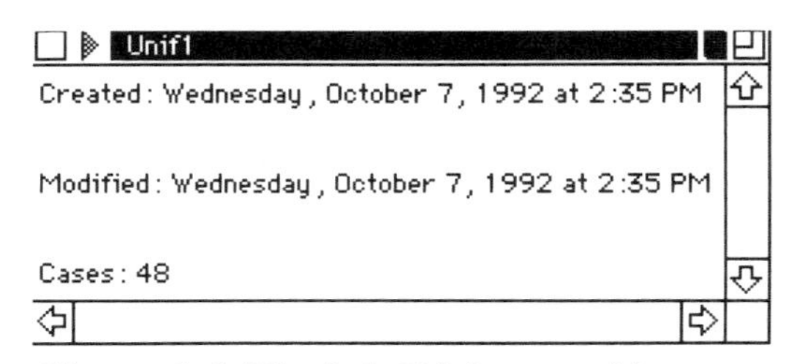

Figure 4-4. *The Info Dialog provides information about an icon in a window associated with that icon.*

Each icon has an *information record* that holds additional information about the icon. To see or alter the information record, select the icon and choose **Info** from the **Data** menu. The information record displays information such as the date and time the icon was created, and the date and time its contents were last modified. (The date and time information help document the icon's history, but they depend upon the clock built into the Macintosh. You should be sure that the clock and calendar in your computer are set correctly.)

Some icons, such as variables, also offer the opportunity to save comments about the icon. The icon window's HyperView may hold either a **Make Comments** command or, if comments have already been saved, a **Show Comments** command. (Section 4.6 discusses HyperViews.) Comments are a good way to document a variable for future reference, leave a note to yourself or to others about changes, or just jot down a few reminders. Many of the variables in the sample datasets have descriptions in their comment records that document their source and units of measurement and provide some idea of how they might be used.

You can view the information for several variables at once by selecting all of their icons before choosing the **Info** command.

4.9 Sliders

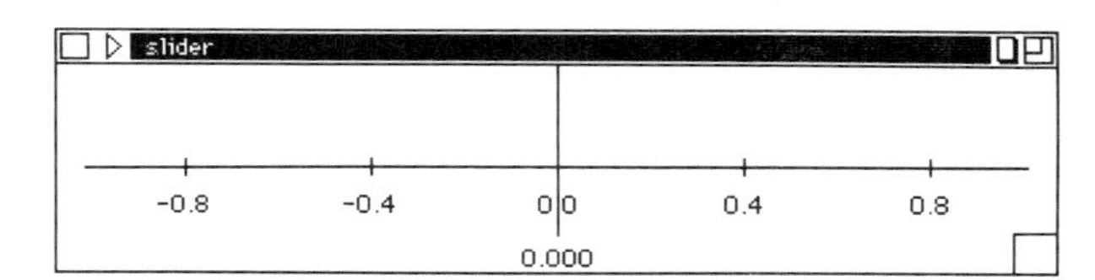

Figure 4-5. *Sliders are building blocks for customized dynamic displays and analyses.*

Sliders are tools with which you can design your own dynamic displays and computations. A slider window holds a horizontal axis intersected by a hairline. You can grab the axis with the ☝ tool and slide it side-to-side. As it slides, it displays the value at which the hairline crosses the axis and publishes it as the value associated with the slider's name. If that name is used in a derived variable expression (See Chapter 11), the derived variable takes on a new value whenever the slider is moved.

Section 11.13 discusses sliders in more detail and provides an example.

4.10 Dependencies

Data Desk icons often use other icons. For example, plots use the variables they display. In general, results are computed from some underlying variables. In a scatterplot of Income *vs.* Age, both Income and Age are used by the scatterplot. If either changes (for example, because you open the variable and change a value), the scatterplot will show the ! symbol and offer to update or redo in a new window.

> REMEMBER
>
> The {Special ▶ Locate} **Users of** command finds and selects the icons that directly use the currently selected icons.

Data Desk keeps track of which icons use which other icons so that all windows know if the icons they use have changed. The {Special ▶ Locate} **Users Of** command selects and locates all icons that use the selected icons, opening and scrolling icon windows if necessary to make them visible. Issue the command again to see the next level of use.

Similarly, {Special ▶ Locate} **Arguments of** selects and locates the icons of objects used by the selected icons. Issue the command again to see the next level of dependence.

4.11 *Datafile Details*

Data Desk opens your datafile only to read from it. All changes you make and all results you create during an analysis are saved in a temporary file. If something goes wrong while you are using Data Desk — even if the power plug is pulled from the wall — your original datafile should be safe. The changes are copied into the main datafile when you Save or Close the datafile or Quit Data Desk.

REMEMBER

Your datafile is vulnerable to errors and power failure only during a **Save** or **Quit.**

The **Save Datafile** command in the **File** menu copies all changes into your datafile immediately, thereby protecting them from future failures. *During a Save, your datafile is vulnerable to errors and power failures.* The **Save Datafile As...** command saves the current version of the data, including any changes not yet recorded in the datafile, under a new name. The original datafile (under its original name) remains unchanged. For absolute datafile security, use the **Save Datafile As...** command to create a copy of your datafile, and work with the copy. **Save Datafile As...** requires more disk space because you must have two full copies of your data, but it protects your original datafile from damage.

TIP

When closing a datafile, a dialogue asks if you want to save changes before closing. **⌘N is a shortcut for No.**

No changes you make to your data or results you produce are saved permanently until you issue a **Save Datafile** *or* **Save Datafile As...** *command, or Quit and Save changes in Quit's exit dialog.* (The **Store** and **Revert** commands in the **Data** menu, on the other hand, operate only on Data Desk icons and not on the datafile.)

Remember, the student version is limited to 15 variables and 1000 cases. The program will not allow you to exceed these limits when using either the **Save Datafile** or **Save Datafile As...** commands.

The **Compact** command in the **Special** menu compacts icon contents by discarding information that Data Desk ordinarily keeps in the datafile but that could be reconstructed. Compacting can shrink a datafile by as much as 50%, but slows down the next computation using compacted icons because the information must be reconstructed. You may want to compact datafiles for storage or backup, or when you are low on disk space.

REMEMBER

***Always* keep a backup copy of your data.**

Despite the safety features built into Data Desk, *you should **always** keep a backup copy of any datafile that would take substantial time to recreate.* The safest backup copies are those preserved on another disk that is not in the computer. A disk placed safely on a shelf cannot be harmed by any computer failure. This is the *only* full guarantee against disk damage resulting from power failures, physical shocks, or unanticipated interactions among programs and desk accessories.

The **Revert To Saved...** command in the **File** menu discards any changes you may have made since the last **Save Datafile** or **Save Datafile As...** and restores your datafile to what was last saved. If you have not saved the datafile, Revert To Saved... restores it to the form it

REMEMBER

No changes you make are saved permanently until you **Save Datafile**, **Save Datafile As...**, or **Quit** (and Save changes).

was in when first opened. The Revert To Saved... command is in the *File* menu and thus it operates on Data Desk datafile. (The *Data* menu has a **Revert** command that discards changes made to individual data values in variable editing windows during the session, but does not directly affect the state of the datafile.)

4.12 *Memory Requirements*

Data Desk does not need to fit all of your data into the Macintosh random access memory (RAM). It only needs room in memory for the variables required by the current computation. Some computations have even lower memory requirements. When running alone, Data Desk should have ample room in memory for most analyses. When run under MultiFinder or System 7, Data Desk requests a partition of memory that should be large enough for most operations.

Data Desk does need room in memory for all of the data in a text file being Imported. It needs room in memory for all variables exported together in a single Export or Copy command.

Some Data Desk calculations, such as ANOVA, can generate large intermediate results that may require larger amounts of memory.

Data Desk makes efficient use of internal memory and disk storage. If it needs more memory for any operation, Data Desk saves data and parts of the program on the disk to make more room. If you are using a floppy disk, such *swapping* of information to disk can slow down the operation of the program. It is always a good idea to give Data Desk as much memory as possible to improve its speed and efficiency. Nevertheless, Data Desk can analyze a datafile too large to fit in the available memory if there is enough disk space to hold the data and analysis results.

On the rare occasions when Data Desk runs too short of memory, it puts a special alert on the screen and tries to continue. You should immediately close any windows you can and quit the program as soon as possible. For greatest safety, use **Save As...** rather than **Save** to preserve your data. During the memory crisis, Data Desk plots work very slowly and may not perform all functions. If you continue working during a memory crisis, Data Desk may run out of memory entirely. In such a case, the program would stop working. Your original data would remain unharmed, but any changes you made since the last Save could be lost.

OK
Data Desk is almost out of memory. Please close excess windows. If possible, please Quit Data Desk and restart with a larger memory allocation. The previous operation may not have completed successfully. Until the memory crisis is over, plots may not work well.

Figure 4-6. *When Data Desk runs short of memory it aborts the operation and posts this alert.*

MultiFinder and System 7 let you ask for a larger memory partition for Data Desk. (In Finder, ask for Info about the Data Desk icon.) If Data Desk lacks the memory to complete your analysis while running under Multifinder or System 7 expand the memory allocation and try again.

If you need more memory but you are not using an environment that allocates a specific memory partition, check that your RAM cache is turned off and remove any RAM disk you may be using. (The Control

Panel desk accessory offers controls for the RAM cache and for some RAM disks.) If there are no other ways to provide more memory for Data Desk, then consider working with fewer cases. You can also provide as much memory as possible for your analyses by closing windows as soon as you are done with them.

4.13 *Multiple Window Modifier*

Data Desk commands ordinarily operate on the icons selected or on the frontmost window. However, some commands can operate on several windows. To apply a command to several windows, first select the icons of those windows. (Click on the icon aliases or on the original icons.) Then hold the Shift key down while issuing the command.

For example, to close selected windows, select the icons of the windows you wish to close, hold the Shift key, and choose {Data} **Close.**

Only some commands can be extended to several windows in this way. However, it can do no harm to select several window icons, depress the Shift key, and try a command to see if it works on all selected windows.

Appendix 4A
Hints and Shortcuts

This appendix discusses ways in which you can improve your efficiency using Data Desk.

Command Keys

Many commands in Data Desk can be issued by holding down the command key (marked with a ⌘) and pressing another key. These "command-key equivalents" appear next to the commands in the menu. For example, the **Undo**, **Cut**, **Copy**, and **Paste** commands in the **Edit** menu can be issued with ⌘-Z, ⌘-X, ⌘-C, and ⌘-V, respectively.

Tab and Enter

Whenever a dialog has an **OK** button or an outlined button, pressing Enter or Return is equivalent to pressing that button.

In dialogs that ask you to type something, Tab skips to the next item to be typed, so you need not remove your hands from the keyboard. Tab also advances to the next icon to the right when you are editing icon names on the Data Desktop.

Dragging several icons at once

To drag several icons at once, select them, pick up any of them, and drag. A gray outline shows the icons gathered together under the mouse arrow. You can drag several icons into another icon window, the icon of another window, or the trash icon, even if they come from different windows. You can only drag icons into a relation window if they have the right number of cases to conform with that relation.

Keeping disk files small

Disk operations slow down as datafiles grow and as free space on the disk is used up. It is always a good idea to discard unused or unnecessary icons and unneeded files. You must empty the Trash to recover the disk space.

Progress reporting

When commands show a rotating cursor while they compute, you can press ⌘-period to abort the operation. ⌘-period also aborts some other operations even when no rotating cursor is visible.

Several of the more time-consuming commands provide visual feedback in the form of a count-down "clock" or "thermometer" that appears below the datafile icon on the Data Desk Desktop.

Important warnings:

Never work with the original Data Desk disk. *Always* copy the disk first and work with the copy. It is very tempting to "just try out the program once", but please protect your investment by making a backup copy.

Never work with the original copy of any important datafile. *Always* copy the datafile first and work with the copy. Despite the precautions taken by Data Desk, there is always a risk, as with any program, that a failure could damage your datafile. The only safe datafile is one that is on a disk in a safe location (i.e., not in the computer). For additional safety, you should keep two or three copies of any important datafile (the latest version, the one before that, and the one before *that*) on separate disks.

APPENDIX 4B
Hot, Warm and Cold Objects

Some Data Desk objects and windows change instantly to reflect changes in other windows, such as changes in data values, changes in expressions, or changes in a plot. Others offer to update, but do not change until you specifically request the change. Still other objects change only when you change them directly. In common computer interface parlance, these are *hot*, *warm*, and *cold* responses to changes, respectively.

Data Desk balances among hot, warm, and cold responses to give you the greatest flexibility in analyzing your data while minimizing the chance that you will inadvertently analyze the wrong version of the data. This appendix discusses the specific assignment of updating methods.

COLD OBJECTS

Cold objects are the bedrock of your analysis. In many ways they define what you are working with. In Data Desk, the fundamental cold objects are variables and expressions. The only way to change your data is to explicitly edit it.

Similarly, the expressions in derived variables that define their structure can only be changed by explicitly editing them. There is a distinction between the defining *expression*, which is an algebraic formula, and the *values* generated by evaluating the expression, which are numbers or text for each case.

HOT OBJECTS

Hot updating is common in spreadsheets, where a change in one cell instantly changes others. It has the advantage that the relationship between two objects with a hot connection appears to be almost a physical link. It has the disadvantage that the updating can occur before you have a chance to look at it, so you may miss seeing the change and lose the ability to compare "before" and "after" views or the freedom to decide not to make the change.

In Data Desk, derived variable *values* are always hot. Any change in the values of variables used in a derived variable expression immediately changes the values generated by the derived variable. This reflects the basic concept of a derived variable as a function of other variables.

By maintaining a hot link between the basic underlying data values and the values generated by evaluating derived variable expressions, Data Desk reduces the chance that you might correct a data value but forget to correct calculations that depend upon it.

Several operations on plots are hot. For example, the selection of data points immediately selects those points in all other plots and editing windows. This makes brushing and slicing possible. (Chapter 9 discusses ways to work with plots.) The assignment of plot symbols or colors immediately changes the plot symbol or color of the corresponding points in all plots. Hot links such as these help to preserve the impression that different plots are showing different views of the same data.

WARM OBJECTS

Data Desk extends the usual dichotomy of hot or cold links to include warm links. Warm links *offer* to recompute as soon as the underlying data or expressions are changed, but actually recompute only when you tell them to. Warm links are primarily part of the HyperView links among windows. Changing data values in a variable or the formula in a derived variable immediately notifies all analyses and plots that depend upon those values or formulas that they may want to recompute. The analyses and plots then display a ! symbol in their window to alert you that they no longer reflect the current state of the data.

Click on the ! symbol to see a HyperView menu that offers to update the window in place or to redo the calculation or drawing in another window. Redo is offered so that you can easily compare the "before" and "after" views of the analysis, and is often the best choice.

Warm links have the advantage that you can observe the consequences of editing the data or formulas, and that you need not suffer those consequences if you do not choose to. In fact, you may choose to update only some of your plots and tables.

Many windows offer a global HyperView **Automatic Update** command to change from warm linking to hot linking. When a window is hot, its global HyperView button is gray.

APPENDIX 4C
Data Desk Limits

Data Desk adapts dynamically to the size and configuration of the computer on which it is running. As a result, it is difficult to define the limits of the program exactly. This technical appendix describes some of the limits as precisely as possible. The Student Version of Data Desk has limits imposed on it that to not apply to the full Professional Version. The guides given here should be taken as "rules of thumb" rather than as absolute limits, unless a specific limit is specified.

> TIP
> To alter the memory allocated to Data Desk by MultiFinder or System 7, select the Data Desk icon on the Finder desktop, Choose **Info** from the **File** menu, and edit the Application Memory Size. Then restart Data Desk.

MEMORY
Data Desk requires at least 512K of RAM, and uses all the memory provided. Data Desk manages memory dynamically, trading space among program, data, and results. The standard memory allocation should be adequate for all student needs.

VARIABLES
Each Data Desk Student Version datafile is limited to 15 variables. The limit is imposed upon Saving the file. During a session you may generate and use any number of variables.

CASES
Variables in the Data Desk Student Version can have up to 1,000 cases.

WINDOWS
Data Desk can open as many windows as memory allows. However, it is a good idea to clean up your desktop occasionally.

ANALYSIS LIMITS
Most Data Desk analyses have no fixed limits. Regressions can have as many predictors as memory space allows. Contingency Tables can use variables with up to 50 categories. Data Desk warns you if a requested analysis is too large for your system.

CHAPTER 5

Entering and Editing Data

5.1	Entering Data for One Variable	*53*
5.2	Editing a Variable	*54*
5.3	Extended and Discontinuous Selection	*55*
5.4	Editing Several Variables	*56*
5.5	The Editing Sequence	*56*
5.6	Moving Around	*57*
5.7	Cutting, Copying, and Pasting	*58*
5.8	Undo	*58*
5.9	Updating Relations	*58*
5.10	Scrolling	*59*
5.11	Managing Windows	*59*
5.12	Details of Editing	*59*
5.13	Shifting Cases	*60*
5.14	More on Linking	*60*
5.15	Finding and Replacing Cases	*61*
5.16	Store and Revert	*62*
5.17	Printing Variables	*63*
5.18	Example	*64*
5.19	Example: Editing Data	*66*

APPENDIX

5A	Configure Editing	*69*

DATA DESK'S VARIABLES are always immediately available. You can read, type, copy, or paste new data values at any time. This chapter discusses data entry and editing. In Data Desk there is no difference between the two. You view, edit, alter, and correct data in exactly the same way as you type in new data, and these methods closely resemble the methods you probably already know for entering and editing text in a text processor or numbers and equations in a spreadsheet.

As a student, you will usually work with the datasets provided with Data Desk or with datasets provided by your professor. Thus, you may not need to read this chapter until later on in the course.

While Data Desk's full data editing capabilities are very powerful, basic data entry and editing are quite simple. If you are just beginning to learn Data Desk, you may want to read only selected sections of this chapter and then work with data imported from other sources (see Chapter 6) or with the example datasets. You can learn basic skills easily and add to them as you go.

The beginning of this chapter describes data entry and editing operations. The final sections show an example. You may want to turn to the example as you read the descriptive sections, and even work through parts of the example to see how the operations work.

5.1 Entering Data for One Variable

To create a variable, choose {Data ▶New} **Blank Variable** from the **Data** menu. If you have just launched Data Desk, press the **Enter Data From Keyboard** button.

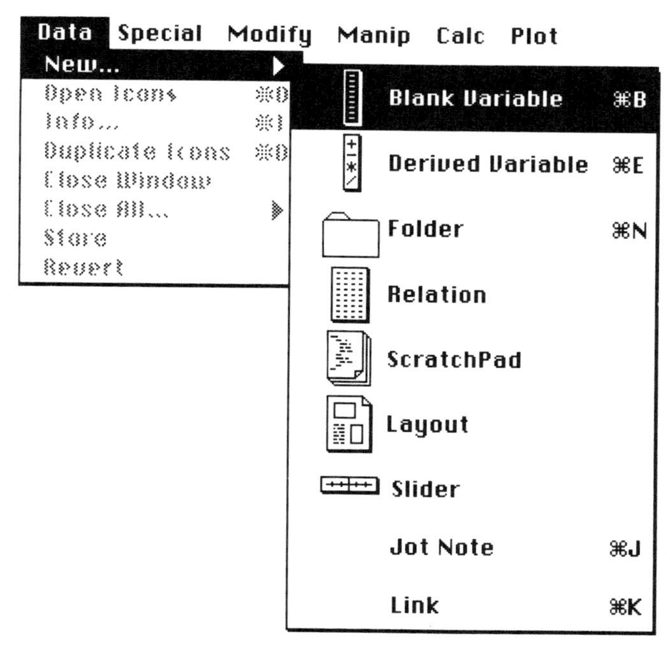

Figure 5-1. *To create a variable choose {Data ▶New...}* ***Blank Variable****.*

Data Desk displays a dialog requesting a name for the variable.

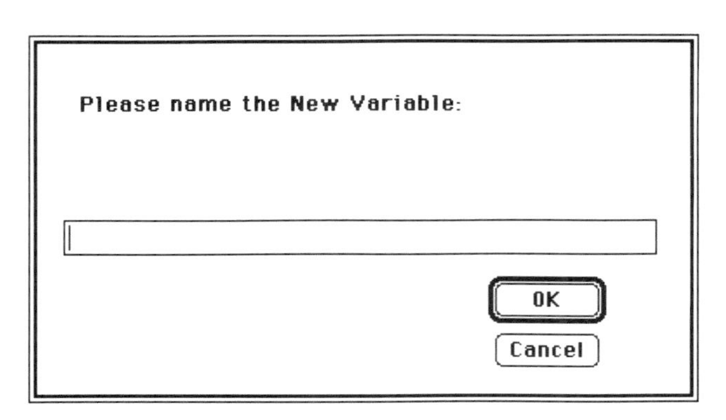

Figure 5-2. *Data Desk asks you to name a new variable.*

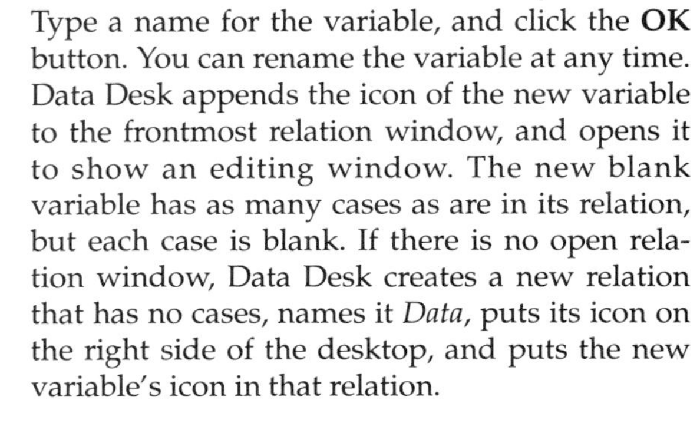

Type a name for the variable, and click the **OK** button. You can rename the variable at any time. Data Desk appends the icon of the new variable to the frontmost relation window, and opens it to show an editing window. The new blank variable has as many cases as are in its relation, but each case is blank. If there is no open relation window, Data Desk creates a new relation that has no cases, names it *Data*, puts its icon on the right side of the desktop, and puts the new variable's icon in that relation.

To enter data, type values one row at a time, ending each row by pressing the Return key. You can make the window wider by dragging the size box to the right, or make it longer by dragging the size box downward. As you type, you replace the old blank cases or, if you are at the bottom of the variable window, you append new cases to the relation.

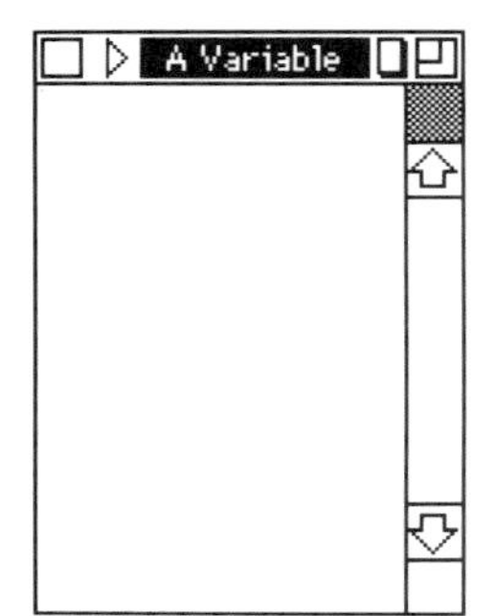

Figure 5-3. *A new variable editing window.*

When you are done typing, click the close box in the upper left corner of the window or choose {Data} **Close Window**. The editing window closes into the variable's icon.

Most statistics are computed on numbers, but variables may hold ordinary text as well. For example, a variable might hold the name of each respondent in a survey, or the name of each sales region in a set of sales data. Some data are best represented as words; for example "male" and "female" for a variable recording gender. Unlike many other programs you may have used, Data Desk lets you mix numbers and text freely in a variable.

Many of Data Desk's commands operate only on numbers, and expect variables to hold numbers. For example, it is impossible to average "rich", "middle class", and "poor", but it is natural to average incomes recorded in dollars. Other commands recognize text — usually to identify different groups or categories of cases. Commands that expect numbers treat non-numeric data values as if they were missing observations.

5.2 *Editing a Variable*

text insertion point

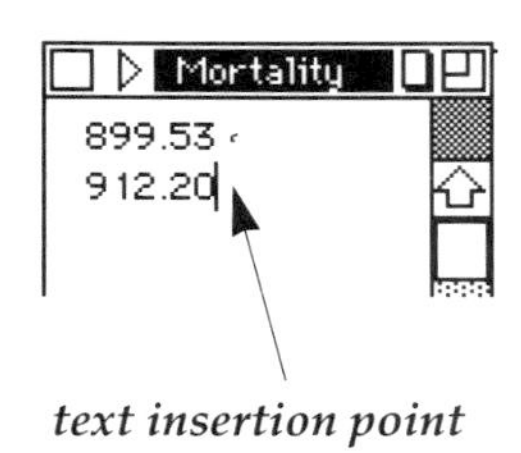

Text editing works in essentially the same way in all Macintosh programs. Whenever a text editing window is frontmost, anything typed is either inserted in the window at the vertical blinking *text insertion point* or replaces text that is selected. The Backspace key deletes the character before the insertion point or the entire highlighted selection.

To alter a single value in a variable, edit it with the standard Macintosh methods. Click between two characters in the text of a case to place a text insertion point between them, or drag across characters to select them. Type or paste to insert text at the blinking cursor or to replace the selected text.

case insertion point

Data Desk extends these Macintosh conventions to data cases. Thus, to insert a case between two other cases, click between the cases to place a horizontal blinking *case insertion point* between those cases. Type or paste text to start a new case at that point.

You can drag up or down across several cases to select them. If you drag off the top or bottom of a window, it scrolls automatically and selects cases as they become visible. Type or paste text to replace all of the selected cases and begin inserting new cases at that point. Press Return to begin a new case.

You can tell whether a click will place a vertical or horizontal insertion point by the orientation of the mouse cursor. A vertical "I-beam" places the vertical text insertion point, while a horizontal "cross-beam" places a horizontal case insertion point. As you move the mouse up and down along the cases in a variable, the cursor alternates between these two shapes.

5.3 Extended and Discontinuous Selection

To select a range of text, click at one end of the text and then hold down the Shift key and click at the other end of the text. Shift-clicks extend text selections in editing windows in the same way that they extend icon selection in an icon window.

You can extend a selection of cases with a shift-click as well. Click at one end of the range of cases and then shift-click at the other end. All the cases between the two clicks will be selected.

discontinuous selection

Data Desk offers an additional case editing feature that is not commonly available for text but is available in some spreadsheets. If instead of holding down the Shift key you depress ⌘-key, the cursor changes to a cross: . A ⌘-click on a case adds it to the set of selected cases, but does not select the intervening cases. A *discontinuous selection* such as this provides many powerful capabilities. Discontinuous selections can also arise from a **Find** command (see Section 5.16) or from selecting cases in plots. (See Section 9.4.)

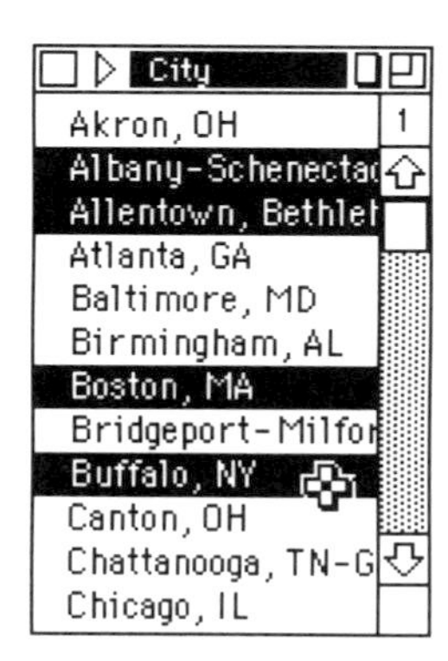

Figure 5-4. Discontinuous selection.

Discontinuous case selections do not correspond directly to standard Macintosh text editing. When you edit text, there can be only one continuous selection, so you always know which parts of the text will change when you type or paste. When cases are selected discontinuously, you no longer know where typed or pasted text should go. Consequently, you may not type or paste text into a discontinuous selection. If you cannot type or paste into your variable editing windows, it might be because what appears to be a simple continuous case selection is in fact part of a discontinuous selection with other selected cases scrolled beyond the screen. Try again to select the cases you wish to replace.

You can Cut, Copy, or Clear discontinuous cases, *but you cannot Undo the operation.* Cut or copied cases form a data table on the Clipboard and can be pasted elsewhere in the variable windows, pasted directly into an icon window as variables, or moved to another program from the Clipboard. (See Chapter 6 for information about importing, exporting, and data tables.)

If you are editing text (there is a blinking vertical text insertion point or some text in one case selected) and you shift-click with a cross-beam cursor to select several cases, the entire case in which you were editing text becomes one end of the selection. However, if you ⌘-click another case, the case in which you were editing text *remains* selected, and the case you clicked on is selected also. Data Desk behaves in this way to reduce the chance of your accidentally including a case in a discontinuous selection.

5.4 *Editing Several Variables*

You may wish to edit several variables together. While you can open variables from different relations at the same time, it only makes sense to edit together variables from the same relation. Because all the variables in a relation refer to the same cases, if your editing deletes, adds, or changes the order of any cases, those changes affect the entire relation — both the variables that are open and those that are not.

TIP

Windows "remember" where they were before.

Hold the Option key down while opening to make variable windows align.

Usually, when you open two or more variables they are arrayed neatly across the screen. You can force variable windows to align in such a table by holding down the Option key while opening them. Editing windows for variables in the same relation scroll together so that the same case number is at the top of each window. As long as the windows are aligned, each *row* across all open windows in the relation represents a *case* in that relation. You can align data editing windows that are already open with the {Special ▶Arrange} **Align Editing Windows** command, but the command only shifts windows up and down, so you must first position them across the screen as you wish.

5.5 *The Editing Sequence*

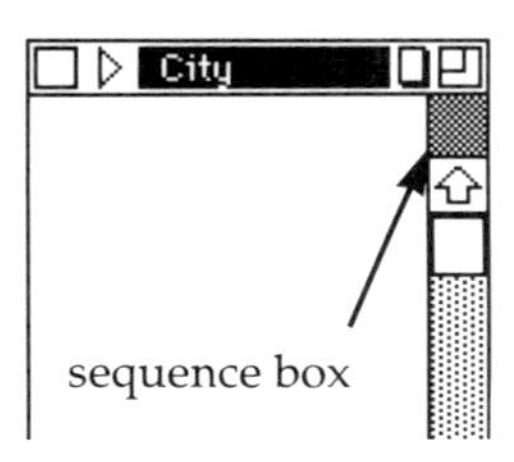

To enter data values one variable at a time, create each variable in turn, type data into it, following each case with a Return, and close the variable when you finish.

You may prefer to enter the values for each case moving along a row extending across several variables. You can create or open as many variables as you wish. Their windows may overlap and you may resize them as you wish.

editing sequence

The Tab key moves the text insertion point to the next variable. The *sequence box* in the upper right corner of the window, at the top of the scroll bar, specifies the order of the variables in the *editing sequence*. A variable whose sequence box is gray is not in the editing sequence, and will be skipped over by the Tab key. If the sequence box holds a number, then this number specifies the place of the window in the editing sequence. (See Figure 5-5.)

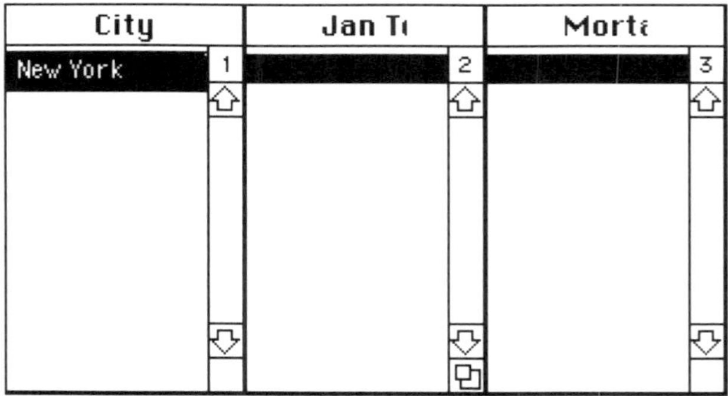

Figure 5-5. *Sequenced variable editing windows.*

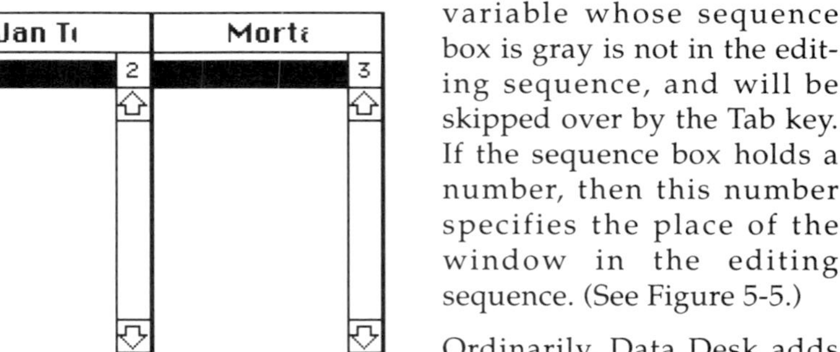

Ordinarily, Data Desk adds newly opened editing win-

HOW-TO

To Move to the Next Variable in the Editing Sequence:
- Press Tab

To Start a New Case:
- Press Return.

TIP

If variables do not join the editing sequence, look for an open variable from another relation, establishing *that* relation as the one with the editing sequence. Close the other variable to make a new sequence in another relation.

dows to the end of the editing sequence. To remove a window from the editing sequence, click on the sequence box; it will turn gray. The {Special ▶ Sequence} **Clear Editing Sequence** command clears the entire editing sequence. The {Special ▶ Sequence} **Remove From Sequence** command removes the frontmost window from the editing sequence. To add an editing window to the end of the sequence, click the sequence box or bring the window to the front and choose {Special ▶ Sequence} **Add to Editing Sequence**.

All variables in the editing sequence must be in the same relation. If a newly opened variable is not added to the editing sequence, check to be sure that it is in the same relation as the variables in the editing sequence. Data Desk objects if you try to add a variable from a different relation to the existing editing sequence.

When variable editing windows are sequenced, the Tab key advances to the next window in the sequence at the current case. You can always tell which variable you are editing because its window is active. Thus, you can type a value in the first variable, press Tab, type the value for the same case in the second variable, press Tab, and type the value for the third variable, and so on.

Variable editing windows that are not in the editing sequence are skipped over by the Tab key, but if they are in the same relation, they still scroll with the other variable windows and are still affected by copying, cutting, clearing, and pasting cases.

5.6 *Moving Around*

Pressing the Return key returns the cursor to the first variable of the editing sequence and makes its window active. The Return key either selects the next case for editing or places a case insertion point after the current case. The **Preferences...** command in the **Edit** menu lets you choose which it should do (See Appendix 5A about **Preferences...**).

The Tab key always moves to the next variable in the editing sequence. Usually, it selects the value at the same case for editing. However, if the text entry point is in the last variable of the editing sequence or in an unsequenced variable window, Tab "beeps" and then behaves like Return.

In addition to Tab and Return, you may want to use the Enter key, which is on the numeric keypad. The **Preferences...** command lets you specify whether the Enter key should work like the Tab key or like the Return key.

In addition to using Tab, Return, and Enter as editing keys, Data Desk supports the cursor control (arrow) keys found on many Macintosh keyboards. The ↑ and ↓ keys move to the previous and subsequent *case* in the current editing window, respectively, selecting the entire case for editing. The ← and → keys move to the previous and subsequent *variable* in the editing sequence, respectively, and stays at the same case.

Of course, you can always place an insertion point anywhere by simply pointing and clicking. Or you can select an entire value by pointing and double-clicking.

5.7 *Cutting, Copying, and Pasting*

HOW-TO

To move cases from one place to another:

- Select the cases to be moved. They may be discontinuous.
- **Copy cases**.
- Click between two cases.
- **Paste Cases**.

The **Cut**, **Copy**, and **Paste** commands in the **Edit** menu work the way they do in most Macintosh applications, except that they also operate on selected cases in variable windows. When operating on cases rather than on characters in text, these commands read **Cut Cases**, **Copy Cases**, and **Paste Cases**.

The Cut Cases and Paste Cases commands operate on the entire relation. If you cut cases from one or more variables, those cases are removed from the entire relation. If you paste cases into one or more variables, extra cases (filled with the missing value indicator, •) are created in the other variables in the relation. It is easy to move cases around in a relation by cutting them from one place and pasting them into another. You can work in a single variable; cases in the entire relation will be re-ordered.

The Cut Cases and Copy Cases commands can also be used to place selected cases from Data Desk onto the Clipboard. Only cases from variables in the Editing Sequence are placed on the Clipboard. The Editing Sequence serves to specify the order of the cases. The case values form a data table and are separated by the *data table delimiter*. Section 6.4 discusses this delimiter and shows how to set it.

5.8 *Undo*

Most of Data Desk's editing operations can be reversed with the {Edit} **Undo** command. This includes reversing the deletion of cases across several variables. You can only undo the previous operation. The Undo command changes to indicate the kind of operation that can be undone or to report that it **Can't Undo**.

If you choose Undo, you can then choose to "Redo". The menu will specify the action that can be taken. Undo works even if some windows have been closed since the original operation.

5.9 *Updating Relations*

Although Data Desk behaves as if any operation that alters the number of cases or their order in a variable operates on the variable's entire relation, in fact the relation is only changed when necessary. Until the relation has been updated, you have the option of reverting all variables in the relation to their previous state with the **Revert** command in the Data menu.

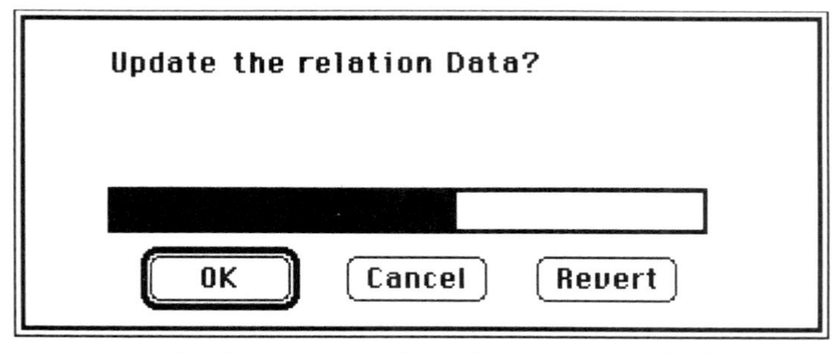

Figure 5-6. *You must confirm changes to a relation before they become permanent.*

Data Desk will not update a relation without your approval. However, Data Desk will do most of the work of updating the relation while you respond to the request. The thermometer indicator in the dialog shows the progress of the update; for large relations updating can be time-consuming. Cancelling the update does not revert the relation to its previous state, but merely aborts the permanent changing of the relation. The Revert button returns the relation to

> **REMEMBER**
> Even after you update a relation, you must use the Save command to store the result permanently in the datafile. Until you Save, you can revert to the last saved version of the file.

the way it was when you first opened variables, or to its state the last time changes were stored.

Ordinarily Data Desk updates a relation when an edited variable is closed. Data Desk also updates relations when it needs to use a variable in the relation or needs to know the size of the relation.

5.10 *Scrolling*

The vertical scroll bar in a variable window moves the text up and down to reveal other cases. Data Desk keeps rows aligned across all variables in the relation. Whenever one variable scrolls up or down, all variables in the relation scroll together. Even if some windows are short and wide and others are tall and narrow, the *top* line of every variable window in the relation always displays data for the same case in each variable.

5.11 *Managing Windows*

active window

When many variables are open at once, the screen can become cluttered. The *active window* — that is, the one that will be affected by typing — is the one with a darkened title bar. It is in front of any other windows that may overlap it (except palette windows), and usually has an insertion point in it or some text highlighted for replacement.

Often variable editing windows are positioned neatly on the screen and the top-left one is the first in the editing sequence. However, you may reposition and resize editing windows as you like. For example, you may have variable editing windows large enough to cover most of the screen. You might then position them as a stack of windows and use Tab to move to the next window in the stack.

> **REMEMBER**
> Variable editing windows must be in the editing sequence for the Tab key to move from one to the next. To add a window to the editing sequence click its sequence box.

No matter what the actual position of variable editing windows on the screen, the editing sequence, and not the windows' positions, defines the actions of Tab and Return.

A window that is in the current relation but is not part of the editing sequence can be very useful. Such a window might hold case identifiers such as the names of the individuals in the dataset or simply the sequence numbers from one to the total number of data values. Case identification windows help you to be certain that the data you are typing is correct for the case in which you are typing. Because Tab skips over windows not in the editing sequence, you can enter or edit data without having to work past the identification window.

5.12 *Details of Editing*

> **TIP**
> Three ways to find a case:
> - Search for it with the **Find** commands in the **Edit** menu
> - Use {Edit}**Go To Case #**...
> - Select it in a plot and use {Edit} **Go to Next Selected Case.**

To edit variables, you must first open them. To correct a case, scroll the variable windows until the case appears. (If you know the case number, the {Edit} **Go To Case #...** command will get you there quickly. If you selected the case in a plot, the {Edit} **Go To Top Selected Case** command scrolls the window to show the case. If you know the data value, the {Edit} **Find** commands may help.) Select the value you want to change, and type the correct value. To alter only a few characters, backspace over them or drag across them and type their replacement. To insert new characters, click where they should be inserted and type. To correct more than one data item in the case, press Tab to move the insertion point to the next variable at the same case.

If you select text in one variable, the corresponding cases in all of the other opened variables are selected for replacement. To replace them, type the new values. To deselect them, click anywhere in a variable window.

To delete one or more cases, select them by dragging the cross-beam cursor in any variable editing window; the cases highlight in all linked variable editing windows. Press Backspace to delete the cases.

To insert a new case in the middle of the data, click in the first variable editing window in the editing sequence between the two cases where the new case should be inserted. Type the first value of the new case and press the Tab key. Type the new value for the second variable and press the Tab key to proceed to the third variable.

To continue inserting new cases conveniently, use the {Edit} **Preferences...** command to set the Return key to start a new case rather than moving to the next case. Then pressing Return (or Tab at the end of the case) will start a new case.

5.13 *Shifting Cases*

TIP

To make a lagged variable:
- Duplicate a variable.
- Open the duplicate.
- Select same number of cases as you wish to lag.
- Choose **Shift Cases Down** from the **Edit** menu.

You can alter the case alignment across variables with the **Shift Cases Up** and **Shift Cases Down** commands in the **Edit** menu. Shift Cases Up and Shift Cases Down shift the cases in the variable editing windows in the editing sequence. If the frontmost window is not in the editing sequence, then it alone is shifted. Shifting realigns variables that may have become misaligned and can create "lagged" variables. Both shift commands expect a continuous range of cases to be selected.

Shift Cases Up deletes the selected cases in the frontmost window and shifts the remaining cases up to fill the hole. Blank cases are appended to the end of the variable to preserve its length.

TIP

You can also lag a variable with the lag function in a derived variable expression.

See Chapter 11 for information about derived variables.

Shift Cases Down shifts downward all cases from the first selected case through the last case in the data, opening up a gap of blank cases in the variable. You indicate how large a gap you want by *the number of cases that you select*. Thus, to open a gap of two blank cases, select two cases and choose **Shift Cases Down**. The two blank cases are inserted just above the selected cases. Cases shifted past the end of the variable are deleted from the bottom to preserve its length. (To make a variable longer relative to other open variables, put it in its own new relation and insert new cases by typing or pasting.)

5.14 *More on Linking*

Data Desk is a highly integrated data analysis environment. One important aspect of this integration is that case selection operates simultaneously in all plot and variable editing windows in a relation. That is, selecting a case in a variable editing window highlights corresponding parts of plots. Selecting a point in a plot selects the corresponding case in any variable editing windows. You can work with both plots and variable editing windows together to identify and edit cases.

For example, to find out more about a plotted point, select it in the plot and open an appropriate variable. The case will be selected for editing. ({Edit ▶ Go To...} **Next Selected Case** gets you there quickly.) If one of your variables holds names or other case identifiers, you can open it, select cases in a plot, and **Copy Cases** from the identifying variable to place a list of the case names on the Clipboard. You can then Paste this list into a ScratchPad or into a report you may be writing with another program.

For example, you might select the best performers on some measure in a plot, open the variable holding case names, Copy Cases, type in a ScratchPad "The best performers are:", and Paste their names into your text.

You can use variable windows to identify points selected in a plot. Select a group of points in a scatterplot and see them highlighted in a variable containing case identifiers. You can also use the **Find** command to select cases that match particular text in a variable; they will highlight in all plots as well.

5.15 *Finding and Replacing Cases*

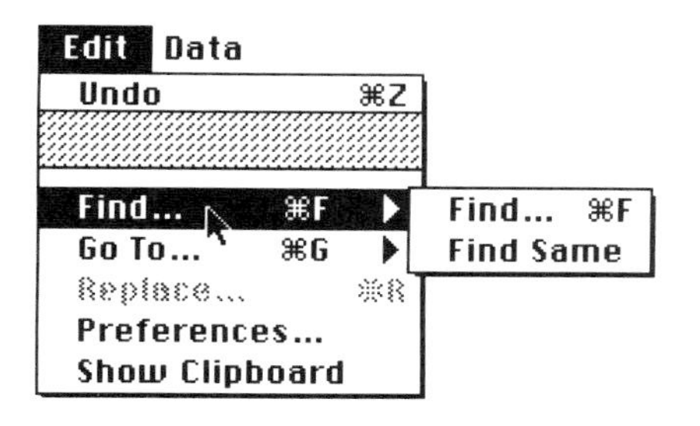

Figure 5-7. The ***Find*** *commands are in the Find submenu*

The **Find...** and **Go To...** commands in the **Edit** menu (Figure 5-7) locate cases in the frontmost open variable editing window. The **Find...** and **Go To...** commands are grouped together into submenus of related commands. Together they offer elementary data management and editing functions to help you work with your data.

Find... prompts for the text to find. According to the setting in the dialog, it either finds all occurrences of the specified text or finds the first occurrence of the text *after the current insertion point.* It selects either the text itself or the entire case for replacement according to choices made in the dialog.

The search does not "wrap" to the first case after checking the last one, so be sure to scroll to the top of the variable and click before the first case if you want to search the entire variable. The **Find next nonnumeric case** option locates cases that are not numbers and would be treated as missing values in a calculation. It is particularly helpful for finding typographical errors.

Find Same moves to the next case that matches the search criteria most recently

Find ☒ all occurrences of OK Cancel
◉ the string:
○ Whole word ◉ Partial word
○ Respect capitalization ◉ Ignore capitalization
○ next nonnumeric case

Figure 5-8. *The* ***Find...*** *dialog offers several searching options.*

specified, but does not prompt for new text or settings. It remains active only while there is text to find.

> **REMEMBER**
> Remember to click before the first case in an open variable window if you want to search the entire window with a **Find** command. Find searches down from the current insertion point or selection.

The **Go To...** submenu contains commands that help you step through cases.

Because cases can be selected easily in any plot or editing window, it is common to have many cases selected that are not continuous. Most of the **Go To...** commands help you to look through the selected cases. The Go To... commands are:

- **Go To Next Selected Case**
- **Go To Previous Selected Case**
- **Go To Top Selected Case**
- **Go To Bottom Selected Case**
- **Go To Case #...**

The **Go To Case #...** command locates a case by its case or row number. The other Go To... commands step through selected cases either forward or backward.

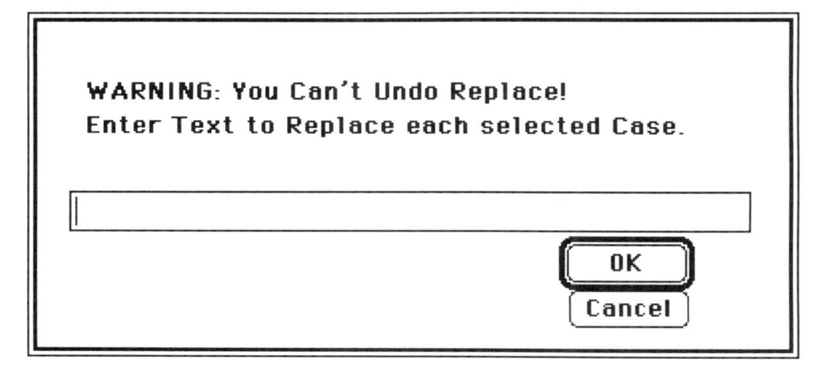

Figure 5-9. The Replace dialog asks for text that will replace each selected case in the front editing window.

The {Edit} **Replace** command prompts for text and replaces *each* selected case in the frontmost window with that text.

Replace overwrites entire cases, not parts of the text in cases. The Replace command is not active unless at least one case is selected. It is not active if you have selected only some of the text of a case. Drag vertically across a case to be sure it is selected.

recoding

Because the Replace command replaces all selected cases in the frontmost window, it provides a way to *recode* data. You can find all cases satisfying some criteria and then replace them with the new value. You can also recode by identifying cases in a suitable plot and replacing them in a variable. Alternatively, you can create a new variable and use **Replace** to recode its (empty) cases to the codes you choose. Another alternative is to recode data with the IF, THEN, ELSE commands and logical operators of derived variables, discussed in Chapter 11.

5.16 *Store and Revert*

As you enter or edit data in a variable window, your changes are held in that window. Any changes you make are immediately available. The variable's icon is active and can be used in most Data Desk operations even though it is gray (indicating that the window is open). Windows that display analyses or plots using a variable immediately offer to update their views when the variable is changed.

The {Data} **Store** command records the current values of the variable and updates the relation. As long as the variable editing window is open, you can discard all changes you have made to a variable since it

REMEMBER

- The {Data}**Store** command records the current values of a variable.
- The {Data}**Revert** command restores the frontmost variable window to its original state.

was opened or last stored and revert to its original values. The {Data} **Revert** command restores the relation to its original state or to the state it was in at the last **Store** command.

Important Note:

Because Data Desk provides its own desktop on top of the standard Macintosh system, there are two levels at which data are saved. The **Store** and **Revert** commands in the **Data** menu record or restore data from the Data Desk datafile. They protect your variables from accidental changes made while variable editing windows are open. You can revert to a previously stored version of a relation as long as it has not been updated.

REMEMBER

The **File** menu commands work with the Macintosh file system. They control the saving, copying, and reverting of Data Desk datafiles, but cannot alter Data Desk icons.

Variables that have been saved with the {Data} **Store** command and those in relations that have been updated are *not* protected from system failures. In fact, even after storing a variable, you can still choose not to "Save changes before closing" when you quit, and return your datafile to its original condition when opened or last saved with a **Save Datafile...** command.

REMEMBER

Only the **File** menu commands can permanently record changes to a datafile on the disk.

The datafile is physically recorded on your disk. The **Save Datafile** and **Save Datafile As...** commands in the **File** menu record the Data Desk datafile on the disk for permanent storage. These commands protect your entire datafile from accidental damage due to system failure.

> To ensure the safety of your editing changes in event of a system failure you must issue the **Save Datafile...** or **Save Datafile As...** commands.

REMEMBER

The **Data** and **Edit** menu commands work with Data Desk icons and windows. They control the creating, saving, copying, editing, and reverting of icons within Data Desk.

Similarly, the **Revert To Saved...** command in the **File** menu works like the **Revert** command in the **Data** menu except that it restores the *entire datafile* to the state it was in at the last save or when opened.

5.17 *Printing Variables*

TIP

The **Text Format** command from the variable window's HyperView menu, lets you specify the font and text size for both printing and editing variables. Select a LaserWriter font for best results when printing on a LaserWriter

To print a variable's contents, select its icon and choose **Print Variables** from the **File** menu. The Print Variables command prints all selected variables as a table with each variable's name at the top of its column. Print Variables reformats tables that are too wide or too long for one page so that they print on several pages, and supports all Macintosh printers. Choose landscape orientation with the {File} **Page Setup** command to fit more columns on a page.

You must have a printer driver file in the System folder for these commands to work. (Refer to the Apple manuals for your printer.)

REMEMBER

- {File} **Print Variables** only prints variables.
- {File} **Print Front Window** prints a display.

Data Desk's {File} **Print Variables** command only prints variables. To print a display, make its window frontmost and use the {File} **Print Front Window** command. This command prints the contents of any Data Desk window.

Chapter 6 of this book discusses other ways of printing and saving windows in Data Desk.

5.18 *Example*

Data entry and editing are easy to learn by example. If you know how to edit text on the Macintosh, you will find that editing data in Data Desk uses all of the standard Macintosh text editing conventions and extends them to include the editing of cases across several variable editing windows and the manipulation of discontinuous case selections.

The step-by-step example given here shows a data entry and editing session. You can work along with the example doing each of the things suggested and matching the figures against how your screen looks.

At any step in the example you can save your work. The first time you save (with the {File} **Save Datafile...** command) Data Desk will ask you to name the datafile. To resume work at a later time, open the saved datafile. The windows on the desktop will reopen automatically to the places they occupied when you saved the file.

(1) To start the example, **Open** Data Desk from the Finder. The initial screen offers a choice of the most common ways to put data into Data Desk. Choose **Enter data from keyboard**.

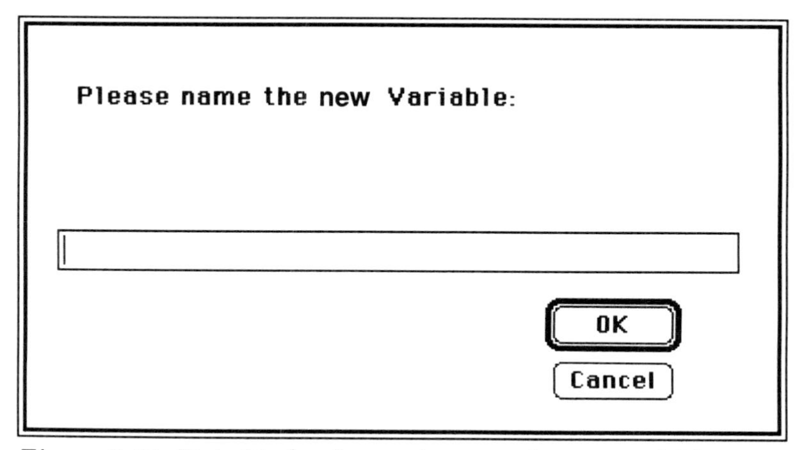

Figure 5-11. *Data Desk asks you to name the new variable.*

Data Desk asks you to name the first variable (Figure 5-11). Name it *City*.

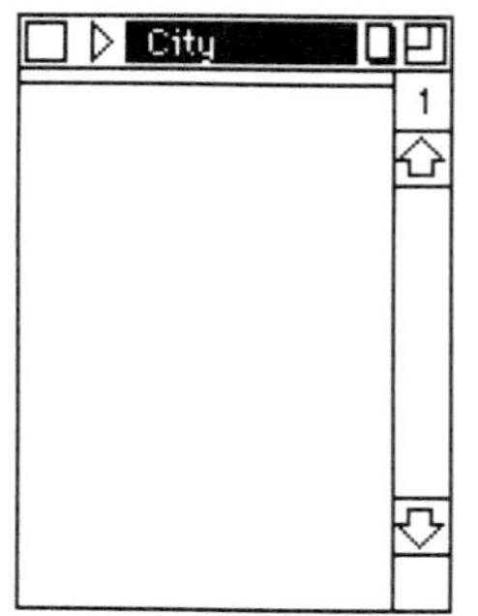

Figure 5-12. *The variable "City" is opened and becomes the frontmost window.*

Data Desk creates a new relation, names it Data, places the icon of the new variable in the relation, and opens it for editing.

(2) Create more variables to hold the data. Choose **New Blank Variable** from the **File** menu and name it *Jan Temp*. Data Desk places *Jan Temp* in the same relation, next to *City*.

(3) Press ⌘–B to create another new blank variable. (Hold the ⌘-key down and press the B key.) Name this variable *Mortality*.

The editing windows should now look like Figure 5-13.

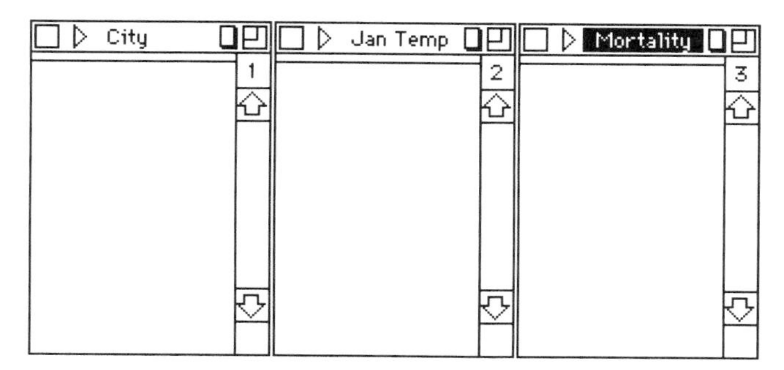

Figure 5-13. The Jan Temp and Mortality variables.

As each variable is created it is automatically added to the editing sequence. The editing sequence numbers at the top of the vertical scroll bars specify the order of the windows for the Tab key and for Copying or Pasting cases.

(4) Point to *City* and click anywhere in the window. The title bar of the City window highlights, and a flashing horizontal case insertion point appears at the top of the window to indicate that this is where the new case will appear.

(5) Type the name of the first city: Akron. If you make a mistake, backspace and correct it. As soon as you start typing, Data Desk creates a new case in these three variables and makes room for it in the windows. The new case is highlighted across all three windows and the cursor changes to a vertical blinking bar where you are typing.

(6) Press Tab to move the insertion point to the second window, which now becomes active. Its title bar highlights and the flashing bar indicates that you can type its first case.

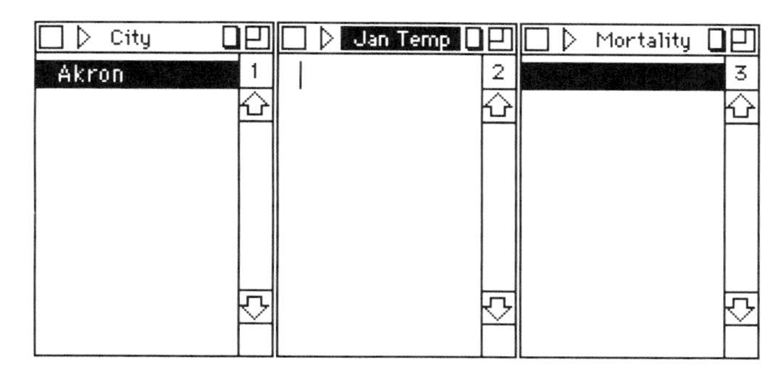

Figure 5-14. Akron highlighted with insertion point at second window.

(7) Type Akron's January temperature: -2.78. (The temperatures are in degrees Celsius.)

(8) Press Tab to move to the *Mortality* window, and type Akron's age-adjusted mortality rate: 921.87.

(9) Press Return to return to the *City* window to place a case insertion point after the case just typed. When you start typing the name of the second city, Albany, Data Desk inserts the new case. Continue entering data for the first 10 cities until the windows look like Figure 5-15.

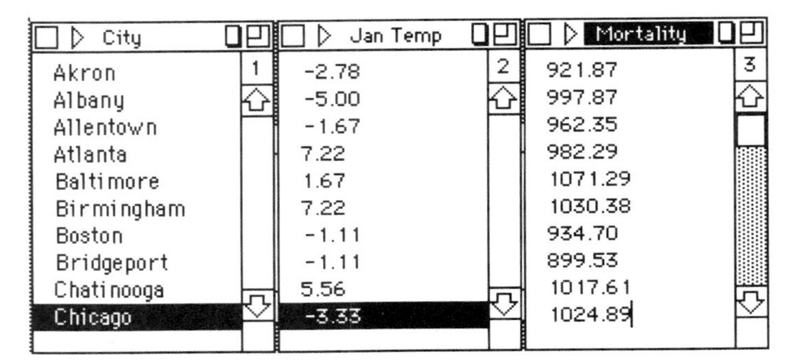

Figure 5-15. *Data for 10 cities.*

(10) Now when you press Return and type, the cases scroll up to make room for the new case. You can use the scroll bar to move up or down through the rows of data. Continue entering the next several rows of data. The scroll bar moves with you so that it is always in the active window. The next seven cases are shown in Figure 5-16.

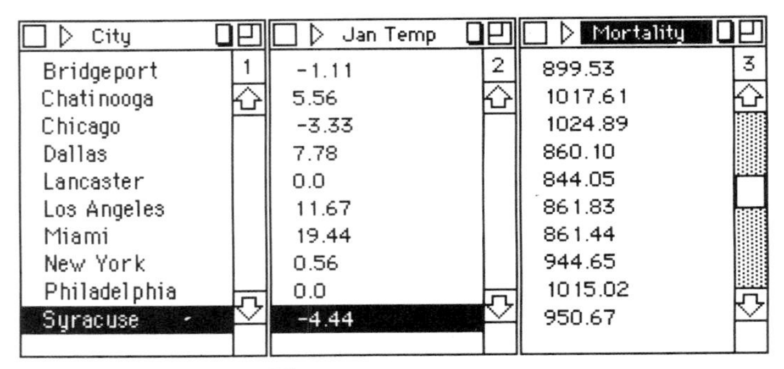

Figure 5-16. *The next seven cases.*

(11) Scroll the windows up and down.

5.19 *Example: Editing Data*

(12) If you have followed along *exactly,* then you copied the misspelling of Chattanooga. Click in the City window and drag across the incorrect characters.

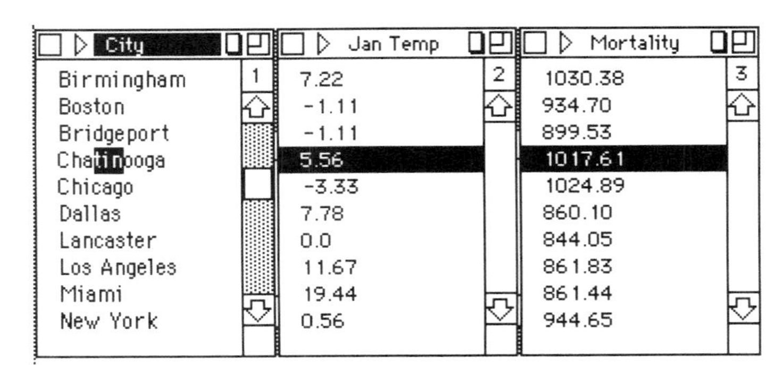

Figure 5-17. *Chattanooga misspelled.*

(13) Type the correct characters. They will replace the highlighted ones.

(14) Insert a new case after Miami.

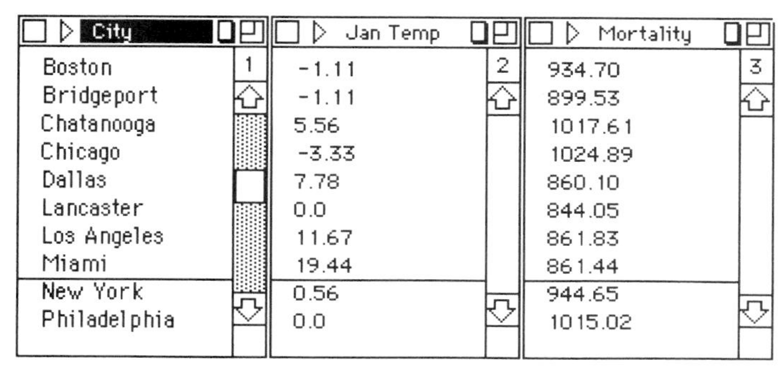

Figure 5-18. *A case insertion point after Miami.*

- Click the City window to make it frontmost.
- Move the mouse between "Miami" and "New York" until it shows a cross-bar cursor: ⟷.
- Click between the Miami case and the New York case line to identify the place to insert a new case. You should now see a blinking horizontal case insertion point.

(15) Type "New Orleans". As soon as you start to type, Data Desk makes room for the new case and selects it for editing. Inserting cases in variable editing windows is analogous to inserting characters in text; click where you want to insert, and type.

(16) Press Tab to move to Jan Temp and type 12.22. Press Tab again to enter New Orleans' mortality rate, which is 1113.16.

(17) Press Return to start another case, and enter the data for Minneapolis (Jan temp = -11.1, Mortality 857.62).

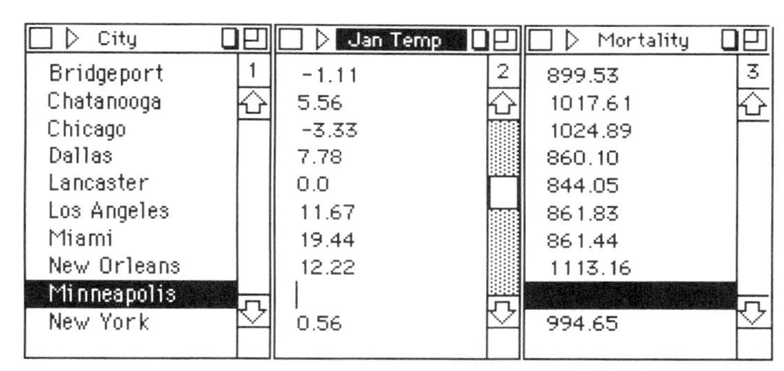

Figure 5-19. *New cases being added.*

(18) Select Minneapolis. Click above "Minneapolis" with a cross-bar cursor and drag down across the case. The entire case is selected and highlights.

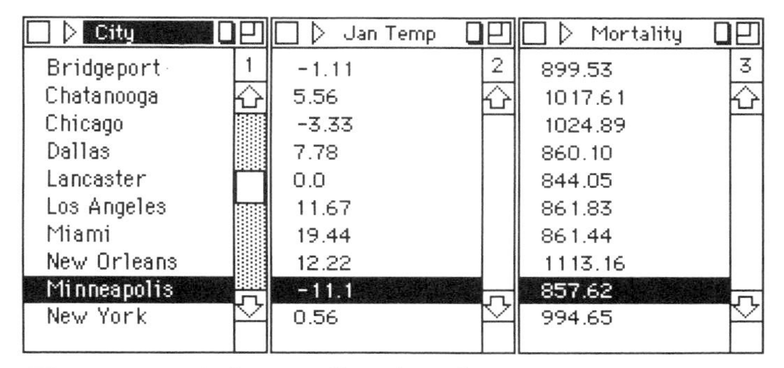

Figure 5-20. *Minneapolis selected.*

(19) The **Cut** command in the **Edit** menu now reads **Cut Cases** indicating that you have selected an entire case rather than characters in text. Choose Cut Cases to remove Minneapolis from the open variables.

(20) Most editing operations can be Undone. The **Undo** command in the **Edit** menu changes to indicate the previous operation. Choose Undo to restore Minneapolis to all three windows.

(21) Cut Minneapolis again to move it to its proper alphabetical location before New Orleans:

- Repeat steps 18 and 19.
- Click above New Orleans with the cross-bar cursor and choose **Paste Cases** from the **Edit** menu. The entire Minneapolis case appears at the case insertion point.

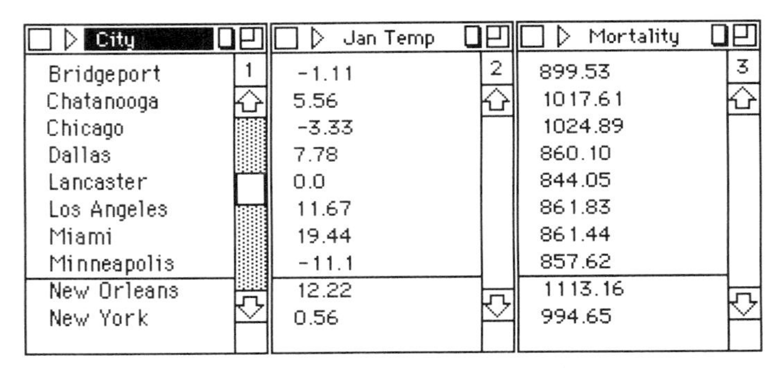

Figure 5-21. *Minneapolis case moved.*

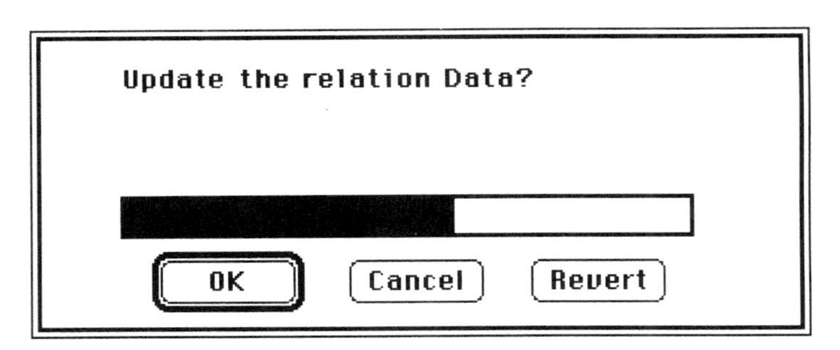

Figure 5-22. *OK to save changes dialog.*

(22) Close Mortality. To do this, click in the Mortality window to make it the active window, and then click the close box in the upper left corner. The window zooms back into its icon, and Data Desk asks you to confirm that you wish to keep the changes you have made to the Data relation (Figure 5-22). Click **OK.**

(23) Choose {Data ▶ Close All...} **Variables**. Alternatively, select the icons for each window (for example, by selecting their icon aliases) hold the Shift key, and Choose {Data} **Close**. The datafile window should now look like Figure 5-23.

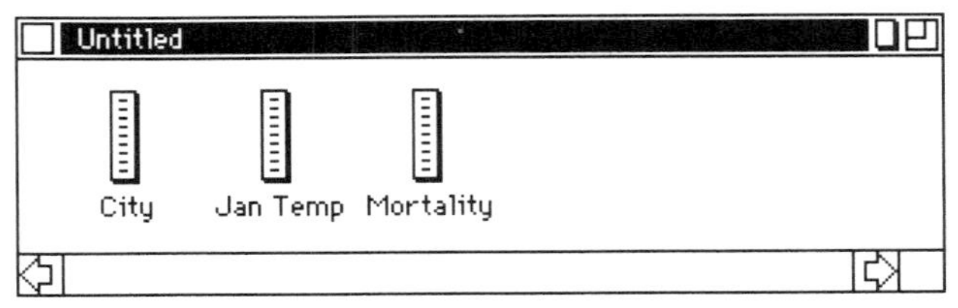

Figure 5-23. *Variables closed into their icons and arranged in the Untitled datafile window.*

APPENDIX 5A

Configure Editing

In Data Desk, the Return key always puts the entry point in the first variable of the editing sequence and moves to a new case. If the window is not in the editing sequence, the entry point moves to the next line. Ordinarily, Return moves down to edit the next case, or inserts a new case if you were at the last case in the variable. The **Preferences...** command in the **Edit** menu brings up a dialog that lets you change the operation of Return so that instead it inserts a new case.

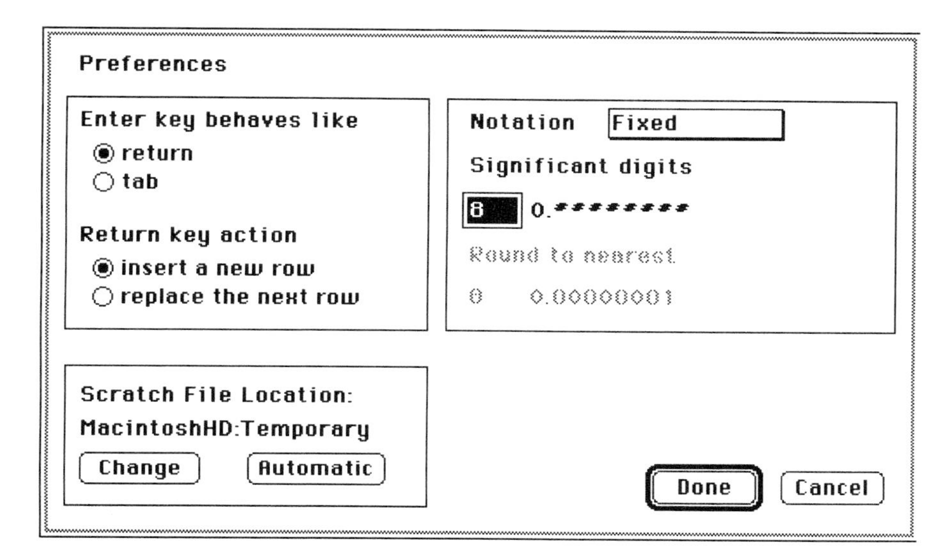

Figure 5-24. *The upper left part of the Preferences dialog lets you specify the operation of the Enter and Return Keys.*

The Tab key always moves to the next variable in the editing sequence. Usually, it stays at the same case. However, if the entry point is in the last variable of the editing sequence or in an unsequenced variable window, Tab behaves like Return, starting a new case and placing the insertion point in the first variable of the editing sequence.

In addition to the Tab and Return keys, it can be useful to use the Enter key. The Enter key is especially useful when you are entering numbers with the numeric keypad because it is a part of that pad. The Preferences... dialog lets you specify whether the Enter key should work like the Tab key or like the Return key.

CHAPTER 6

Importing and Exporting

6.1	Data Tables	*73*
6.2	Copying Variables to a Data Table	*74*
6.3	Pasting Variables from a Data Table to the Data Desk Desktop	*74*
6.4	Alternative Delimiters	*76*
6.5	Writing a Data Table to a Text File	*76*
6.6	Reading Text Files	*76*
6.7	Importing from Several Files	*77*
6.8	Appending Cases in Data Desk	*77*
6.9	Copying Data Desk Results	*78*
6.10	Printing from Data Desk	*79*
6.11	Layout Windows	*80*
6.12	Saving Screen Images	*80*

DATA DESK IS DESIGNED to work along with the other programs you use. You can import data into Data Desk from spreadsheets, databases, other statistics and graphics programs, disk files in text format, and even from other computers (by using your favorite communications software). You can export data and numeric results from Data Desk to other programs for further work. You can also export plots and tables to word processors, page composition programs, and graphic editing programs.

It is easy to import and export data to and from Data Desk, moving entire variables or selected cases. Generally, you can select whatever you wish to export, and copy it to the Clipboard or save it in a text file. To move specific cases, open the variables whose values you want to move and select the cases you want. Alternatively, you can use a selector variable to identify the cases you wish to export. To move entire variables, select their icons and **Copy** them. It is simpler and more efficient to export a variable in this way rather than by opening it and selecting all of its cases.

The Student Version of Data Desk will not allow any **Import** or **Copy** commands that will result in a datafile holding more than 15 variables or 1000 cases. If you raech this limit during one of these commands, Data Desk will display a message explaining that it cannot complete the last command.

6.1 *Data Tables*

data table

Data Desk ordinarily transports variables by placing them in a *data table*. A data table is a table of data values with Tab marks delimiting each successive data value in a case and a Return delimiting the cases themselves. A data table looks like this:

case 1:	value	*tab*	value	*tab*	value	*tab*	value	*return*
case 2:	value	*tab*	value	*tab*	value	*tab*	value	*return*
case 3:	value	*tab*	value	*tab*	value	*tab*	value	*return*

rectangular datasets

Many programs read and write data tables either from text files saved on a disk or from the Clipboard. In Data Desk, each column of a data table is a variable and each row is a case. Data tables are always rectangular. That is, all variables have the same number of cases and every case must have a value (numeric, text, or missing) for each variable. Relations have exactly this structure, and usually it makes the most sense to export variables from the same relation together. However, Data Desk does not impose this restriction, so you may export variables from different relations with different numbers of cases together. When Data Desk creates a data table, it adds missing data values to the end of shorter variables to make the table rectangular.

The first row of a data table may hold the names of the variables. You can think of these names as column labels — indeed, they may be column headings on a spreadsheet or on a table in a text document. Data Desk offers to place variable names in the first row of data tables that it exports, and can read variable names from the first row of data tables that it imports.

The {Manip} **Make Data Table** command combines variables into a data table in a window on the desktop so you can scroll up and down in it to view variables together. (See Section 12.9 for details.)

6.2 *Copying Variables to a Data Table*

HOW-TO

How to copy variables:

- Select variables left-to-right.
- Choose **Copy Variables** from the **Edit** menu.

To copy variables to a data table, select the variables *in the order in which they should appear in the table (left-to-right).* Recall that when you drag across icons to select them, they are ordered left-to-right, and that when you select a folder or relation icon, its contents are selected left-to-right. When you have selected variables, the **Copy** command in the **Edit** menu changes to **Copy Variables**. Choose it to copy the entire contents of the selected variables onto the Clipboard.

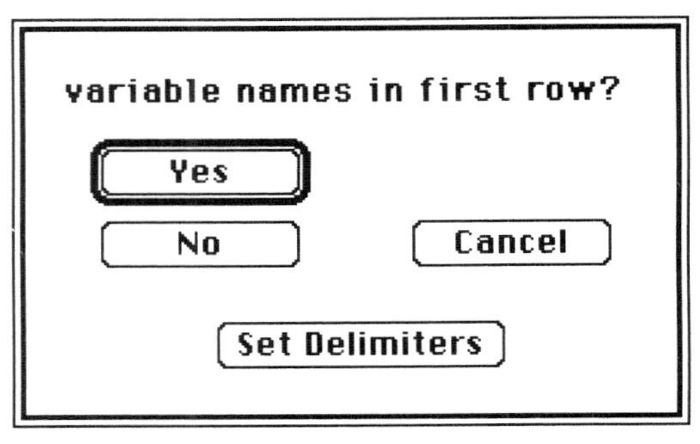

Figure 6-1. *The Copy Variables Dialog.*

Data Desk asks if you wish to place variable names in the first row of the data table. If you are moving the data to a word processor or a spreadsheet, you may want to have the variable names at the top of the columns. Other programs may be confused by nonnumeric information in the first row, so you need to know something about the program to which you are moving the data. You can also substitute a different character for the Tab delimiter. (See Section 6.4.)

You now have a data table on the Clipboard. You can move to another program, and paste the data table into that program.

While the Clipboard is a convenient way to move data, it is not designed to handle large amounts of data. If the Clipboard cannot do the job, try moving smaller amounts of data in several operations by switching between applications with MultiFinder or System 7, or export the data in a text file.

6.3 *Pasting Variables from a Data Table to the Data Desk Desktop*

When the Clipboard contains text and the frontmost window is an icon window, the **Paste** command in the **Edit** menu changes to **Paste Variables**. Choose it to paste each column of the data table into the datafile as a variable. Data Desk creates a new relation named *Clipboard* to hold the new variables. You may drag them into other existing relations if they have the correct number of cases, but you must then be sure that the new variables are indeed measured on the same individuals as those in the old relation.

REMEMBER

The Student Version of Data Desk will not allow any **Import** or **Copy** commands that will result in a datafile holding more than 15 variables or 1000 cases.

Data Desk shows you the first row of the data table and offers a choice among six alternatives:

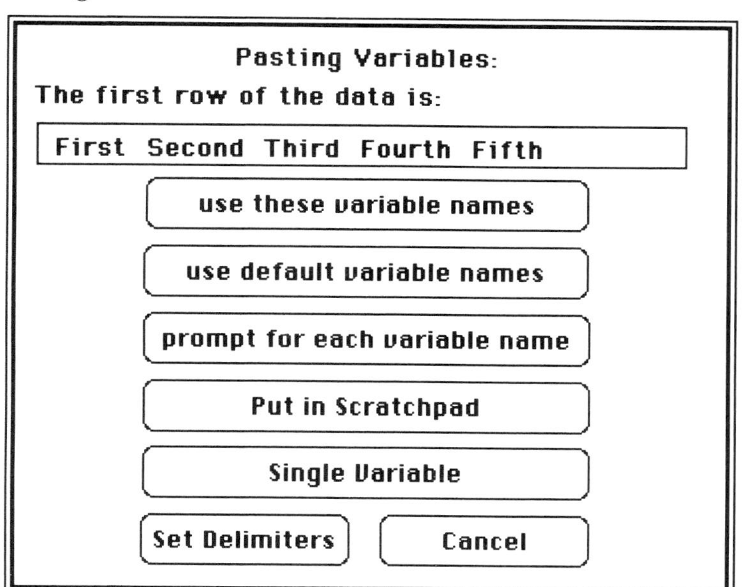

Figure 6-2. *Options for pasting variables. The Set Delimiters button is discussed in Section 6.4.*

- You may elect to use the displayed text as variable names. The variables' values then start with the second row of the data table.
- You may elect to have Data Desk generate the default variable names *Var1, Var2, ...* . You can rename them later.
- You may ask to be prompted for a name as each variable is created. The prompt displays the first case in each variable to remind you of its contents, but this case remains part of the data. The prompt also allows you to enter background information for each variable.
- You may paste the contents of the Clipboard into a new Scratchpad. This alternative is particularly useful if you wish to examine the contents of the Clipboard first or if you have imported an ordinary text file (for example, a description of the data) rather than a data table.
- You may paste the entire data table into a single variable. This alternative can be useful if you wish to edit the data table further and want to work with the case-conscious capabilities of Data Desk's data editing commands.
- You may Cancel the Paste command and return to the Data Desk Desktop.

TIP

To join variables side-by-side:
- Select the variables in order.
- **Copy Variables**.
- **Paste Variables**, specifying Single Variable in the Options Dialog.

6.4 Alternative Delimiters

While data tables ordinarily have a Tab between each column, you can specify another choice of column delimiter. Press the Set Delimiters button in either the Paste Variables dialog or the Copy Variables dialog to display another dialog that lets you specify alternative column delimiters.

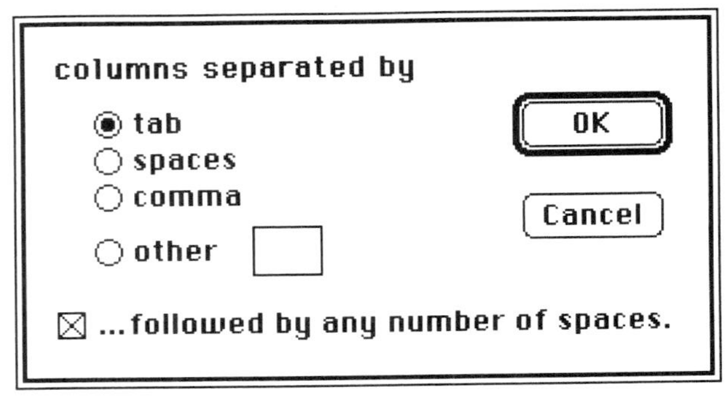

Figure 6-3. *You can specify an alternative delimiter for data tables.*

You can substitute another character, such as a space, comma, or anything you can type, for the Tab character in the standard data table. You can also specify whether spaces following the delimiter should be removed or left in the data. Ordinarily, extra spaces are removed.

Changing the delimiter affects both copying and pasting variables (and thus affects both importing and exporting data.) The delimiter you specify is inserted between columns in output tables exported to a text file or placed on the Clipboard with the **Copy Variables** and **Copy Cases** commands.

If you change the delimiter, the new delimiter remains in effect until you change it again or Quit Data Desk.

6.5 Writing a Data Table to a Text File

HOW-TO

To write data in variables to a text file:

- Select the variables in order.
- Choose {File}**Export**

You write a data table as a text file (sometimes called an ASCII file) in much the same way you copy variables to the Clipboard. Select the variables you want in left-to-right order. Then choose **Export** from the **File** menu. Data Desk offers the option of placing variable names in the top row. It then asks you to name the new file and writes the data into that file in data table form. Many Macintosh programs read text files. Programs on other computers can often read text files that have been created on the Macintosh.

6.6 Reading Text Files

To read from a text file, choose {File} **Import** and select the file from the dialog it presents. Text files typically contain either a data table with each row holding a case and columns (variables) separated by a delimiting character, or a single column of data. You will see the same sequence of dialogs as for pasting variables. Data Desk creates a new relation and gives it the same name as the datafile from which the text was imported. You can drag imported variables from this new relation into existing relations, but you must insure that the new variables are measured on the same individuals as those in the old relation; Data Desk can only check that the variables have the right number of cases.

If the file does not contain a data table, you can paste the file's contents into a scratchpad or a single variable. In this way you can import a text file and edit it in a scratchpad, or use it to document a datafile.

6.7 *Importing from Several Files*

TIP

You can keep text files with one variable per file and combine them as needed by importing.

The Import command provides a simple way to combine variables from two or more datafiles, either by combining all the variables from the files, or by selectively exporting variables from one file and then importing them into another.

If you invoke Data Desk by opening a Data Desk text file, Data Desk creates a relation with the same name as the file and imports the columns of the text file as variables. Only text files created by Data Desk can launch Data Desk automatically when opened. You can open several text files at once or open foreign text files by selecting them on the Finder desktop along with the Data Desk icon, and choosing **Open** from the Finder's **File** menu. Data Desk opens and automatically imports variables from each of the selected files, combining their contents into a single datafile with several relations. If one of the selected files is a Data Desk datafile, it is opened first, and the contents of the other files are imported into it, but additional Data Desk datafiles are ignored.

HOW-TO

To open several files at once:

- Select the file icons on the Finder Desktop.
- Shift-Click to select the Data Desk icon.
- Choose **Open** from the **File** menu.

6.8 *Appending Cases in Data Desk*

You can merge cases from different datafiles or from different relations in Data Desk datafile into a single relation using the **Parallel Append** command. This command allows you to combine data from two separate relations by appending the content of variables in the second relation to the content of variables in the first relation.

For example, part of your data might be stored in a Data Desk datafile and the rest kept in another application. You can import the new data into the Data Desk datafile using the **Import** command from the **File** menu. This new data will be placed in a new relation in the existing Data Desk datafile.

You can now combine the data from the two relations into one. First make sure that you have the same number of variables in the two relations and that they are positioned in the same order within their relation. Select the variables in relation one as *y's*, the variables in relation two as *x's*, and choose {Manip} **Parallel Append**. Data Desk will create a new relation labeled *Data* with the variables labeled as they were in relation one. It will also hold an additional variable labeled *Group*, which will provide the names of the two relations from which the data came.

This operation is very memory-efficient; Data Desk does not require that all variables selected be stored in memory at one time. Therefore if the file you wish to bring into Data Desk is too large to be imported at one time, the **Parallel Append** command provides an efficient way to bring in data in smaller sections and combine them in Data Desk.

6.9 *Copying Data Desk Results*

REMEMBER

Copy tables as pictures to preserve special symbols and formatting.

Copy tables as text to be able to reformat and edit.

Data Desk presents statistical results (summary reports, regression summary tables, correlation tables, ANOVA tables, contingency tables, etc.) in text form using boldface, italics, and special symbols to enhance readability. You can copy these tables as pictures to preserve this formatting.

Data Desk can also copy statistical output as text with a tab mark separating each of the columns. This lets you move tables to a word processor or page formatting program, insert your own tab-stops, and reformat the text however you would like. For example, you might change the column headings to another font or to a bigger size.

To copy output such as a regression summary table, make it the frontmost window on the Data Desk Desktop, then choose **Copy Window** from the **Edit** menu. (The Copy Command in the Edit menu says Copy Window when the frontmost window is a result window.) Data Desk asks if you want a picture-format copy (which preserves text faces and special symbols, but cannot be edited) or a text-format copy. If you choose text-format copying, Data Desk places a tab-delimited table of text on the Clipboard. You can then paste this into the Scrapbook for later use, or move it immediately to another program.

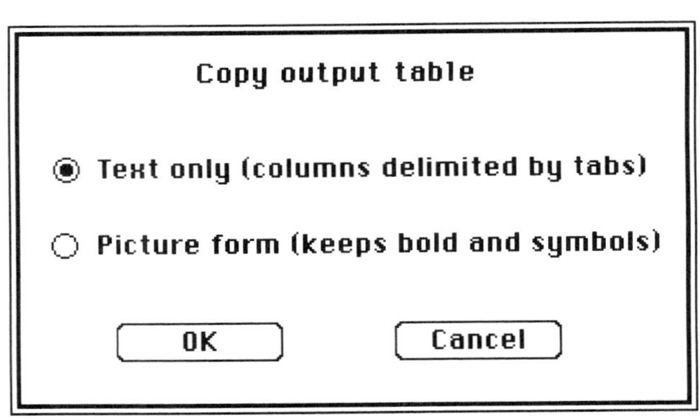

Figure 6-4. *Data Desk offers to copy tables as tab-delimited text or as a bit-mapped picture.*

You can copy plots in exactly the same way. Plots copied onto the Clipboard are in Macintosh PICT2 format. Each element of the plot is recorded as a graphic object. You can paste the picture of the plot into virtually any Macintosh graphics editing program and modify any element of the plot easily. If you print the plot from a graphics editor, a word processor, or from a page composition program it will print with the full precision of your printer rather than with the limited precision of the screen.

6.10 *Printing from Data Desk*

Data Desk supports all standard Macintosh printers. The **Page Setup** command in the **File** menu presents printer setup options according to the printer you have selected using Apple's Chooser accessory. (See your Macintosh manuals for information about Chooser.) You must have the appropriate printer file in your System folder to be able to print or to perform a page setup.

> HOW-TO
>
> To print a table of data:
> - Select the icons of the variables to be printed in the order they are to appear.
> - Choose **Print Variables** from the **File** menu.

The **Print Variables** command in the **File** menu gathers all selected variables together into a data table and prints them in side-by-side columns. For tables of many variables, you may prefer the Page Setup option that prints sideways because it puts more columns on a page.

Print Variables only prints variables. You can print variables while they are open, but you must select their *icons* before choosing the command.

You can print any window with the **Print Front Window** command in the **File** menu. It prints the contents of the frontmost window in PICT form to take greatest advantage of the resolution of your printer. Plots usually look better printed black-on-white rather than the screen standard white-on-black, and Data Desk defaults to this choice. In fact, some plots will not print correctly on the LaserWriter in white-on-black form because they use pattern transfer modes that are not supported by the LaserWriter. You can change the default with the {Plot ▶ Plot Options} **Print White-on-Black** command.

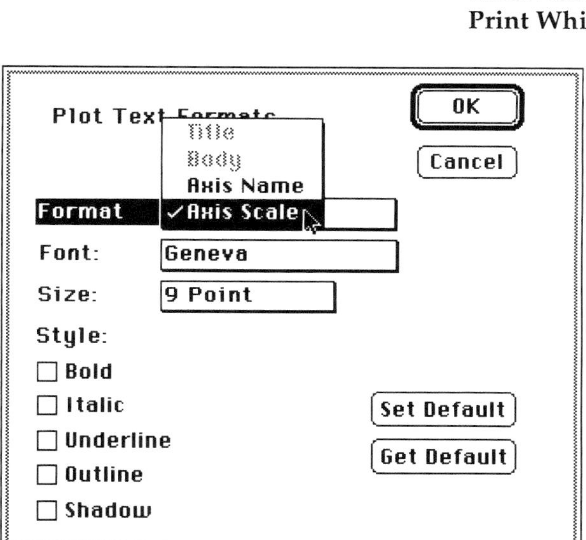

Figure 6-5. *Text format dialog lets you specify text format for each part of the plot or summary table.*

You can specify the font and size of the text in any plot or summary table by choosing **Text Format...** from the window's Global HyperView. The window shown in Figure 6-5 allows you to specify the format for each part of the plot or table. From the *Format* pop-up menu, select the part of the plot or table you want to modify. If the window is a plot, you can change text formats for the **Axis Name** and **Axis Scale.** If the window is a summary table, you can change the text formats of the table's **Title** and **Body.** You can also record and apply this text format setting to other windows by clicking on the *Set Default* button. The *Get Default* button will recall the previous text format settings

6.11 *Layout Windows*

Layout windows are output formatting windows into which you can paste or drag pictures of other Data Desk windows. You can then position the pictures as you please and then print it with **Print Front Window**.

To make a new layout window select {Data ▶ New} **Layout**. Data Desk creates a new layout window and opens it. To place pictures of plots or tables in the layout window use the {Edit} **Copy Window** command to copy a picture of the window, click on the Layout window, and Paste. Alternatively, drag the icon (or icon alias) of the window into the layout window. (You can only drag icons of open windows into layout windows.) To reposition a picture in the layout window, click on it and drag it where you would like it. Plots in layout windows are transparent so you can overlay several plots.

To remove a picture from a layout window hold the Backspace key down and click on the picture. It is a good idea to click first to be sure which picture you are about to delete. This deletion has no Undo.

To add comments or titles to a layout window open a new ScratchPad and type them there. You cannot copy a ScratchPad as a picture (Copy will copy selected text), so you must drag its icon or icon alias onto the Layout.

Plots and analyses in a layout window are pictures. Thus they reflect the state of a window when it was copied. Future changes to data or updates of the window are not reflected in the layout window. Thus you can copy a window into a layout window, alter the plot or analysis, and copy the new version next to the old.

New layout windows consult the Page Setup information to learn how big they should be. You can use the {File} **Page Setup** command to change this information, but the change will only affect future layout windows.

6.12 *Saving Screen Images*

The **Save Desktop Picture** command in the **File** menu saves a MacPaint-format picture of the Data Desk Desktop on the Hard Drive and names its icon Screen 0 or Picture 1 depending on which operating system you have on your computer. Before saving the picture, it turns the desktop white because the standard gray background usually gets in the way of picture editing. Because this command uses the built-in screen saving features of the Macintosh system, it may work differently on different Macintosh computers and with different versions of the System file. For example, in System 7, the screen image is saved as a TeachText document called "Picture #" and is in full color. With older system files color may not be available, the size of the saved screen may be restricted, or the saved picture may be rotated 90°.

CHAPTER 7

Simple Summaries

7.1	Measures of Center	*83*
7.2	An Example	*84*
7.3	Measures of Spread	*84*
7.4	Order Statistics	*85*
7.5	General Summaries	*86*
7.6	Moments	*87*
7.7	HyperViews	*87*
7.8	Summaries As Variables	*88*
7.9	Summaries As Variables by Group	*88*

APPENDIX

7A	The Biweight, a Robust Center	*90*
7B	Coefficient of Skewness and Kurtosis	*91*

SUMMARY STATISTICS PROVIDE concise descriptions of the numbers in a variable and permit simple comparisons of the variable to others or to external standards. Many different summary statistics are common, but they describe only a few different characteristics of data. Because summary statistics summarize a single variable at a time, they are sometimes called *univariate* statistics.

univariate

Data Desk computes a wide variety of summary statistics. The {Calc ▶ Calculation Options} **Select Summary Statistics** command displays a dialog offering a choice of statistics, each with a check box next to it. Any statistic marked with an **X** in its check box will be computed by any subsequent **Summary Reports**, **Summaries as Variables**, or **Sammaries As Variables By Group** command. To put an **X** in a box, click the box. To remove the X, click the box again.

☐ kth %ile, k= ☐

Some of the statistics (for example, the kth *%ile*) require a number as well. For these, click the check box to the left of the statistic and type the number into a text box. If you don't specify summary statistics options, Data Desk computes the sample mean, the standard deviation, and the number of numeric values by default. However, you can set new defaults by clicking the boxes for the desired statistics and then clicking on the **Set Defaults** box.

To compute summary statistics, select the variables to summarize and choose **Summary Reports** from the **Calc** menu. The resulting report shows each requested summary statistic computed for each selected variable.

```
Summary statistics for    Weight
Mean 2.8629
Numeric 38
StdDev 0.70687
```

Figure 7-1. *The default summary report of Weight variable in the Cars data.*

7.1 Measures of Center

Centers
☒ Mean
☐ Median
☐ Midrange
☐ Biweight, c= ☐

The center is the most common single numeric description of a batch of values. Measures of center are so common in ordinary speech that it is easy to forget that they have precise mathematical definitions.

The center goes by many names. It is often called the *level*, the *middle*, or the *average*. Many statistics texts refer to centers as *measures of location* or *measures of central tendency*. Whatever the name, all centers obey two rules:

- If every value in a variable is incremented by a value, a, then the center increments by the value a.
- If every value in a variable is multiplied by a value, b, then the center is also multiplied by the value b.

7.2 An Example

The Cars dataset contains information about 38 automobiles. A summary report of four centers computed for car weight is shown in Figure 7-2.

```
Summary statistics for    Weight
Mean 2.8629
Median 2.6850
Midrange 3.1375
Biweight 2.8637
```

Figure 7-2. *Four centers computed for Weight.*

While each of the centers satisfies the two rules, they differ in value because they are computed differently.

mean
average

The sample *mean* or *average* (in the example, 2.8629) is the most commonly used measure of center. It sums the numbers and divides by the total number of values summed.

The sample mean is the basis for many statistical methods. It is easy to compute, but it can be misleading if the variable contains any extraordinarily large or small numbers. For example, consider the student who tries to salvage his grade average following a zero exam grade.

median

The *median* (2.6850) is the middle value. That is, half of the numbers in the variable are less than or equal to the median and half are greater than or equal to it. The median is used less often than the mean because it is harder to compute and more difficult to deal with mathematically. Nevertheless, it has some advantages. For example, the median is not affected by occasional extraordinary data values.

midrange
range

The *midrange* (3.1375) is the mean of the largest and smallest data values. It is useful primarily when the overall extent, or *range*, of the data is of particular interest.

biweight

The *biweight* (2.818) is a *robust* center, which means that it is not unduly affected when the data have extreme values. The biweight is not as common a measure of center as the others discussed here. It is defined in detail in Appendix 7A.

7.3 Measures of Spread

Spreads
☒ **Standard Deviation**
☐ **Interquartile Range**
☐ **Range**
☐ **Variance**

Measures of spread describe the extent to which individual values cluster around a particular center. Like centers, measures of spread go by several names. Terms such as *variability, variation,* and *dispersion* are common synonyms for *spread.*

Measures of spread obey two rules:

- If every value in a variable is incremented by the value, c, then the spread remains unchanged.
- If every value in a variable is multiplied by a value, d, the spread is multiplied by the absolute value of d.

Measures of spread computed for car weight are shown in Figure 7-3.

Summary statistics for Weight
StdDev 0.70687
Interquart 1.2613
Range 2.4450
Variance 0.49967

Figure 7-3. *Spreads computed for car weight.*

While each of these spreads (except the Variance) satisfies the two rules, they differ in value because each is defined and calculated differently.

standard deviation

The *standard deviation* (0.70687) is the most frequently used measure of spread. It is a natural companion to the sample mean because it describes the extent to which the collection of data values scatter around the sample mean. Like the sample mean, the standard deviation can be affected by extreme data values.

Standard deviations can be calculated in several different ways. For n values, each denoted $y_1, y_2, \ldots, y_n$, Data Desk first computes the sample mean, $\bar{y}$, and then calculates the standard deviation as

$$s = \sqrt{\frac{\sum (y_i - \bar{y})^2}{n-1}}$$

variance

The *variance* (0.49967), which is the square of the standard deviation, is often thought of as a measure of spread. Although the variance grows with increasing dispersion and shrinks when the data values cluster closely about the mean, it is not a true measure of spread because it does not obey the second rule for spreads. Specifically, multiplying a variable by a value d has the effect of multiplying its variance by d^2.

range

The *range* (2.4450) is the absolute difference between the largest and smallest data values. It summarizes the overall extent of the data, and is a natural companion to the midrange.

interquartile range

The *interquartile range* (1.2613) measures the range of the middle half of the data. It is the absolute difference between the data *quartiles*. (Quartiles are discussed along with other percentiles in the Section 7.4.) The interquartile range is related to the median in the sense that both are found by ordering the data values and then counting in from the ends.

7.4 Order Statistics

Order Statistics
☐ Minimum
☐ Maximum
☐ kth %ile, k=
☐ kth %ile, k=
☐ kth Largest, k=
☐ kth Smallest, k=
☐ Mid k%, k=
☐ kth %ile Diff, k=

Many summary statistics, including some of those discussed so far, order the data values from lowest to highest and then select values based on their position in the ordered list. For example, the *minimum* and *maximum* values are common order statistics.

minimum, maximum

Because order-based statistics depend only upon the relative ranking of values, they resist being unduly influenced by extraordinary values. An extreme value that might render the mean or standard deviation misleading will have only a slight effect on the median, mid-25th percentile, or interquartile range. Because order statistics deal with infinities gracefully, Data Desk does not ignore infinities when computing order statistics. Other summary statistics treat infinities as missing values. (As with any value, you can search for infinities (∞ is Option-5) with {Edit > Find} **Find** and replace them with a nonnumeric value such as * to omit them from the ranking.)

percentile

Percentiles specify relative position in an ordering of the values in a variable. A percentile is the ordered data value falling a specified fraction of the distance between the minimum and the maximum. Generally, the kth percentile (where k is between 0 and 100) in a variable with n cases is the value that is the (k/100) x *n* smallest. The minimum is the 0th percentile and the maximum is the 100th percentile. The median is the 50th percentile and the lower and upper quartiles are the 25th and 75th percentiles, respectively. If the k% value falls between two data values, Data Desk interpolates between the adjacent data values.

quartile

To specify which percentiles to compute, click the check box next to "k^{th} %ile, k = ☐", and type a number between 0 and 100 in the box. Data Desk offers you the choice of two different percentiles in each summary report because it is often useful to compare symmetric percentiles at k% and (100-k)%.

rank

Another way to specify relative position in a variable is by the ordered position or *rank*. Percentiles do not depend directly on the number of numbers in a variable. Ranks count individual values from 1 to n — either counting up from the minimum (to obtain the kth smallest) or counting down from the maximum (to obtain the kth largest). The 1st largest value is the maximum. The nth largest value is the minimum. The 1st smallest value is the minimum. The nth smallest value is the maximum.

Mid k %

The *Mid k %* defines a general family of order-based centers. It is the average of the value at the specified k^{th} percentile and the value at the symmetrically placed $(100 - k)^{th}$ percentile. For example, the Mid 0% is the midrange. The Mid 25% is the average of the two quartiles.

k^{th} %ile difference

The *k^{th} %ile difference* defines a general family of order-based spreads. It is the difference between the value at the specified k^{th} percentile and the value at the symmetrically placed $(100 - k)^{th}$ percentile. Thus, the 0%ile difference is the range. The 25%ile difference is the interquartile range. The 50%ile difference is always zero because it is the difference between the median and itself.

7.5 General Summaries

General
- ☒ **Numeric**
- ☐ **Non Numeric**
- ☐ **Total Cases**
- ☐ **Sum**

Data Desk provides information fundamental to interpreting other statistics.

 # numeric cases

The # *Numeric Cases*, usually denoted by *n* in formulas, counts up those cases in a variable that are numbers. It does not count missing values, infinities, or cases with nonnumeric text.

 # nonnumeric cases

The # *NonNumeric Cases* is the number of cases that do not contain numbers. Any case that is not numeric is treated as missing by Data Desk in any operation requiring numbers, so the number of nonnumeric cases is also the number of missing values. (You can make a numeric case missing simply by editing its value to be nonnumeric — for example, by putting a "*" or a ">" next to it.)

Total # Cases

The *Total # Cases* is the sum of the # Numeric Cases and the # Non-Numeric Cases. Because Data Desk permits both text and numbers in variables, the count of numeric values may be different from the total number of cases. Most formulas in statistics texts are written for an ideal world in which a numeric value is recorded for every case. As a result, many texts use n to denote the number of cases rather than the number of numeric values.

Sum

The *Sum* is the sum of all values in the selected variable. Nonnumeric or missing cases are not included in the calculation.

7.6 Moments

Moments
☐ **Skewness**
☐ **Kurtosis**

Moments summarize numerically the characteristics of data distributions observed in histograms by summarizing the shape of a data distribution. Moments should be used with caution, however, because they are easily affected by extraordinary values.

coefficient of skewness

The *Coefficient of Skewness* is a moment-based summary that describes deviation of a distribution from symmetry. A symmetric distribution has a skewness coefficient of zero. Positive skewness indicates a longer tail stretching into higher values. Negative skewness indicates that the longer tail stretches into lower values.

coefficient of kurtosis

The *Coefficient of Kurtosis* describes the degree of peakedness in the distribution's shape. Distributions with positive kurtosis have long tails and a narrow, peaked, central hump. Distributions with negative kurtosis have short tails and a wide, flat, central hump. Data Desk adjusts the kurtosis so that the Normal (or Gaussian) distribution has a kurtosis coefficient of zero. This adjusted form is sometimes called the *Coefficient of Excess*.

Appendix 7.B gives the fromulas for skewness and kurtosis.

7.7 HyperViews

Summary reports offer an excellent opportunity to use HyperViews. Data Desk HyperViews suggest other analyses or plots that might tell you more about your data. For example, summary statistics treat each variable individually, summarizing the center of the values, their spread, and other aspects of how they are distributed. Histograms (see Chapter 8) convey much of the same information graphically, so it is often helpful to see a histogram along with a summary report.

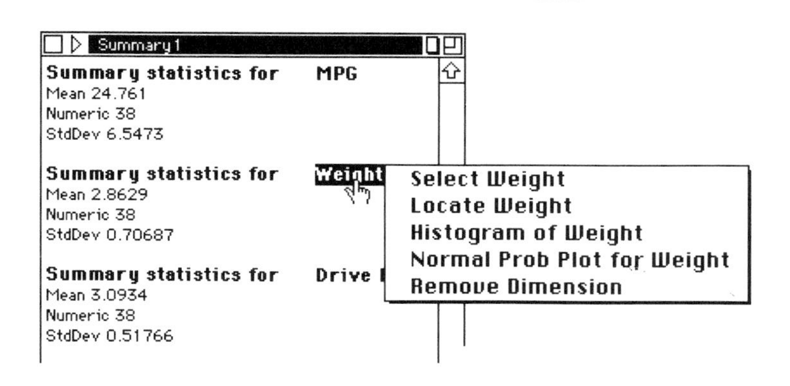

Figure 7-4. *The HyperView menu on a summary report.*

When you move the cursor over the name of the variable in a summary report, the cursor changes to a cursor to indicate that the name can behave like a button. Click at that point to pop up a short HyperView menu of suggested plots and analyses. In this

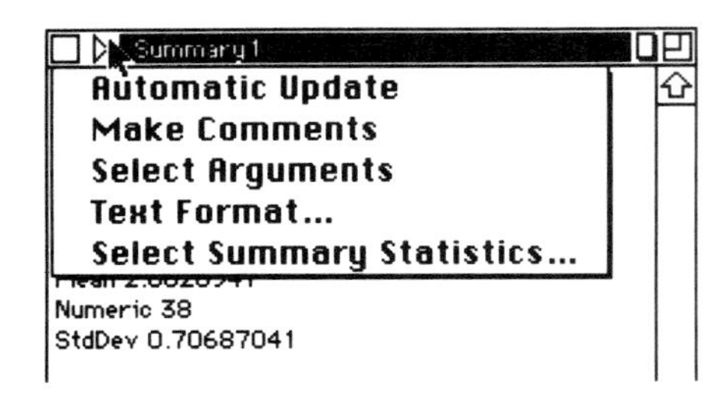

the menu suggestion of a histogram. Another HyperView command locates the icon of the summarized variable so that you can work with it.

In the upper left of the window's title bar is an ˘ much like the > that ordinarily indicates a submenu. Pressing the arrow drops down a HyperView menu. The commands in this *HyperView* refer to the entire summary report rather than to a particular variable result.

7.8 Summaries As Variables

Data Desk offers a second way to compute summary statistics that makes it easy to compare the values of each summary statistic across several variables. This command generates a variable for each summary statistic selected from the **Select Summary Statistics** dialogue. Select the variable(s) for which you want to compute statistics and choose {Calc > Summaries} **As Variables**. Data Desk creates a new relation named "Summary" containing a variable for each summary statistic chosen. Data Desk also creates a variable named "Identities" that names the variables. In this new relation, each case (row) holds the statistic values for one of the originally selected variables and each variable (column) holds values for a particular statistic. You can thus read down the column of means (for example) to compare them easily.

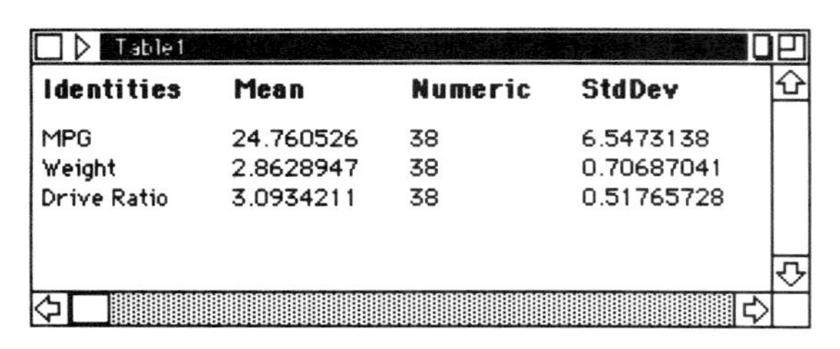

Table1

Identities	Mean	Numeric	StdDev
MPG	24.760526	38	6.5473138
Weight	2.8628947	38	0.70687041
Drive Ratio	3.0934211	38	0.51765728

To display these results conveniently, select the summary statistics variables and choose {Manip} **Make Data Table**. The summary statistic variables can also be graphed or used in other calculations. For example, a scatterplot of standard deviations *vs* means is an effective way to look for increasing variance with increasing level — a violation of assumptions for some analyses.

7.9 Summaries As Variables By Groups

Data Desk also computes summary statistics for cases categorized into groups, saving the results in variables as for the Summary Statistics As Variables command. Choose the desired summaries from the {Calc > Calculation Options} **Select Summary Statistics** dialogue. Next, select the variables to be analyzed as y and the variable(s) holding the categories as x and choose {Calc > Summaries} **As Variables By Groups**. Data Desk creates a new relation and places in it a folder for each of the summarized (y-) variables. Each folder holds a variable for each summary statistic requested. Each of these variables has a case for each category in the grouping (x-) variable. In addition, Data Desk places a variable named with the name of the group variable in the relation but outside the folders. This variable simply names the categories, providing a label for the rows of the summary statistic variables.

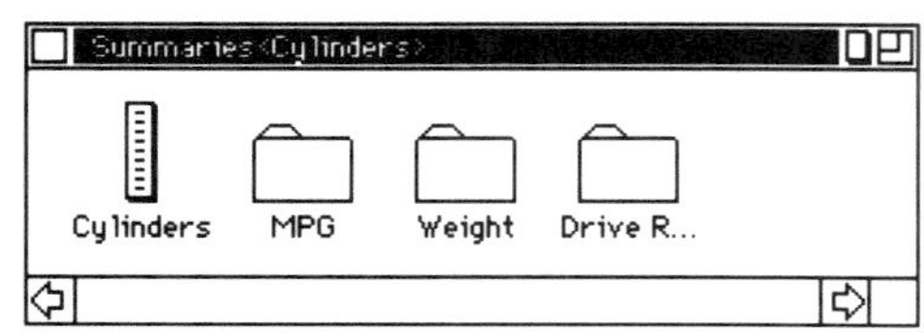

If you select more than one group variable, Data Desk computes summaries for each cell of the table defined by the group variables. Data Desk first determines every possible combination of categories in all of the category (x-) variables. It then computes the selected summary statistics for all the cases in each combined category. Outside the folders, Data Desk creates a variable for each of the group variables so that the set of these variables together labels each case (row) of the summary statistic variables with the combination of categories defining that case.

Summaries for multiple group variables are a valuable adjunct to factorial Analyses of Variance (See Chapter 19). Select the ANOVA dependent variable as y and the factors as x-variables. The summary statistics will report summaries for each cell of the experiment design.

You can create a convenient display of the results by selecting the variables that name the categories and any of the folders and choosing **Make Data Table** from the **Manip** menu. You can also use any of these variables in future plots or analyses.

Let's look at an example from the Cars dataset. Suppose we want to see the mean, sum and standard deviation for MPG and Weight broken down by each combination of Country and Cylinder. That is, we are interested not only in differences by Country and Cylinders, but in differences among 4, 6, and 8 cylinder cars made in the U.S. Choose the summaries Mean, Sum and Standard Deviation from the **Calculation Options** > **Select Summary Statistics** dialogue. Select MPG and Weight as *y* and Country and Cylinders as *x* and choose {Calc > Summaries} **As Variables By Groups**. Data Desk creates a new relation holding variables named Country and Cylinders, and two folders named MPG and Weight. Select all of the icons in the relation window and choose **Make Data Table** from the **Manip** menu. The resulting table displays the Means, Sums and Standard Deviations of MPG and Weight for each combination of countries and cylinders.

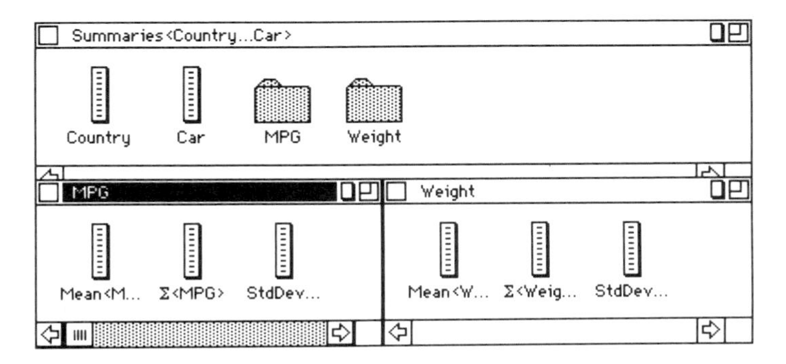

Figure 7-5. *When you choose **Summaries** > **As Variables by Group,** Data Desk creates a new relation holding variables named Country and Cylinders and two folders named MPG andWeight.*

APPENDIX 7A
The Biweight, A Robust Center

The Biweight

The biweight is a center that is said to be *robust* because it is relatively unaffected by extreme values and yet behaves like the mean for well-behaved data. The biweight is a *weighted mean* in which the weights are computed *adaptively* depending on the data values themselves. Specifically, the biweight is computed as follows:

1) For each data value, x_i, compute

$$u_i = \frac{x_i - median(x)}{cMAD(x)}$$

where median(x) and MAD(x) are the median and median absolute deviation from the median of the x-values, respectively, and *c* is a tuning constant that you may specify. Constants smaller than 4.0 are dangerous; larger constants give less robust results. 7.0 is often a good choice, and is the default choice given in the **Select Summary Statistics...** dialog.

2) For each u_i, compute an associated value, w_i:

$$w_i = (1 - u_i^2)^2 \qquad u_i^2 \leq 1$$

$$w_i = 0 \qquad u_i^2 > 1$$

3) Compute the biweight by taking the weighted average of the data values, using the collection of w's as the weights:

$$biweight = \frac{\sum w_i x_i}{\sum w_i}$$

See Mosteller and Tukey (1977) for more information about the biweight.

APPENDIX 7B
Coefficients of Skewness and Kurtosis

The coefficient of skewness describes deviation of a distribution from symmetry. It is calculated using the following formula:

Skewness = $M_3 / M_2^{3/2}$, where:

$$M_2 = \sum (y_i - \bar{y})^2 / n$$

$$M_3 = \sum (y_i - \bar{y})^3 / n$$

The coefficient of kurtosis describes the degree of peakedness in the distribution's shape. It is calculated using the following formula:

Kurtosis = $(M_4 / M_2^2) - 3$, where:

$$M_2 = \sum (y_i - \bar{y})^2 / n$$

$$M_4 = \sum (y_i - \bar{y})^4 / n$$

Subtracting 3 makes the kurtosis equal 0 for the Normal distribution. This form is sometimes called the *Coefficient of Excess.*

CHAPTER 8

Displaying Data

8.1	Data Analysis Displays	*95*
8.2	Plotting Conventions	*96*
8.3	Histograms	*99*
8.4	Recentering and Rescaling Histograms	*100*
8.5	Dotplots	*102*
8.6	Dotplots of Separate Variables	*103*
8.7	Boxplots	*104*
8.8	Bar Charts	*106*
8.9	Pie Charts	*107*
8.10	Scatterplots	*109*
8.11	Lineplots	*110*
8.12	Normal Probability Plots	*112*

APPENDIX

8A	Boxplot Definitions	*113*
8B	Probability Plots	*114*

DATA ANALYSIS DISPLAYS are powerful tools for finding patterns in data. Innocent looking data can reveal remarkable and unanticipated structure when displayed in the right way. Pictures naturally convey general trends and patterns and let extraordinary values or unexpected behavior stand out. When plots work together even complex relationships among several variables are easy to see and understand.

interactive

- *DATA DESK'S DISPLAYS ARE INTERACTIVE.*

To get the full benefit of a display, you need to be able to work with it while analyzing your data. Data Desk does not limit you to simply reading a static display. Instead, you can identify subgroups and create new variables based on them, learn about special cases, record projections, or trace sequences. You can even drag a new variable onto an axis and watch the display change.

dynamic

- *DATA DESK'S DISPLAYS ARE DYNAMIC.*

Data Desk displays use animation to reveal aspects of your data that static plots cannot show. You can build your own dynamic displays with sliders. And all Data Desk displays are fully integrated with other analysis methods, so you can save projections and newly found structure for further analysis.

linked

- *DATA DESK'S DISPLAYS LINK TOGETHER.*

Linking highlights points that are selected in one display in *all other* displays—including variable editing windows. This helps you to see relationships among several variables at once. Data Desk also links choice of plot symbol and color so that the representation of a case is consistent across plots.

HyperViews

- *DATA DESK'S DISPLAYS ACTIVELY GUIDE YOU TO NEW ANALYSES.*

Data Desk's displays feature HyperView menus, which suggest additional analyses that build on your progress so far. HyperViews put new ideas for analyzing data at your fingertips.

update

- *DATA DESK'S DISPLAYS INCORPORATE NEW INFORMATION TO KEEP UP-TO-DATE.*

When one of the underlying variables or functions in a display is modified, Data Desk offers to *update* the display to incorporate the new information. This feature significantly simplifies the analysis process by keeping track of when displays could be updated — and then doing it when you wish. "Before" and "after" comparisons are easy when you Redo a plot and place the versions side-by-side.

This chapter describes the displays available in Data Desk. Chapter 9 discusses tools and commands for working with displays and shows how to use them. Chapter 10 discusses dynamic graphics in detail.

8.1 Data Analysis Displays

area principle

Data Desk displays obey the *area principle,* which says that the visual impact of a part of a display is proportional to the area it occupies. Displays that observe the area principle are more likely to reveal patterns in data and less likely to create optical illusions that might lead to false conclusions. Data Desk does not use a false impression of depth for decoration, because that can leave the illusion that equal *volumes* represent equal amounts.[1]

Data displays can be grouped into two broad classes:

data analysis displays

vs.

presentation displays

- *Data analysis displays* address the question, "What do I want to *know* about the data?" They help you discover patterns and avoid unwarranted conclusions. Data analysis displays are usually as plain as possible because excess clutter and irrelevant changes of pattern or color distract the eye from the data.
- *Presentation displays* address the question, "What do I want to *show* about the data?" They depict conclusions already reached and try to convince the viewer to believe them. They often incorporate shading, patterns, or color to improve the artistic appeal of the display.

Both kinds of displays are useful, but it is best to use them in proper order, first understanding your data with data analysis displays and then explaining your insights to others with presentation displays. Data Desk is designed to analyze data, and thus concentrates on data analysis displays.

Of course, many of Data Desk's plots are excellent presentation displays. You can combine and overlay displays in a Layout window. For greater artistic control, you should use one of the many graphics programs available for the Macintosh. To transfer a Data Desk display to another program, copy the display onto the Clipboard with the **Copy Window** command in the **Edit** menu, (active when the frontmost window is an output window) or use the **Save Desktop Picture** command in the **File** menu to save a MacPaint picture of the screen on the disk.

8.2 *Plotting Conventions*

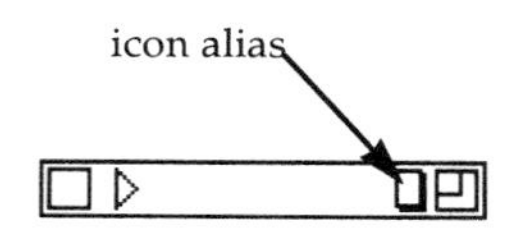

Most of Data Desk's displays are requested from the **Plot** menu, or from HyperViews. Each plot command creates an icon that holds the plot, then opens it to show the plot. Data Desk constructs a name for the plot by abbreviating the names of the variables depicted. To rename a display, select the name under its icon and type the new name. (You can locate the icon quickly with a double-click on the icon alias in the window's title bar).

Data Desk initially arranges displays so that four or more are visible at once on the screen. You can often see more in your data by comparing several plots to each other. Each shows a different view of the data. By integrating these views into a coherent whole (perhaps with the aid of selecting parts of one display to see the corresponding parts of the other highlight), you can see far more in the data than any one display can show.

---

[1]For an amusing and informative discussion of these concepts, see the classic book *How to Lie with Statistics* by Darrell Huff (1954) W.W. Norton & Co., New York.

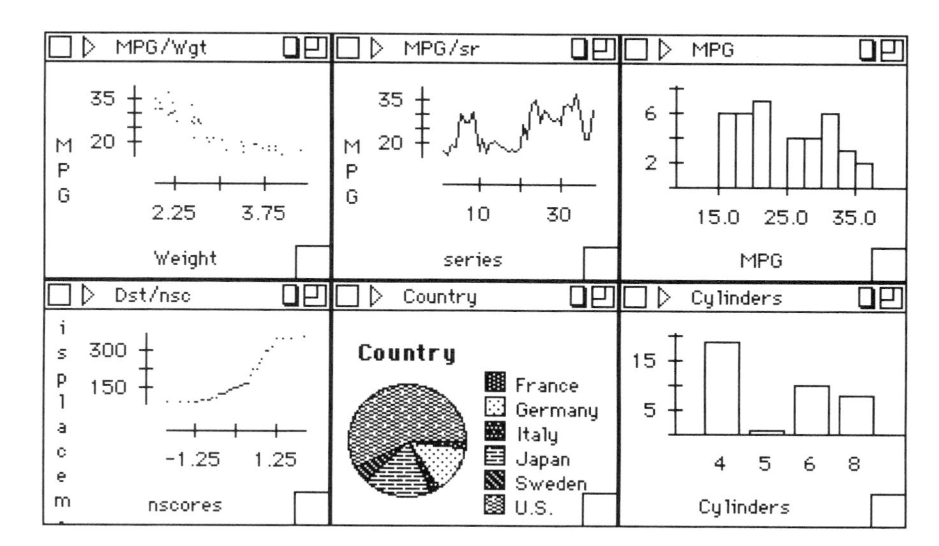

Figure 8-1. Tiled windows cover the screen. Click a plot's zoom box to see it full-size.

You can arrange and resize display windows to suit your particular needs, or use the {Special ▶ Arrange}**Tile Windows** or **Stack Windows** commands to reshape and reposition all displays. **Tile Windows** arranges all open displays so that they are all visible at once. **Stack windows** arranges the displays in large format in a stack, fanning their title bars so you can select any display with a single mouse click.

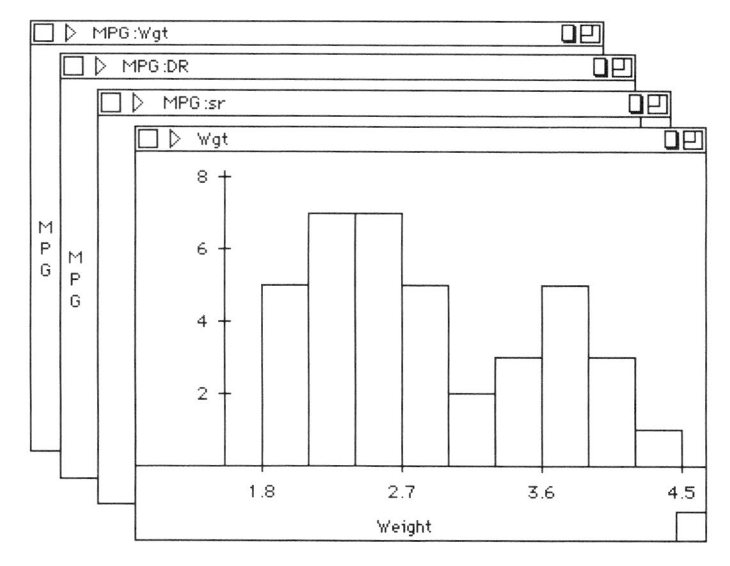

Figure 8-2. Stacked windows.

The **Close All ▶ Displays** command in the **Data** menu provides a convenient way to close all open display windows into their icons. To close selected displays select their icons (or icon aliases), hold the Shift key, and choose **Close.**

Most **Plot** commands can generate several plots at once. Commands that are plural (for example, **Histograms**, **Scatterplots**, etc.) generate a separate display for each icon selected or for each pair of x- and y-variables.

Data Desk draws most plots in white on a black background. Points plotted white-on-black look like stars on a black background and seem to glow on the screen, making them easier to see. You can change displays to plot black-on-white with the {Plot ▶ Plot Options} **Black on White** command. Black on white plotting is generally better for printing (Data Desk uses it by default for printing), but is slower for interactive displays.

If the drawing of a plot is aborted (for example, because Data Desk alerts you that a variable may have the wrong structure and you Cancel the command, or because you aborted plotting by pressing ⌘-period), Data Desk may create an empty plot. You can discard the plot (drag its icon alias to the trash), or substitute new variables in it as described in Section 9.11.

PLOTS TO DEPICT DISTRIBUTIONS

The plots discussed in the following section depict the distribution of a variable, or compare the distributions of several variables or groups. By understanding a variable's distribution we can be aware of extraordinary cases, of clumping, or of asymmetries that might alter our conclusions about the data.

8.3 Histograms

histograms

Data Desk makes two kinds of displays that show data in bars. Bar charts (Section 8.8) display variables that contain category names or identifiers. Each bar depicts the number of cases in a category. Histograms display variables that hold numeric values. Each data value is represented by an equal amount of area in the display, and these little bits of area are collected into bars placed side-by-side. Thus, histograms depict the overall distribution of data values.

> TIP
>
> **Histograms** depict the distribution of values in a *numeric* variable.
>
> **Bar Charts** depict the number of cases in each group of a *categorical* variable.

> HOW-TO
>
> How to make histograms:
> - Select the variable(s) to display.
> - Choose **Histograms** from the **Plot** menu.

Histograms show:

- The range of values covered by the data
- Where the values concentrate
- Whether the values are distributed symmetrically around the center, or trail off to one side
- Whether there are gaps where no values were observed
- Whether any values stray markedly from the rest

Histograms show the distribution of numeric values. The division into bars is arbitrary and can be changed to adjust the display. (By contrast, Bar Charts show counts of cases in pre-defined groups.) Often the first questions asked about a numeric variable can be answered by a histogram. For example: How big are the values? Are they spread out or compact? Are they symmetrically distributed about the middle or skewed? Do they cluster into two or more groups?

To make histograms, select the variables to be displayed and choose **Histograms** from the **Plot** menu. The first histogram opens in the upper left corner so you can drag its size box down and right, or click its zoom box to fill the screen if you wish. Histograms take the name of the variable they display.

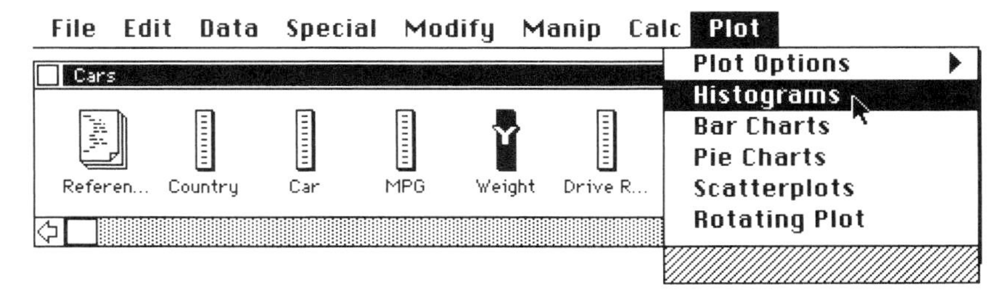

Figure 8-3. *To make a histogram, select the variable to display and choose* ***Histograms*** *from the* ***Plot*** *menu.*

Figure 8-4 shows the weights (in thousands of pounds) of the cars in the Cars dataset. The height of each bar shows the number of cars whose weights fall in a given range. The range covered by a bar starts at the value labeled at its left edge and extends up to, but not including, the value at the next tick mark to the right. Thus, in Figure 8-4 each bar depicts a range of 300 pounds. The cars weigh between 1800 pounds and 4500 pounds. If we look more carefully, we can see that there is one car between 4200 and 4500 pounds, three between 3900 and 4200, and so on[2].

symmetric

skewed

tails

The shape of the histogram reveals whether the data distribution is generally *symmetric*. Symmetrically distributed data are usually easier to work with and are more likely to satisfy the assumptions required by most statistics. If a histogram is not symmetric, it is said to be *skewed* toward the side that is more stretched out. The *tails* of a distribution contain the extreme largest and extreme smallest values in the data.

unimodal

bimodal

multimodal

A histogram may have a single central hump or *mode*, or it may have more than one mode. *Unimodal* (single-hump) data distributions are usually easier to work with and more likely to satisfy the assumptions of common statistical methods. *Bimodal* (two-hump) and *multimodal* (many-hump) distributions may indicate subgroups in the data. For example, a histogram of heights of people will usually have two modes—one consisting mostly of men and one mostly of women. Statisticians generally do not think that ordinary bar-to-bar variation in a histogram identifies modes, but rather they look at the overall shape of the histogram to see if it shows large-scale humps.

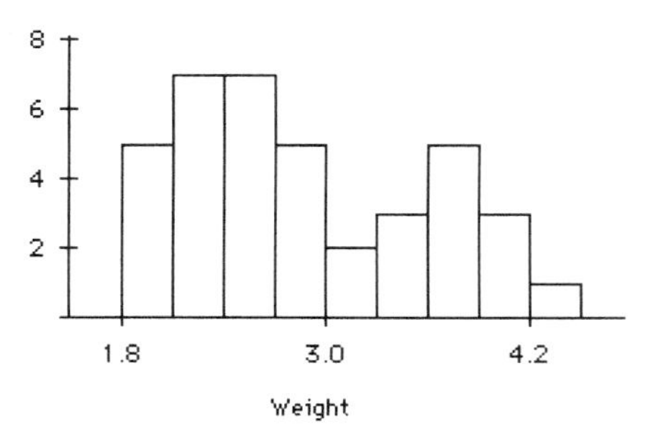

Figure 8-4. *Histogram of car weights*

The histogram of car weights is interesting because it is bimodal. Probably this represents a division between compact cars and larger cars, but it is remarkable to see such a strong split into groups rather than a smooth distribution including both large and small cars.

8.4 *Recentering and Rescaling Histograms*

With all we can learn from a histogram it is important to keep in mind what we *cannot* learn. When a histogram is drawn, the data values are grouped together and each group is graphed with a bar. Histograms of the same data drawn on different scales or with different group boundaries can look quite different. For example, if you drag the histogram window's size box to the right, the histogram gives the impression of an almost flat distribution of weights.

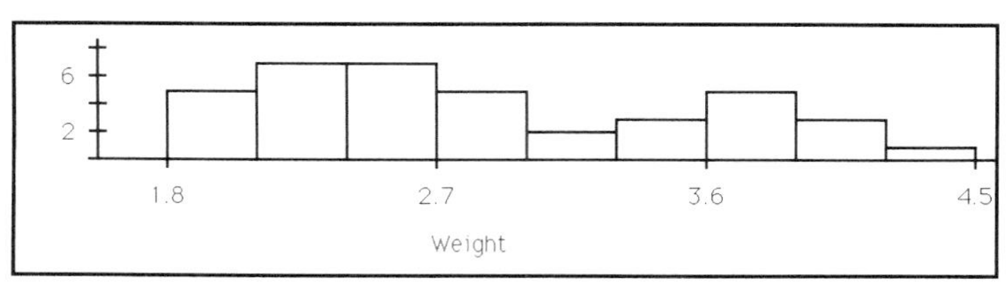

Figure 8-5. *Drag the size box to the right to make a histogram that gives the impression of a flat distribution.*

---

[2]The plot tool, discussed in Chapter 9, will grab the histogram and slide it side-to-side, so you can position any bar directly over the vertical axis to read its height easily.

If you drag the size box down and left to make the histogram tall and thin as in figure 8-6, it gives the impression of greater variation from bar to bar.

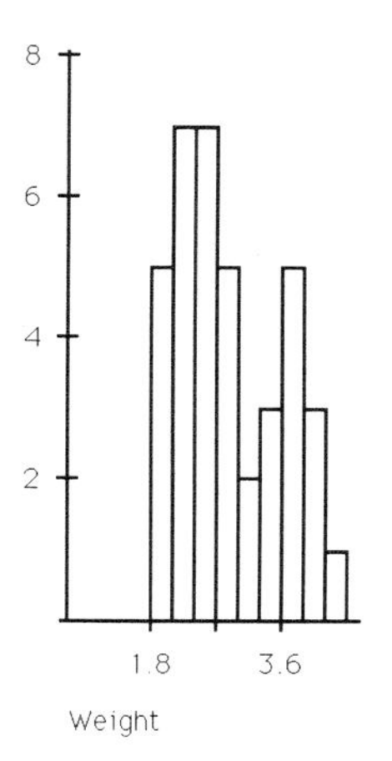

Figure 8-6. A tall histogram gives the impression of greater variability.

The bars of a resized histogram adjust proportionately to the changes in the window dimensions, but ordinarily do not change in number.

If you press the Option key before pressing the mouse button, and hold it down while resizing the window, the histogram bars remain about the same size but the number of bars changes to fill the new window. For example, you can rescale the car weights histogram so that each bar represents a weight range of 1000 pounds instead of 300. Hold down the Option key (the cursor will change to a left-right arrow), and drag the window smaller to leave room for slightly more than four bars:

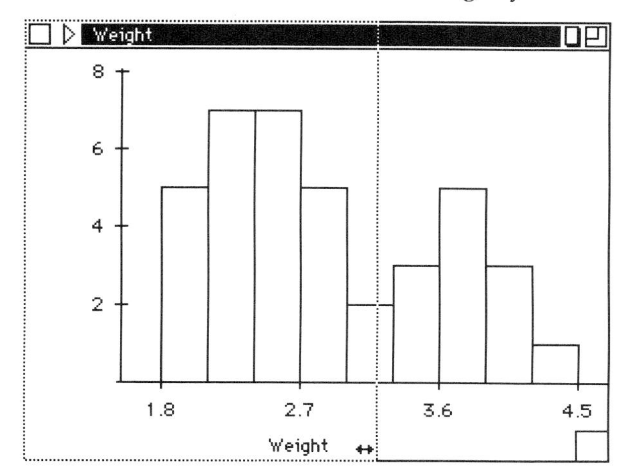

Figure 8-7. Hold the Option key and resize the window to specify more or fewer bars.

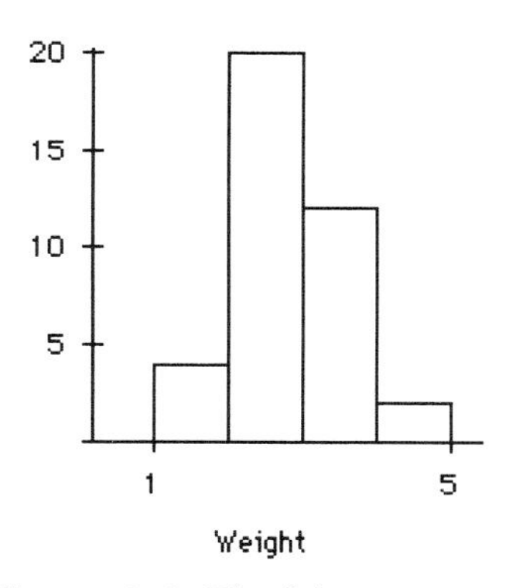

Figure 8-8. The histogram of Figure 8-7 resized to four bars.

When you release the mouse button, the data are redistributed into four bars, as Figure 8-8 shows. This scaling hides the bimodal pattern—a good example of the importance of finding the right scale for a display.

These two rescaling methods make it easy to adjust a histogram to have bars of any size and to have roughly the number of bars that makes the display look best. For example, to make the bars smaller, make the histogram window smaller (without pressing the Option key). To increase the number of bars, hold down Option key and make the window wider. At one extreme, you can make a histogram with very few bars. At the other extreme, many of the bars are empty, so the filled bars appear to be separated. For example, the car weights were recorded to the nearest hundred pounds, so a histogram with bars representing ranges smaller than that may have empty bars.

Because differently scaled histograms of the same data can look remarkably different, it is a good idea to rescale a histogram a few times to get a sense of what it *doesn't* say about the data.

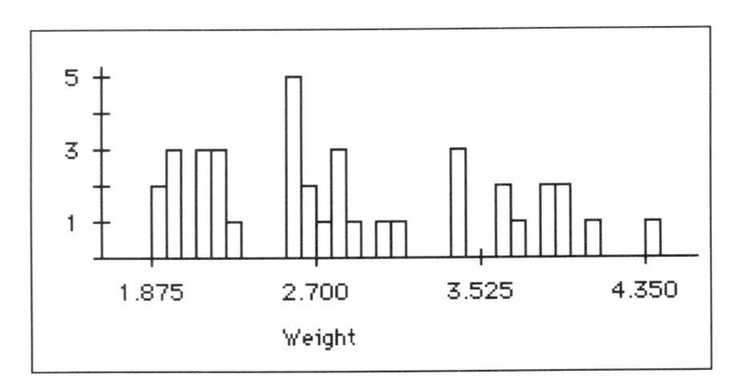

Figure 8-9. Requesting too many bars leaves empty bars.

HOW-TO

Change the number of bars in a histogram:

- Press Option.
- Resize the histogram window to accommodate the desired number of bars.

OR

- With the histogram window frontmost, Choose {Modify ▸ Scale} **Specify Plot Scale**.

The {Modify ▸ Scale} **Set Plot Scale...** command offers a way to rescale a histogram when it is the frontmost window to a specific starting bar and bar width. The **Set Plot Scale...** command is also available from the Global HyperView of the plot.

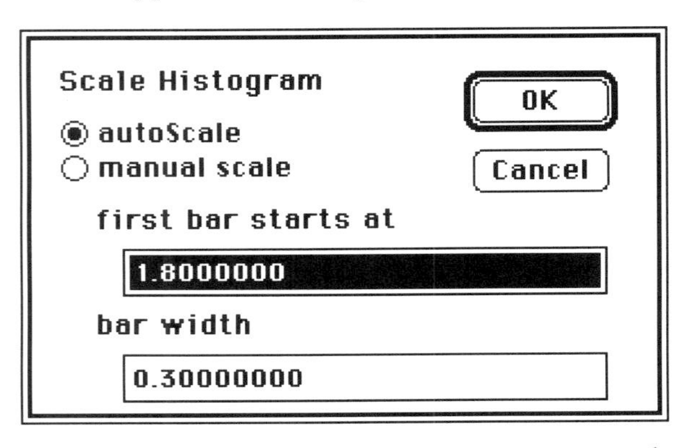

Figure 8-10. Specify Plot Scale lets you control histogram scaling in detail.

The dialog initially displays the current values for the frontmost histogram. Type a new value for the left edge of the first bar and for the width of the bars. Keep in mind that these values are in the same units as the data being plotted, in this case thousands of pounds.

Data Desk sets the histogram to **Manual Scale.** To return to automatic scaling, click the **Auto Scale** radio button in the Plot Scale dialog or choose **Automatic Scale** from the histogram's HyperView menu.

To slide a histogram side-to-side, grab it with the Grabber, ✋, from the Plot Tools palette, and slide it to the desired location (see Chapter 9).

8.5 Dotplots

Histograms display the distribution of all the values in a variable. Dotplots compare the distributions of values in each of several groups. For example, the data in the Cars dataset include the grouping variable *Country*, which specifies the group (country) in which each car was manufactured.

Car	Country	MPG	Weight
Buick Estate Wagon	U.S.	16.9	4.360
Ford Country Squire Wagon	U.S.	15.5	4.054
Chevy Malibu Wagon	U.S.	19.2	3.605
Chrysler LeBaron Wagon	U.S.	18.5	3.940
Chevette	U.S.	30.0	2.155
Toyota Corona	Japan	27.5	2.560
Datsun 510	Japan	27.2	2.300
Dodge Omni	U.S.	30.9	2.230
Audi 5000	Germany	20.3	2.830
Volvo 240 GL	Sweden	17.0	3.140
Saab 99 GLE	Sweden	21.6	2.795
Peugeot 694 SL	France	16.2	3.410
Buick Century Special	U.S.	20.6	3.380

Figure 8-11. *The variable Country can be a group variable.*

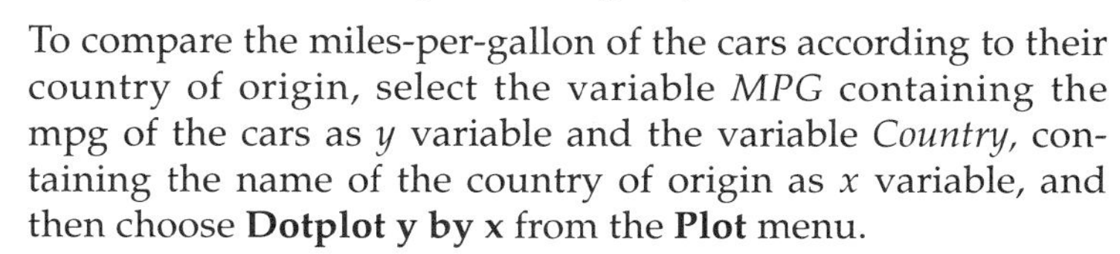

To compare the miles-per-gallon of the cars according to their country of origin, select the variable *MPG* containing the mpg of the cars as *y* variable and the variable *Country*, containing the name of the country of origin as *x* variable, and then choose **Dotplot y by x** from the **Plot** menu.

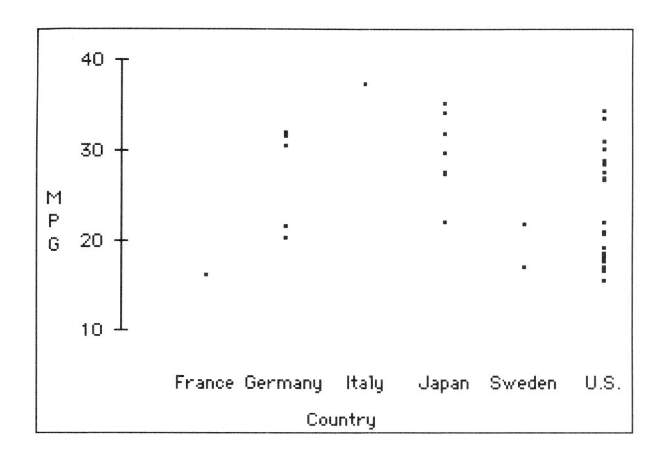

Figure 8-12. *Dotplot y by x of MPG by Country.*

A dotplot displays each group as a thin vertical stripe and each value as a single dot in its stripe. You can see where the dots clump together and how the groups compare in location and range.

For each group, the dotplot shows:

- The overall level of values
- The range of the data
- Any clumping of values in the variable

In addition, the collection of dotplots together shows:

- How the levels of the groups compare
- How the ranges of the groups compare
- Whether clumpings are similar across groups

If there is no room to put the full name of a group under its stripe of dots, Data Desk abbreviates the name, first to three characters then to just the initial letter. If you resize the window quite small, there may be no room for labels at all.

8.6 Dotplots of Separate Variables

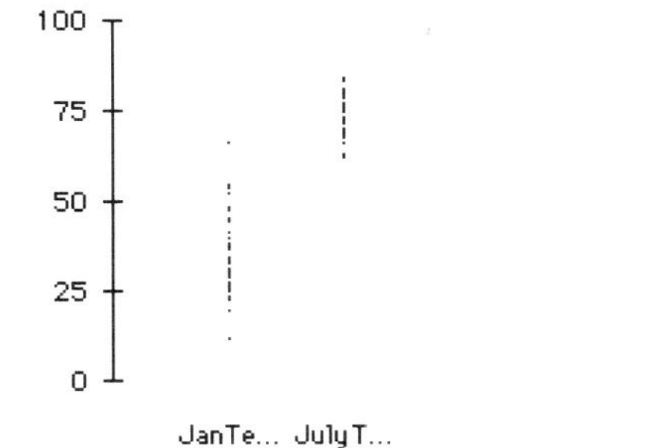

Figure 8-13. *Dotplot Side by Side of January and July temperatures.*

Sometimes you may want to compare several variables in the same relation. They might, for example, be measurements of the same quantity taken at different times or places. The **Dotplot Side by Side** command offers a way to display variables from the same relation side-by-side. Select the variables to be plotted and choose **Dotplot Side by Side** from the **Plot** menu.

Dotplots provide a convenient way to display the relationship between data values and color when you assign colors according to the values in a variable. (See Section 9.13 for details.) A dotplot of the variable from which colors were assigned shows the colors and the values that go with them, and can serve as a key for the color assignment in other displays.

8.7 Boxplots

Boxplots display much the same information as dotplots, but hide value-by-value detail to show more summary information. To make a boxplot, select a variable holding the values to be displayed as *y* and a variable holding the group identities as *x* and choose **Boxplot y by x** from the Plot menu. To compare several variables in a relation, select the variables to display and choose **Boxplot Side by Side**.

Boxplots have three components:

- The outlined central box depicts the middle half of the data between the 25th and the 75th percentiles. The horizontal line across the box marks the median.
- The *whiskers* extend from the top and bottom of the box to depict the extent of the main body of the data.
- Extreme data values are plotted individually, usually with a circle. Very extreme values are plotted with a starburst

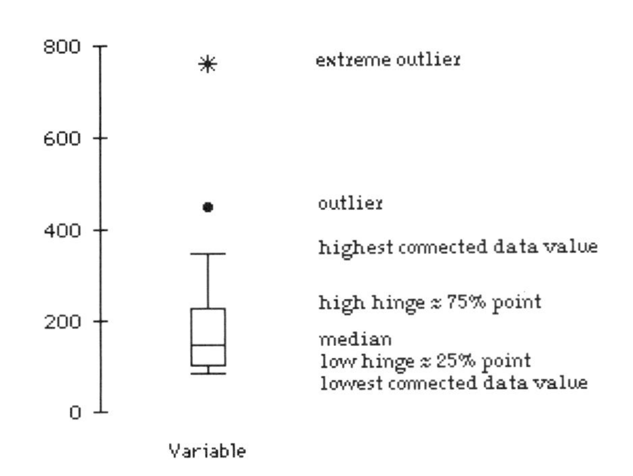

Figure 8-14. *Parts of a Boxplot.*

Appendix 8A defines the components of a boxplot precisely.

Boxplots comparing several variables show patterns in level and spread, and possible outlying values. For each variable selected, the boxplot shows:

- The overall level of values
- The overall variability or spread of the data
- Whether the main body of data values is distributed symmetrically around the median
- Any values that stray markedly from the rest

In addition, the collection of boxplots shows:

- How the levels of the variables compare
- How the spreads of the variables compare
- Relationships between the levels and spread. For example, do variables with a higher overall level tend to be more variable as well?

Figure 8-15 shows that the tenors and basses are generally taller than the

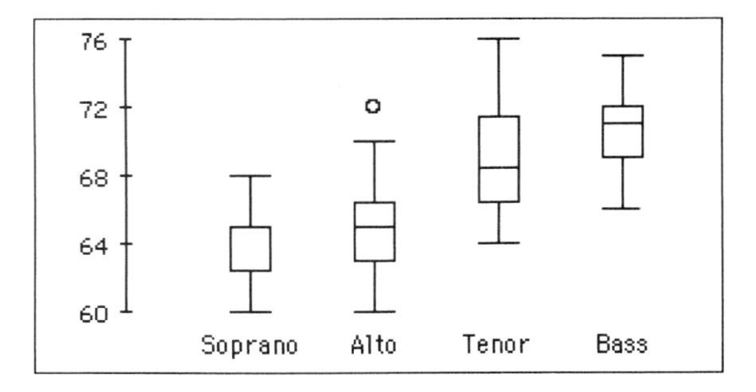

Figure 8-15. *Boxplots of Singers' heights by Part. The median Soprano height overlaps the upper edge of the box.*

sopranos and altos, which is not surprising. All four groups seem to be comparably variable. We also note one unusually tall alto.

Resizing a boxplot window changes the heights and horizontal locations of the boxes, but not their widths. If the window is too narrow the boxes may overlap, and the box labels are abbreviated as for dotplots.

Boxplots and Inference

confidence intervals

It is tempting to think that when two boxplot boxes do not overlap, the difference between their medians is statistically significant. However, while inferential statistics usually take account of sample size, boxplot boxes do not. We can augment boxplots to indicate whether the difference between any two of the group medians is statistically significant. The {Plot ▶ Plot Options}**Set Boxplot Options** command offers such *confidence intervals*. You can also set boxplot options from the Global HyperView of any Boxplot window.

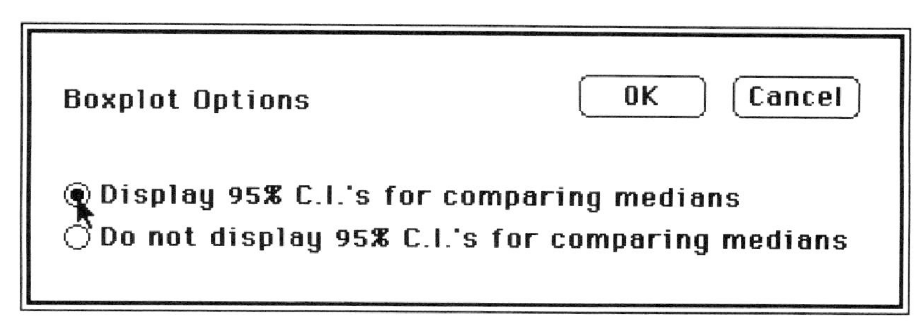

Figure 8-16. *Boxplot options offer to display confidence intervals for comparing medians of two boxes.*

The *Display 95% C.I.'s for comparing medians* option superimposes a shaded area on each box indicating confidence interval bounds around its median. These are not individual confidence intervals, but rather are constructed so that if two gray boxes fail to overlap, the corresponding medians are discernably different at approximately the 5% significance level. (See Appendix 8A for Boxplot definitions.)

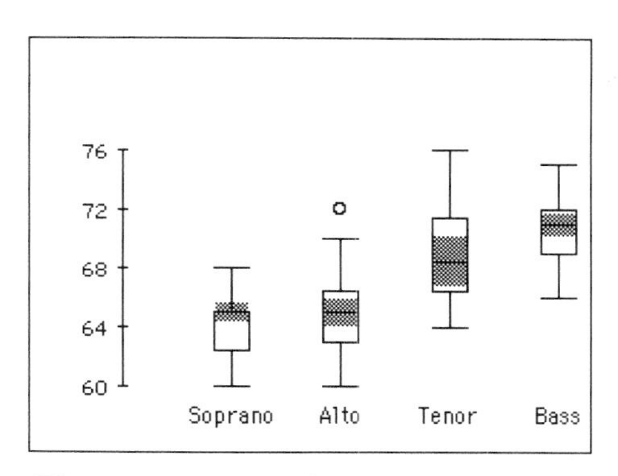

Figure 8-17. *Boxplots shown with confidence intervals.*

For the Singers' heights, the tenors and basses appear to be significantly taller than the sopranos and altos because the shaded areas of the boxes do not overlap (Fig. 8-17). It is not clear whether the basses are significantly taller than the tenors. We may want to look into this question more carefully with more powerful inferential methods. (See Chapters 16 and 17 for a discussion of hypothesis testing and confidence intervals.)

PLOTS TO COMPARE CATEGORIES

category variables

Variables that classify cases into *categories* rather than reporting measurements or values require special displays. Data Desk offers two displays for such *discrete* data: bar charts and pie charts. They are discussed in the next two sections.

8.8 Barcharts

REMEMBER

Histograms show the distribution of a variable that holds numbers.

Bar charts show the counts in each category of a variable that names groups.

Bar charts are sometimes confused with histograms because the two displays look very similar. The difference between them is in the kind of data they display. Bar charts display variables that contain category names or identifiers. Histograms display variables that hold numeric values. A bar chart depicts each category with a bar whose length is proportional to the number of cases in the category. Unlike histogram bars, the left-to-right order of bar chart bars is arbitrary so the overall shape of a bar chart is meaningless. By default, Data Desk places bars in alphabetical order according to the category names.

HOW-TO

How to make a bar chart:

- Select the variable that identifies the groups.
- Choose **Bar Charts** from the **Plot** menu.

To make a bar chart, select the variable that identifies the groups and choose **Bar Charts** from the **Plot** menu. For example, a bar chart of Country from the Cars data is shown in Figure 8-18.

Bar charts make it particularly easy to compare categories because most people can discern differences in the lengths of two bars set on a common baseline quite well, and even (somewhat less precisely) the ratios of the lengths. For example, it is easy to see in Figure 8-18 that there are slightly more Japanese cars than German cars in this sample and that U.S. cars dominate the sample with three or four times as many cars.

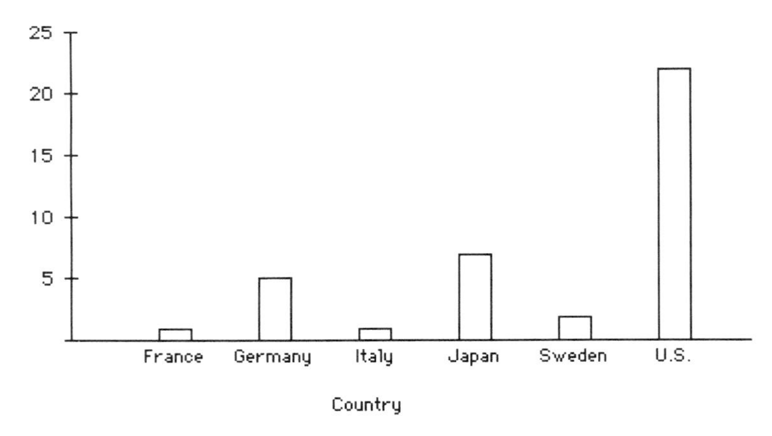

Figure 8-18. *Bar chart of Country for the Cars data.*

However, bar charts do not emphasize the relationship of category size to the whole sample, so they are usually used when the relative sizes of the groups are of greater interest than the division of the whole into subgroups.

A bar chart shows:

- How many categories there are
- The relative size of each category
- Whether any category is particularly dominant or particularly sparse

Group names are abbreviated if necessary. Groups can be identified with numbers as well as names. Data Desk assumes that the variable selected for a bar chart contains names and interprets the contents of that variable according to their *text* values. This means that the values "1" and "1.0" would not be placed in the same category because their text values differ, even though they are equal numerically. If Data Desk suspects that a variable is not a group variable (for example, if it has many categories, each of them with only one case) it checks with you before proceeding.

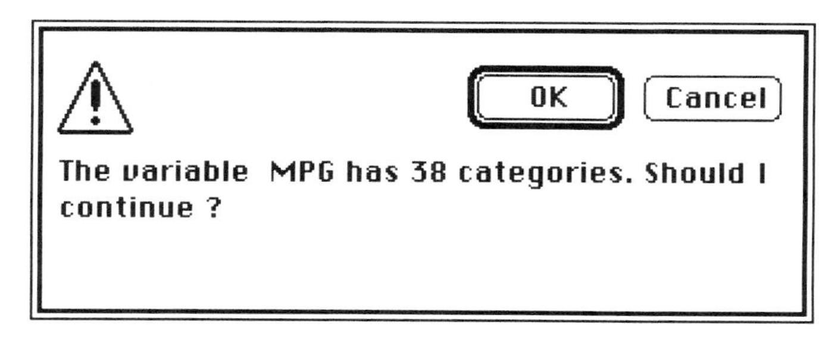

Figure 8-19. *Data Desk posts an Alert if you choose a possibly inappropriate variable for a bar chart.*

8.9 Pie Charts

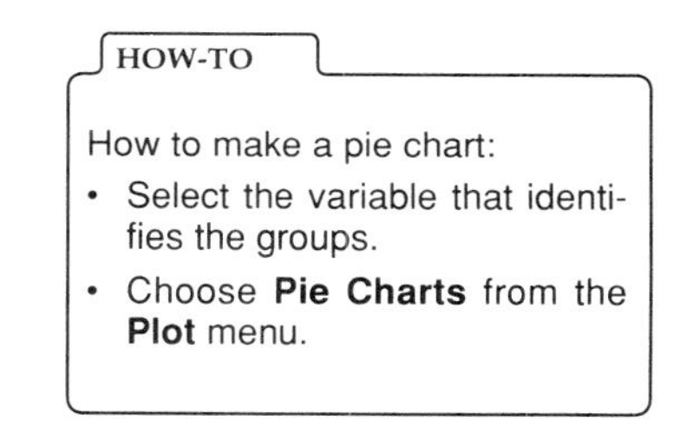
HOW-TO

How to make a pie chart:

- Select the variable that identifies the groups.
- Choose **Pie Charts** from the **Plot** menu.

Traditionally, pie charts have been thought of as presentation displays. A pie chart depicts a variable that contains category names or identifiers. The "pie" represents the whole population, which has been partitioned into categories or groups, each represented by a slice or wedge. The area of each wedge of the pie is proportional to the number of cases in its category. Pie charts are thus particularly suited for displaying the division of a whole into several subgroups.

To make a pie chart, select a variable containing category identifiers and choose **Pie Charts** from the **Plot** menu. To change the variable being displayed in an existing pie chart, drag a new variable containing category identifiers into the pie chart window.

For example, the pie chart of Country from the Cars data in Figure 8-2 helps us think about the *relative fraction* of cars from each country. By contrast, the bar chart in Figure 8-18, emphasizes the *number* of cars from each country.

Although pie charts observe the area principle, they have been criticized because most people cannot discern small differences in the angles of pie wedges, and thus may misjudge what a pie chart displays. Nevertheless, pie charts are intuitively easy to understand and are used widely.

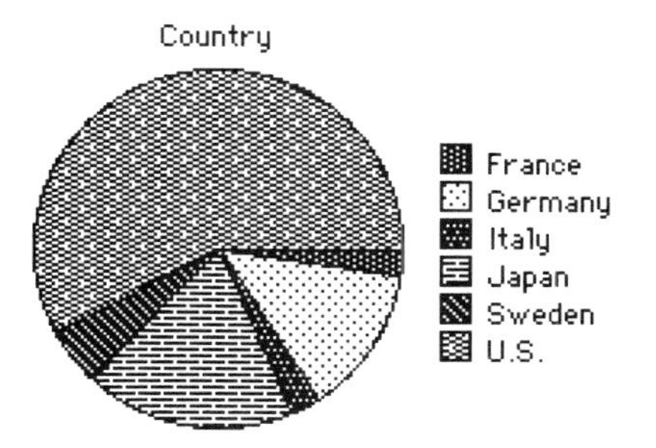

Figure 8-20. *Pie chart of Countries in the Cars data.*

Pie charts show:

- How many categories there are
- The partitioning of the whole into fractions
- Whether any category is particularly dominant or particularly small

On a color display, pie charts can be colored, each slice taking on the color that would be assigned to its category were the displayed variable used to assign colors by group. (See Section 9-13.) To color a pie chart choose **Use Colors** command from its HyperView.

PLOTS TO DEPICT RELATIONSHIPS

Many of Data Desk's displays show the relationships among two or more variables, or the relationship between a single variable and a standard. These plots depict each case as a point in the display. The points can be plotted with different symbols, connected with lines, moved on the display, and otherwise modified by using the plot tools and commands discussed in Chapter 9.

8.10 Scatterplots

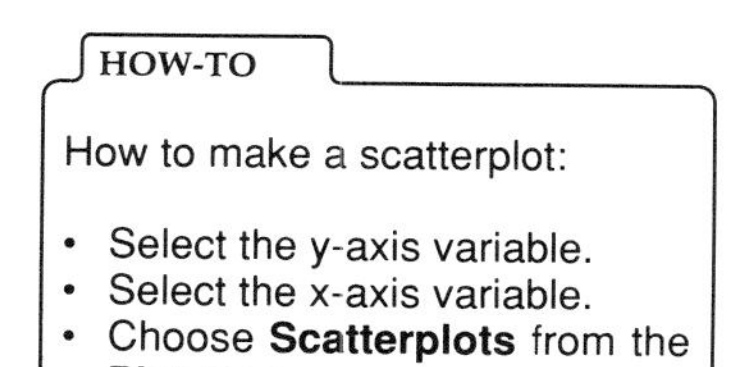

HOW-TO

How to make a scatterplot:

- Select the y-axis variable.
- Select the x-axis variable.
- Choose **Scatterplots** from the **Plot** menu.

Scatterplots show relationships between pairs of variables. In Data Desk, values in the same row of each variable in a relation refer to the same case, and thus are paired suitably for a scatterplot. Every case plotted in a scatterplot has two data values — one for each of the two variables graphed. By convention, the variable whose values are plotted vertically is denoted y, and the variable whose values are plotted horizontally is denoted x. The distinction between y and x variables persists throughout much of statistical analysis, and is reflected in the different ways Data Desk highlights icons designated as y or x.

To make a scatterplot, first select the variable to plot on the vertical (y) axis. Then, holding down the Shift key (to extend the selection and get the ✕ cursor), select the variable to plot on the horizontal (x) axis. Finally, choose **Scatterplots** from the **Plot** menu.

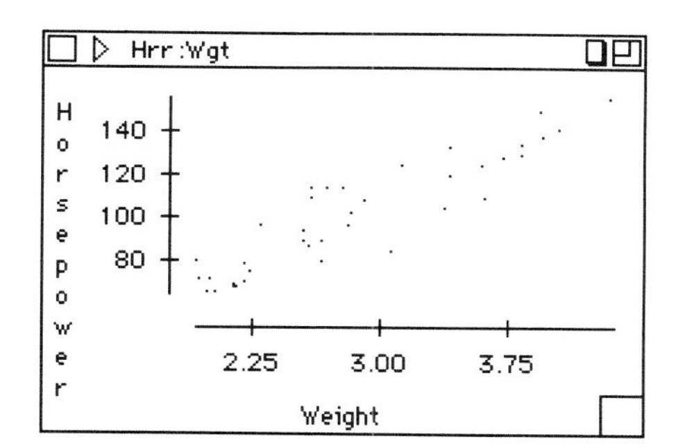

Figure 8-21. Scatterplots show the relationship between two variables.

For example, in the Cars dataset, a scatterplot of Horsepower *vs* Weight shows that heavier cars have more powerful engines, as we might expect:

Scatterplots show:

- Trends between the y-values and the x-values
- Whether a trend is straight or curved
- Clustering of data points
- Changes in the spread of y-values as x-values increase
- Extraordinary data points far from the rest of the data

Trends in a scatterplot are often interesting statistically. Straight-line relationships may be approximated numerically with regression analysis, which is discussed Chapter 20. Simple curved relationships can be described similarly with the use of transformations, such as those discussed in Chapter 11.

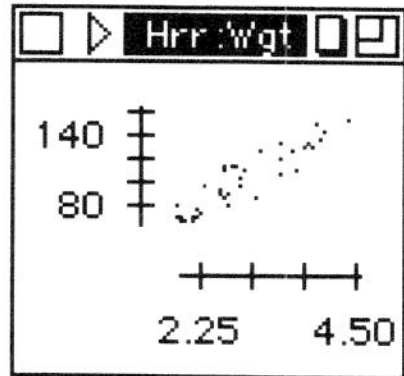

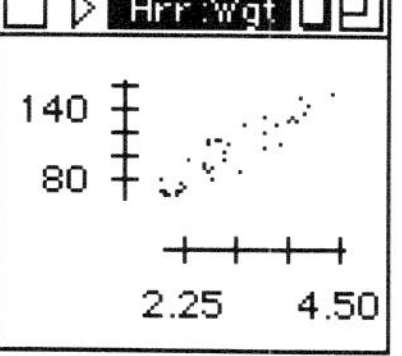

Figure 8-22. A compact version of Figure 8-21 shows much the same information.

As with histograms, the visual impression of trends in a scatterplot can be altered by rescaling its window with the size box. However, a scatterplot can also become too large. A plot with only a small number of data points (around 20 or 30) often is easier to read when drawn in a window about 2 inches square than drawn to fill the entire screen.

You can select several y-variables (press Option to get the **Y** cursor) and one x-variable, or one y-variable and several x-variables to request several plots at once. For example, you might plot the dependent variable in a multiple regression as y against each of the predictors in the regression equation. You can replace either variable in a scatterplot simply by dragging the icon of the new variable on top of the axis label of the variable to be replaced. To replace both variables, select the new y-axis variable as *y* (option-click) and the new x-axis variable as *x* (shift-click) and drag them both into the center of the plot.

8.11 Lineplots

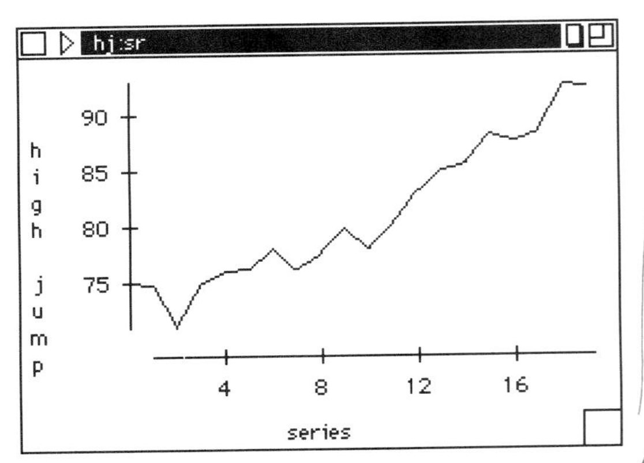

Figure 8-23. *A sequence plot of Olympic high jump gold medal performance since 1900 shows a consistent trend of improvement.*

A lineplot graphs a variable in case order, and connects the successive points with lines. It is like a scatterplot of a variable against another that counts from 1 to n.

To make a lineplot, select the variable to plot and choose **Lineplot** from the **Plot** menu.

The line connecting successive points helps your eye to follow any trends, so lineplots are often used for data recorded over time. However, it is a good idea to make a lineplot of any variable just to check for unexpected trends related to the sequence order in which the data are recorded. This is especially true if the cases are recorded in the order in which they were collected.

Lineplots space the datapoints evenly across the x-axis. If your data were measured in order but at unequal intervals you may get a more appropriate picture of your data by making a scatterplot of the data against an x-variable that specifies the correct spacing and adding lines.

You can turn any scatterplot into a lineplot with the {Modify ▶ Lines ▶ Add} commands. Select the x-axis variable (the **Select** command in the HyperView on the x-axis name is a convenient way) and choose {Manip} **Sort on Y Carry X's**. The Sort command generates a new icon called *Unsort Indices.* Select this variable, click on the scatterplot, and choose {Modify ▶ Lines ▶ Add} **by Series**.

Alternatively, make a variable that is entirely constant, using the {Manip} **Generate Patterned Data…** command (Specify that it should go from 1 to 1 in steps of 1 repeated n times), select the generated variable, click on the scatterplot, and choose {Modify ▶ Lines ▶ Add} **by Group**.

The **Multiple Lineplot** command plots several variables, against a common sequence axis. If you are working without color display, then each variable is plotted with a different symbol (up to eight symbols). If color is available, Data Desk plots each line in a different color. Multiple lineplots show how several sequences move together. They do require, however, that all of the sequences be measured on the same scale or the *y*–axis of the plot is meaningless. Thus, for example, a multiple line plot of the price of Apple Stock (measured in dollars) and the number of Macintosh apples sold annually (in millions of bushels) might show related trends, but would arbitrarily equate 100,000,000 bu. to $100.00. It is important to keep in mind that this equivalence is an artifact of the display with no meaning for the data.

A similar, but more insidious, problem is that when variables displayed together are measured in different units, both their absolute magnitude and their scale are arbitrary. Thus, we could decide to represent the Apple stock prices in pennies. This would multiply all the stock values by 100 but would not make the multiple line plot wrong in any formal

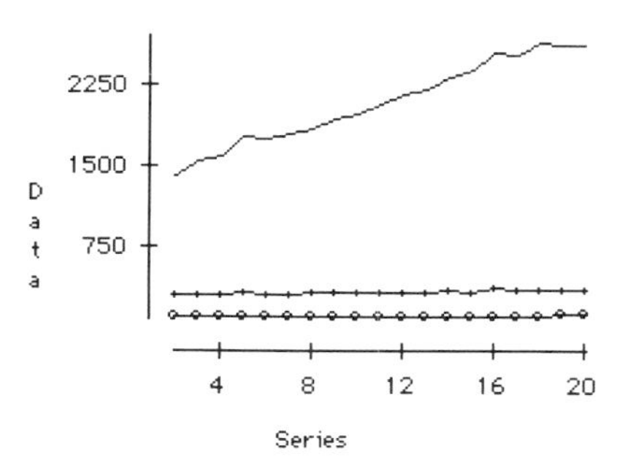

Figure 8-24 *Multiple line plot of the Olympic gold medal performances seems to show that discus gold medal performance has grown far ahead of long jump and high jump. But see Figure 8-25 for another view.*

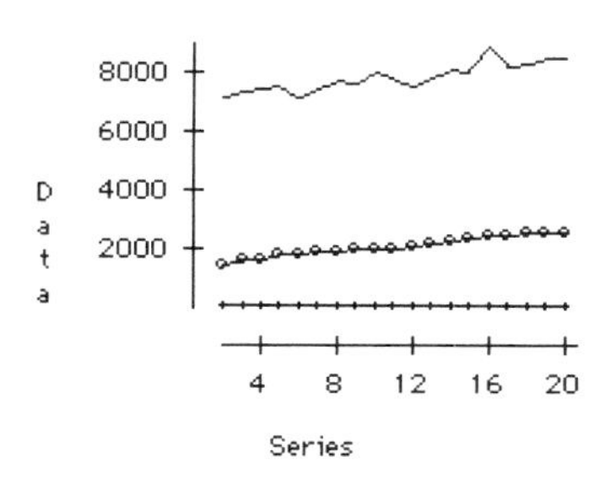

Figure 8-25. *Long Jump performance leaps ahead of other field sports.*

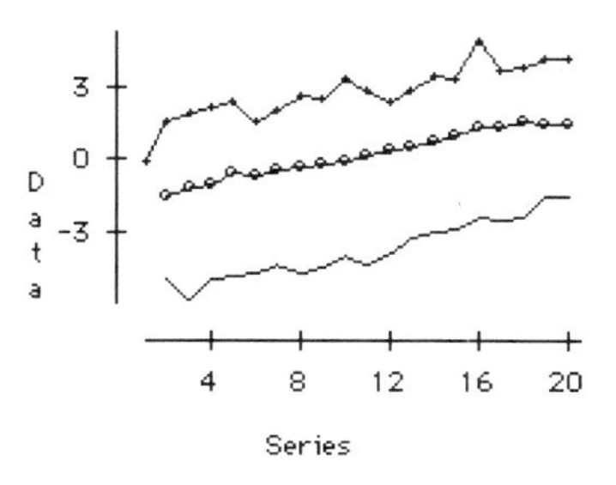

Figure 8-26 *This multiple line plot of the Olympic gold medal performances seems to show that long jump gold medal performance has grown far ahead of discuss throw and high jump. But see Figure 8-24 for another view.*

statistical sense. Multiplication by a large factor will, however, inflate every small fluctuation in stock price into a huge swing, while making the changes in fruit sales seem trivial because they cover only a few million bushels.

For example, the Olympic gold dataset hold three sequences of gold medal performances. While all of these are measured in inches, that does not make them comparable. A typical discus throw is much longer than a typical high jump is high. A simple multiple line plot of the three series looks like figure 8-24.

We could caption it "discus throw performance soars while other track and field sports languish". But a simple rescaling of the long jump results produces the equally valid plot in Figure 8-25, which clearly demonstrates that it is long jump performance that is leaping ahead of the others.

Multiple line plots must thus be used (if at all) with great care. They are most trustworthy when the sequences plotted are directly comparable, measured in the same units, and of the same general magnitude and range.

With these *caveats* in mind, it can be useful to compare sequences with multiple line plots. When our chief concern is whether fluctuations or trends seen in one sequence are reflected in others, rather than the direct comparison of the size or nature of these patterns among sequences, it is often best to standardize the sequences so that they are nearly of the same magnitude and variance. We can do that by converting them to standardized scores, for example by subtracting the mean from each and dividing each by its standard deviation. Thus, for example, the Olympic Gold series could be transformed to:

('high jump' - 100)/5

('Discus Throw' - 2000)/400

('long jump' - 250)/20

Figure 8-26 shows a multiple lineplot of the standardized values. Now we can see how they have grown together, including the effects of the two world wars on all three sequences. We can also look for a possible "Mexico City effect" in 1968 proposed by some to be due to a slightly lower pull of gravity at the high altitude there.

Multiple lineplots create a new relation holding three intermediate variables needed to construct the plot. The first of these appends the selected variables together. The other two provide the x-axis and group names. Because Multiple lineplots work in this separate relation they do not link to the relation of the original data. Initially, the relation is hidden. To make it visible, choose any **Locate** command in a lineplot HyperView.

As with Lineplots, you can make a Multiple Lineplot against an alternative x-axis. This is particularly valuable when the data sequences are not observed at constant intervals. Sort the variables according to the alternative x-axis variable if necessary and make a multiple line plot. Now select the alternative x-axis variable and choose {Manip} **Repeat Variables....** Specify repeating the variable the same number of times as the number of sequences you plan to display. Now drag the repeated variable onto the x-axis of the lineplot.

8.12 Normal Probability Plots

Normal distribution

A Normal probability plot provides a simple way to tell whether the numbers in a variable are approximately Normally distributed. Many statistics assume that data or residuals follow a *Normal distribution*, so checks of normality help to determine the applicability of some methods.

normal score

Data Desk sorts the values in the variable and then, starting with the smallest value, poses the question: "If this were a sample from the standard Normal distribution (that is, the Normal distribution with zero mean and unit variance), what would I expect the smallest value to be?" The answer is the first *normal score*. It depends only on the number of cases in the variable, and can be estimated with the NScores function available in derived variables (see Chapter 11). The probability plot graphs the observed smallest value against the value expected under the assumption of a Normal distribution.

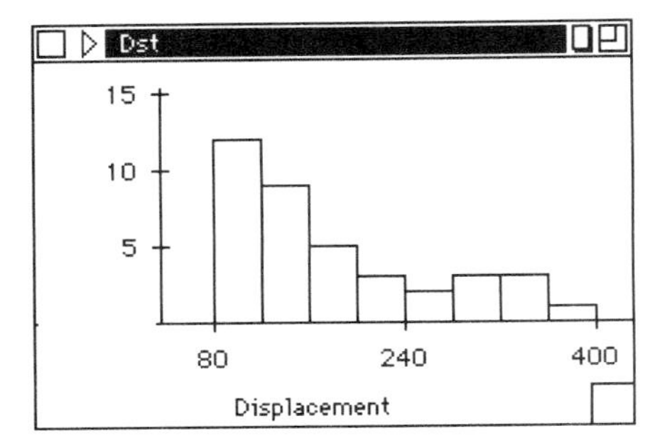

Figure 8-27. *A histogram of car engine displacement does not look Normal.*

The question is repeated for the second smallest value; "If this were a sample from the standard Normal distribution what would I expect the *second* smallest value to be?" The observed second smallest value is plotted against the expected second smallest value.

The corresponding questions are posed and calculations determined for each data value, and the resulting plot graphs the observed value as y *vs* the calculated *normal score* as x for each case.

To make a Normal probability plot, select the variable to plot and choose **Normal Prob Plots** from the **Plot** menu. Normal probability plots are also available from HyperViews in several procedures in which the assumptions of Normality is important.

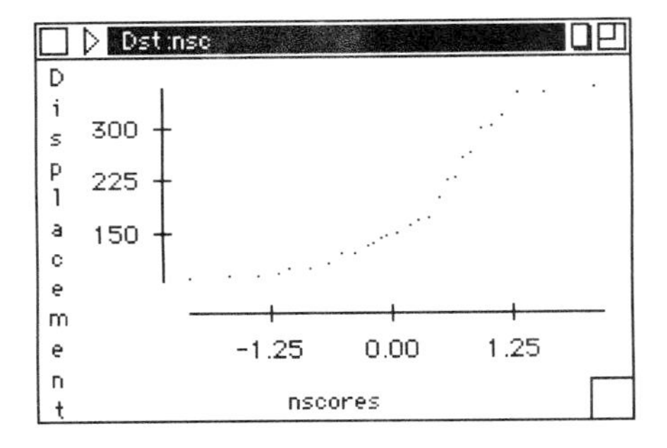

Figure 8-28. *Normal probability plot of Displacement in Cars data.*

For example, the Displacement variable in the Cars dataset is very non-Normal. In fact, the histogram in Figure 8-27 has a dip in the middle. Its Normal probability plot in Figure 8-28 shows the strong deviation from normality.

Probability plots always show a non-decreasing trend from lower left to upper right. If the plot is straight or nearly straight, then the distribution of the variable is nearly Normal, and the slope of the line estimates the standard deviation of the variable. If the plot is S-shaped, with the right and left parts pointing up and down, the variable's distribution is more stretched out, or long-tailed, than the Normal. If the plot is S-shaped, with the right and left parts pointing right and left, then the variable's distribution is more compact, or short-tailed, than the Normal. If only one side of the plot bends away from the line, the variable has a skewed distribution.

The correlation of a variable and its normal scores is sometimes used to measure how near the variable's distribution is to the Normal distribution. The normal scores are available as the NScores function in derived variables. Appendix 8B provides details about the calculation of normal scores.

APPENDIX 8A
Boxplot Definitions

Boxplots are defined in terms of the median and *hinges* of a collection of numbers. The hinges are much like the 25th and 75th percentiles of a variable. More precisely, they are the medians of the data from the minimum to the median (like a 25th percentile) and of the data from the median to the maximum (like a 75th percentile). The median is included in both half-variables if it is one of the data values, but not included if it is the average of two data values.

The "box" in a boxplot extends from the low hinge (roughly, the 25% point) to the high hinge (roughly, the 75% point). The horizontal bar is at the median. The *whiskers* extend from the box to the highest data value not above:

high hinge + 1.5 (high hinge - low hinge)

and from the box to the lowest data value not below:

low hinge - 1.5 (high hinge - low hinge)

Any data value beyond these limits is plotted with a circle unless it exceeds either:

high hinge + 3.0 (high hinge - low hinge)

or:

low hinge - 3.0 (high hinge - low hinge)

in which case, it is plotted with a starburst.

The shaded intervals for comparing medians are placed symmetrically around the median at

$$\text{median} \pm 1.58\,(\text{high hinge} - \text{low hinge})/\sqrt{n}$$

See Velleman and Hoaglin (1981), *Applications, Basics, and Computing of Exploratory Data Analysis,* Boston, Duxbury Press, Chapter 3, for a complete discussion of boxplots and the derivation of the shaded interval formulas.

Alternative definitions of boxplots have been proposed in some publications and implemented in some other statistics programs. The definitions shown here and used by Data Desk are those of the originator of boxplots, Dr. John W. Tukey.

Appendix 8B
Probability Plots

Probability plots (sometimes called Quantile-Quantile or Q-Q plots) provide information about how well the distribution of a variable matches some theoretical distribution by plotting the observed values (on the y-axis) against their normal scores (on the x-axis).

Traditionally, the ith normal score for a sample of size n has been the mean of the *sampling distribution* of the ith order statistic (that is, the ith ordered value) in a sample of n values drawn from a standard Normal distribution. Data Desk instead approximates the *medians* of the sampling distributions of the order statistics of the standard Normal distribution. This has the advantage of being more resistant to the inherent skewness of the sampling distributions belonging to the extreme order statistics. The ith median order statistic of a sample of size n is approximated by

$$Gau^{-1}((i - 1/3) / (n + 1/3)$$

where Gau^{-1} is the inverse Normal (or Gaussian) cumulative distribution function.

This formula is also used by the NScores function in derived variables.

CHAPTER 9

Working with Displays

9.1	Chapter Organization	*117*
9.2	Organization of Features and Commands in Data Desk	*118*
9.3	Basic Plot Actions	*119*
9.4	Select	*120*
9.5	Link	*121*
9.6	Brush and Slice	*122*
9.7	Identify	*123*
9.8	Move	*123*
9.9	Isolate	*124*
9.10	Resize	*126*
9.11	Substitute	*126*
9.12	Symbols	*128*
9.13	Colors	*128*
9.14	Axes	*130*
9.15	Scale	*130*
9.16	Visibility	*133*
9.17	Lines	*133*
9.18	Lines and Color	*135*
9.19	Working with Displays	*135*

APPENDIX

9A	Plot Tool Shortcuts	*137*

DATA DESK'S PLOTS can display your data from many points of view. But the true power of these plots comes from touching them and watching them change dynamically in response to your actions. This chapter discusses the elements of making and modifying data analysis graphics, and shows how various methods and plot tools can help you see more in your data.

A NOTE FOR STUDENTS

The tools and techniques of plot manipulation are simple and intuitive. Nevertheless, they are not central to learning statistics. You can easily proceed with your study of statistics without learning about any of these methods. You have probably encountered this chapter relatively early in your study of statistics. If you are just learning statistics, we recommend that you skip both chapters 9 and 10 for now and return to them as you become more comfortable with Data Desk and with statistics.

If you are studying statistical graphics, or wish to concentrate on getting the most out of your data, then these chapters are just what you need. They require no special knowledge or mathematics and present methods that are visual and intuitive.

Data Desk's plots will help you discover unexpected patterns and relationships, identify groups and clusters, and spot extraordinary cases. Some plots show the distribution of a single variable (for example, histograms), or compare the distributions of several groups (for example, dotplots and boxplots). Some plots show information about category data by depicting relative proportions of the data in each category (for example, bar charts and pie charts). Other displays depict relationships. The relationships can be between a single variable and some standard, such as the sequence order of its cases (for example, lineplots) or expected values from a standard distribution (for example, probability plots), between two variables (for example, scatterplots), or comparing several sequences (multiple line plots).

Please experiment with the plots available in Data Desk. Any collection of methods as rich as these encourages new and creative applications that go beyond the discussions in this book. We hope you will try new combinations of methods if only to see how they work on your data, and that you will find these techniques both effective and enjoyable.

9.1 *Chapter Organization*

Data Desk offers a highly integrated system of data analysis graphics capabilities. Some of the graphics techniques in Data Desk may be new to you, but they are designed to work intuitively and to have consistent controls. An action that modifies a plot in some way will modify other plots in much the same way. For example, the tool always repositions the contents of a plot within its window.

It is easy to start working with Data Desk's plots. As you work with them you will learn more and more ways to combine the tools, commands, and options to see more in your data. The combination of graphic views and ways of working with them is far too rich to be able to document every possible combination. Instead, this chapter presents the basic principles of interactive graphics and discusses the corresponding Data Desk tools and commands.

We divide graphics concepts into *Actions* and *Aspects*. Actions include all ways in which you can move, highlight, inquire, or otherwise act on a plot to see a new view, learn more about the data, or record information for further analysis. Many actions are accomplished through the plot tools, but some work in other ways. Aspects of plots specify how the plot looks and works. Most plot aspects are controlled with commands, although some are controlled by tools in special palettes.

9.2 *Organization of Features and Commands in Data Desk*

In Data Desk, you control displays with the features and commands in the **Plot** and **Modify** menus. Together, these provide many simple and natural ways to work with plots. As with most of Data Desk's features, you should get started with the basic commands, and learn as you work.

REMEMBER

The **Plot** and **Modify** menus hold the features and commands that control displays.

The **Plot** menu holds the basic plot commands and the basic plot options. Plot options are controlled by the **Plot Options** submenu in the **Plot** menu. Plot options usually specify the default operation of plots. As with other options in Data Desk, plot options specify how the plots you are about to make work or look.

Commands in the **Modify** menu alter or work with the frontmost plot. They are grouped into submenus according to the aspect of the plot affected. Thus, commands that alter plot scaling are grouped together under the **Scale** submenu. Commands that affect the axes appear together in the **Axes** submenu.

Figure 9-1. *The Plot Tools palette.*

The **Modify** menu also holds three palettes; the Plot Tools palette, the Plot Symbols palette, and the Plot Colors palette. All are found as menus or submenus. You can select items from these palettes as you would from any submenu, or you can tear them off the menu by dragging down through them and off onto the desktop. Palettes left on the desktop float above other Data Desk windows so that they are always visible. To select an item from a palette when it is on the desktop simply click one of the rectangles in the palette. The **Palettes** (or **⌘M**) command places all four palettes (three if you are not working in color) on the desktop. To hide the palettes choose the **Palettes** command again, type **⌘M**, or close individual palettes by clicking their close box.

The *Plot Tools palette* holds 12 tools for manipulating, moving, and using displays, each of which is discussed in the context of its function throughout this chapter. Because plot characteristics differ, a particular tool may act differently on different plots. However, each tool's basic function and behavior is consistent across all plots.

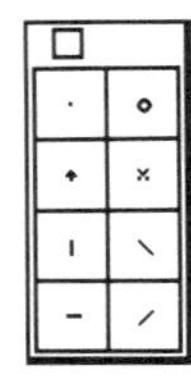

Figure 9-2. *The Plot Symbols palette.*

The *Plot symbols palette* shows the eight available plot symbols. All points are initially represented by a single dot. To change the symbol for a case or a group of cases, select the cases to change and click on a plot symbol in the palette. Those points will then be displayed using the new symbol in all plots that represent individual cases, such as a scatterplot, or dotplot.

The *Plot Colors palette* is active only on systems with color. It has 64 squares showing the 64 colors available in a Data Desk display. Select cases and click on any color to display those cases with that color in any plot that represents individual cases. Data Desk requires that color screens be set to 8-bit depth (256 colors). If your screen is not set to 8-bit color, Data Desk will work without color, and the color palette will not be available.

ACTIONS

9.3 *Basic Plot Actions*

The basic plot actions are:

- *SELECT*
  Selected points or parts of displays highlight to stand out from the rest of the plot and are available for separate analysis or modification. Many commands operate specifically on the selected cases.
- *LINK*
  Data Desk links plots so that parts of plots in the same relation that represent the same cases show them consistently. For example, selecting a point (and, thereby, a case) in one plot highlights the parts of other plots that display that case.
- *BRUSH AND SLICE*
  You can see relationships among several plots at once by brushing or slicing a plot to select points and seeing the corresponding cases highlight in other plots.
- *IDENTIFY*

  Data Desk can tell you the identity of any individual case in plots that display cases individually.
- *MOVE*
  Motion, such as repositioning, turns the abstract concept of a "point cloud" into a physical reality that you can touch and manipulate.
- *ISOLATE*
  You can refocus plots to ignore extreme points or to concentrate on interesting subsets of the data.
- *RESIZE*
  Control the size of the plot contents independently of the size of the plot window.
- *SUBSTITUTE*
  Any plot can serve as a template for new plots. You can substitute a new variable for one in the plot simply by dragging the icon of the new variable over the name of the old one on an axis.

Each of these actions is discussed in its own section below.

9.4 Select

Selecting is a fundamental operation both because selected points stand out from the background of other points, and because a variety of other commands can operate on the selected points. For example, you can choose to display selected points with any of the 8 plot symbols, color them in any of the 64 available colors, or hide them from view to see the remaining points more clearly.

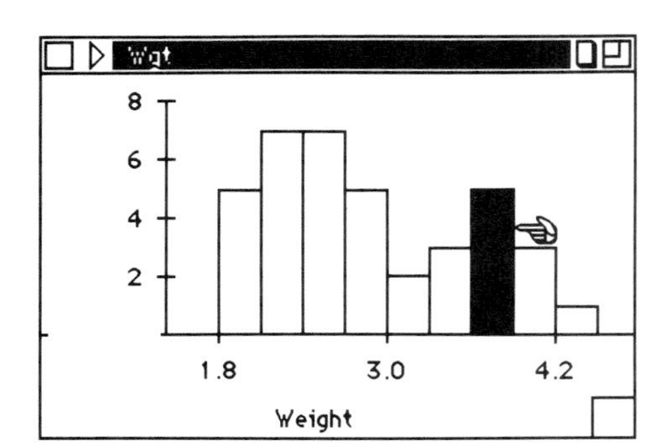

Figure 9-3. Selected parts of a plot highlight.

Selected points and regions highlight by becoming brighter, by becoming slightly larger, by changing color, or by filling in open spaces. In some plots (for example, pie charts, bar charts, and histograms) you cannot select individual cases, but you can select collections of cases (for example, an entire pie chart wedge or histogram bar).

Several plot tools perform selections:

This tool operates on all plots. When the mouse button is pressed, it selects the datapoint or the part of the plot it is pointing to. The tool selects individual points in scatterplots, but selects entire bars in histograms and entire pie wedges in pie charts. Hold down the Shift key to extend the selection, adding unselected points, bars, and wedges to the selection or to de-select parts that have already been selected.

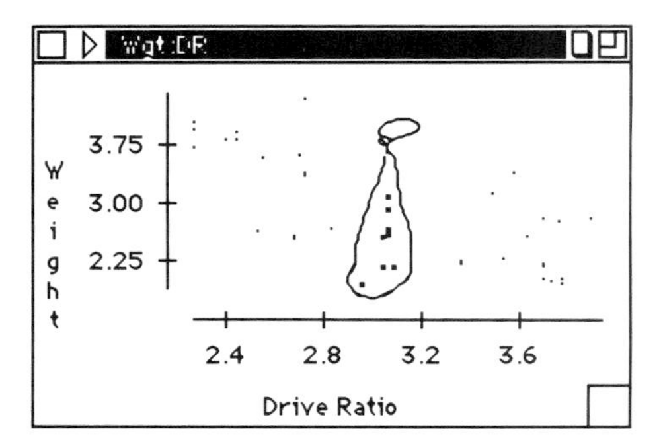

Figure 9-4. The Lasso selects the points it encircles.

You can get the tool immediately in any plot by holding down ⌘-Option.

To select points with the Lasso, draw a line around them. When you release the button, the shape you have drawn is automatically closed, and all enclosed points are selected.

RECTANGLE

To select points with the Rectangle, hold down the mouse button (the cursor is the arrow,) and drag out a rectangle on the plot. When you release the button, all enclosed points are selected.

You can also select with the brush and slice tools discussed in Section 9.6.

EDIT MENU selections

The **Select All, Find...**, and **Go To...** commands in the **Edit** menu operate on plots as well. To select all points in a plot choose **Select All** or use its command-key equivalent, ⌘-A. To select a case by its case number, choose **Go To... Case #** and specify the case number.

The cases selected in data displays are often important in some way. They may be the cases that stand out in each of several displays of different aspects of the data, or an identifiable subgroup of the data. It is natural to select such points with one of the selection tools so that they can be examined more easily and seen in each of several displays. Once this is done, you may want to record the selection for further use. The **Selection** submenu in the **Modify** menu holds commands that record or restore the highlighted selection. The {Modify ▶Selection} **Record** command creates a variable whose values are set to zero for unselected cases

indicator variable

and one for selected cases. Such *indicator variables* have many uses in statistics, and are discussed in Section 11.18. Alternatively, you can use the recorded selection to restore a previous selection by selecting the recorded variable, clicking on a plot, and choosing {Modify ▶Selection} **Add**.

Data Desk 4.0

Selector Variable : Selector

Figure 9-5. *The Selector Button is placed in the lower left corner of the Data Desk Desktop.*

Often the selected cases deserve further attention. You can restrict any Data Desk command to work on a subset of cases by creating a 0/1 indicator variable and setting it as the Selector variable. (Section 12.5 discusses Selector variables). The {Modify ▶Selection} **Set Selector** command does all of this in one command. It first creates a new variable, asks you to name it, sets its values to 0 for unselected cases and 1 for selected cases, places it in the *Selector Button*, and highlights the *Selector Button*. The subsequent Data Desk command operates only on the selected cases.

You can also select cases based on logical criteria. Create a derived variable (See Section 11.2) containing a logical expression and use it as the criterion variable for a {Modify ▶Selection} **Add** command. For example, a derived variable containing the expression

MPG > 30

will select all cars in the CUCars dataset that achieved better than 30 miles per gallon. The more complex expression

Weight < 3.0 AND Horsepower > 100

selects lighter cars with more powerful engines — probably sports cars.

{Modify ▶Selection} **Clear** makes all cases unselected. {Modify ▶ Selection} **Toggle** selects cases currently not selected and de-selects those that are selected.

9.5 *Link*

Each kind of plot shows a different view of your data, but within a relation the underlying cases are the same. A particular case may be a point in a scatterplot or part of a bar in a histogram or bar chart, or all of these at once. And, of course, each case also corresponds to a row of values in

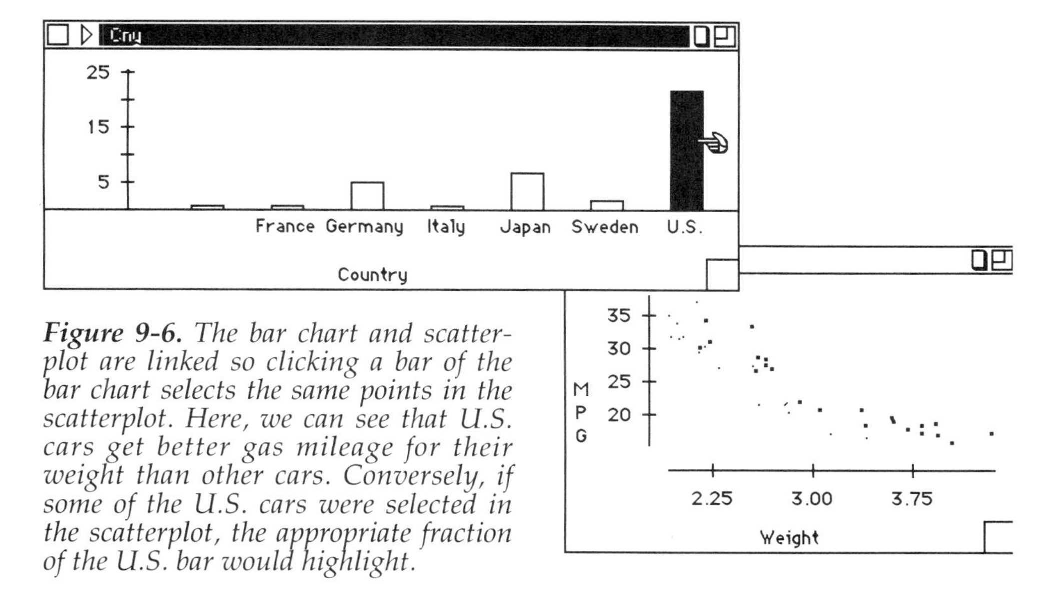

Figure 9-6. *The bar chart and scatterplot are linked so clicking a bar of the bar chart selects the same points in the scatterplot. Here, we can see that U.S. cars get better gas mileage for their weight than other cars. Conversely, if some of the U.S. cars were selected in the scatterplot, the appropriate fraction of the U.S. bar would highlight.*

any open variable editing windows.

Data Desk preserves the individuality of cases so that each view of the data shows each case consistently. For example, when you select a case in one plot or editing window, all views of that case are selected immediately, and highlight so you can see the selection. The case stands out from the other cases in each of these windows, so its relationship to them becomes clearer. You can select groups of points to see whether they appear as a group in other views of the data, or whether the observed grouping is a local feature.

Linking makes it easy to answer questions like:

- Is this extreme point also extreme in any other view of the data?
- Do the points in this part of the histogram cluster together on other variables?
- Is the relationship between these two variables the same for each of the groups in this pie chart?

These questions would require sophisticated and complex statistics calculations to answer numerically, but are easy to investigate with linked plots.

Data Desk displays also link plot symbols and colors for plots that show individual cases (such as scatterplots and dotplots). Each case appears in all of these plots with the same symbol and in the same color.

Only windows displaying data for the same relation can link to each other. If two of your windows will not link with each other, they may be from different relations.

Linking also makes possible the interactive actions *brush* and *slice*.

9.6 *Brush and Slice*

Brushing and slicing can reveal joint patterns and relationships among many variables. Chapter 10 discusses brushing in detail.

BRUSH

The brush tool is a rectangle. As you brush the rectangle across a plot, the points it covers are temporarily highlighted, as are the corresponding points in all open, linked displays. By brushing a part of one display you can see how the corresponding cases look in other displays.

To resize the brush, hold down the ⌘ key.The brush rectangle will resize as you drag it bigger or smaller. For example, you can make the brush a tall, thin rectangle to look at small, local parts of an x-axis variable. The highlighted points in other plots show the patterns and distributions "conditional" on the selected slice of points.

 KNIFE

The Knife tool selects points in vertical or horizontal slices of a plot. The knife slices right-to-left, left-to-right, top-to-bottom, or bottom-to-top according to its initial direction. Points are selected as the mouse passes their position, and remain selected unless you reverse direction and drag the mouse back over them.

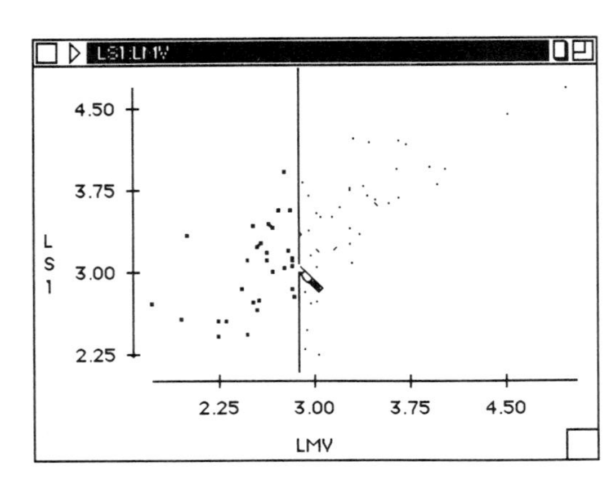

Figure 9-7. *Slicing across a scatterplot. The selection links immediately to other open plots and editing windows.*

Brushing and slicing help you answer questions such as:

- Do the same cases seem to be in roughly the same places in each of these plots?
- Is there any trend in sales from East to West?
- How does the relationship between the gas mileage and weight of cars change as drive ratio increases? (Plot MPG *vs* Weight in a scatterplot, plot Drive Ratio in a dotplot, and brush or slice Drive Ratio while watching the MPG *vs.* Weight display. See which points highlight as you move from low to high Drive Ratio.)

9.7 *Identify*

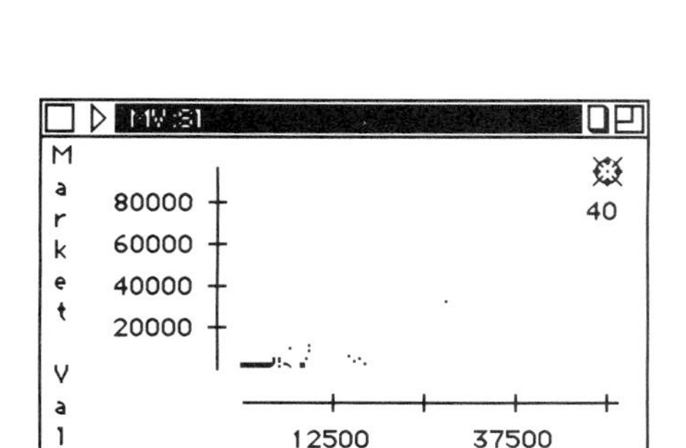

Figure 9-8. *When no variables are open, the* **?** *tool shows the case number of a datapoint.*

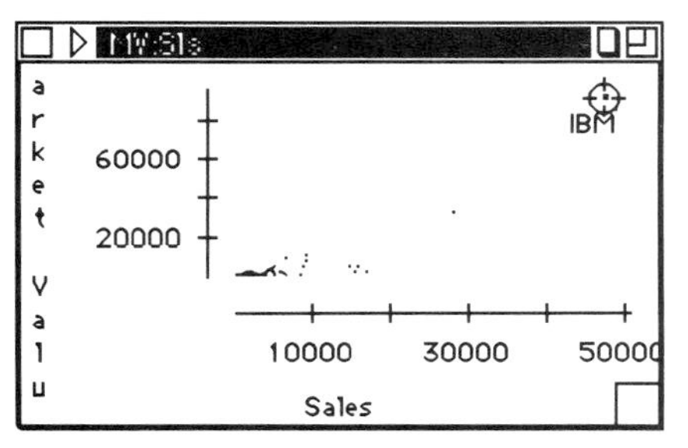

Figure 9-9. *When an identifying variable is open and frontmost among open variable windows, the* **?** *tool displays the case name found in that variable.*

Data analysis displays help you discover patterns and relationships in your data. However, it is not enough to see a general pattern or trend. You usually want to know which cases make up each of the groups, which cases form the heart of the trend, and which cases fail to follow the pattern established by the others. For this, you need to be able to identify datapoints on a plot easily.

? *IDENTIFIER*

The identifier tool provides a crosshair cursor that looks like a bomb sight. Place it over a plotted point and press the mouse button to highlight the point and display its case number.

To display identifying text such as a name rather than a case number, open a variable in the same relation as the plotted variables that contains identifying text for each case. The **?** tool then identifies each point on the plot with the text of the corresponding case in the identifying variable. If two or more variables are open, the frontmost window becomes the identifying variable. The identifying variable's window may be completely covered by other windows, but it must be in the right relation. To see the case number rather than text when a variable window is open, press the Option key.

For example, a scatterplot of *Market Value vs Sales* for the Companies data shows an extreme value. We can identify it with **?**. First open the variable *Company* so that the company names are available as identifying text. Then click on the extreme point with **?**. Note that we could have opened the variable *Sector* instead of *Company* if we had wanted to identify points only by their market sector, or the variable *Profit* to see companies' profits.

Pressing ⌘-Option gives you the ☛ tool to make it convenient to select points in a plot after you have identified them.

9.8 *Move*

Movement provides a natural way to adjust displays. It is easier to push or pull to adjust the way a display looks than to type verbal commands or select menu items. Thus, movements such as repositioning the display contents or bringing other parts of the display into view are natural manipulations.

Grabber

The grabber repositions the contents of a plot within its window. In most plots, the axes adjust as you slide the points so that they are always correct.

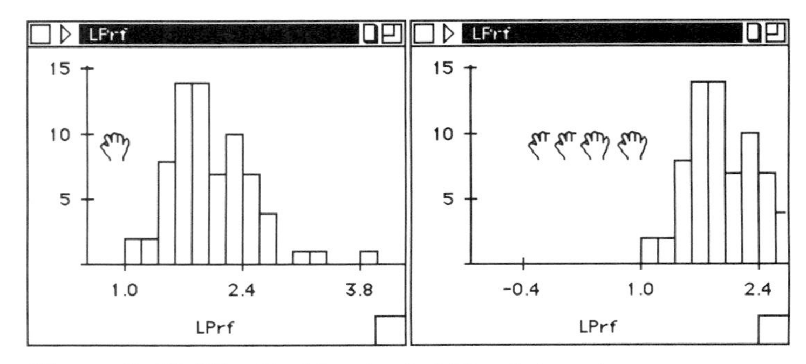

Figure 9-10. *The Grabber tool shifts plot centers.*

Boxplots and dotplots of many groups may have more boxes or stripes than can fit in a window. For such displays, you must slide the display side-to-side with the Grabber to see the remainder of the display. The Grabber also controls the side-to-side movement of the axis in sliders (see Section 11.13). If you hold the Shift key down while sliding many kinds of plots with the Grabber, the plot will continue to slide when pushed. Click again with the Grabber or press the space bar to halt the slide. The {Modify ▶ Scale} **Home** command returns any plot to its original location and scale, and can be a convenient way to retrieve a plot that may have slid too far.

9.9 Isolate

Isolating singles out one part of a display for consideration and hides the other parts of the display. Data that have cases from several groups, and data containing extraordinary cases often call for isolation. We might wish to eliminate a group or outlier from consideration for the moment, or, alternatively, to focus our attention on it.

The commands in the {Modify} **Visibility** submenu isolate points by changing the visibility of points in plots. **Show Only Selected Points** command hides all but the selected points in the front plot. You cannot select or otherwise work with invisible points. However, visibility is local to the plot; the cases may be visible in other plots in the same relation.

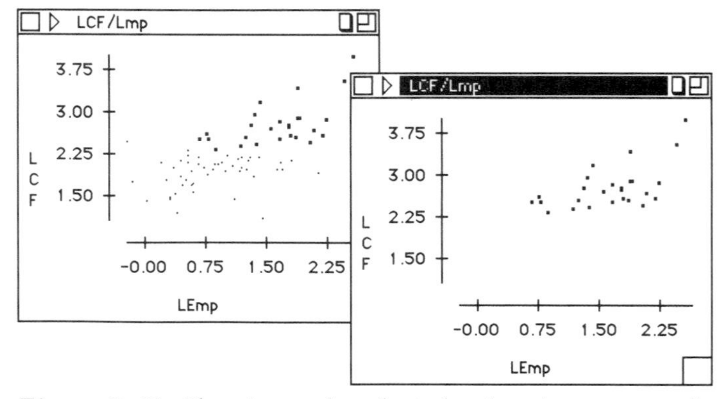

Figure 9-11. *Showing only selected points in a scatterplot.*

The **Show All Points** command makes invisible points visible again. **Toggle Hidden Points** hides the visible points and shows the invisible ones.

Refocus

An alternative to isolating points by making others invisible is to isolate them by focussing the plot on them. Other plotted points are still part of the plot, but fall outside the plot window, so they are not seen.

This tool refocuses a plot to display only the part enclosed by the selected rectangle in the plot window. Drag a rectangle on a plot to define the new area to be plotted. The selected rectangle is rescaled to match the plot window. If the rectangle is too small it will be ignored, so you can start over by collapsing the rectangle down to a point before releasing the mouse button. The {Modify ▶ Scale} **Home** command restores a plot to its original scale. The rectangle can extend beyond the plot window, but you must start dragging within the plot window.

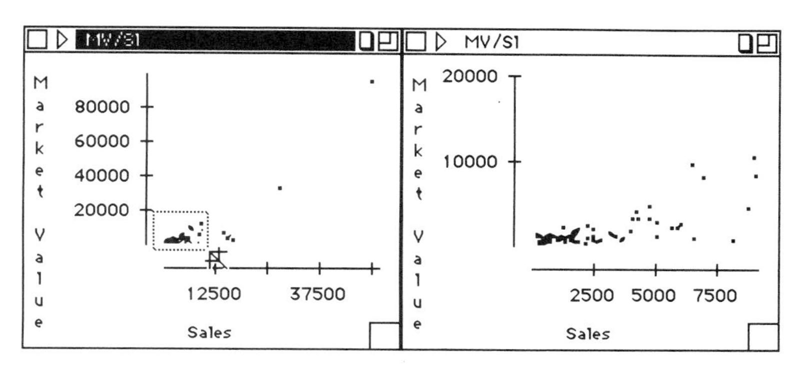

Figure 9-12. *Refocus a scatterplot by selecting the desired part of the plot with the tool. The result of the refocus operation is on the right.*

For example, in a plot of *Market Value vs Sales* for the Companies data, we might isolate the mass of points in the lower left from the others. The rescaled display lets us pick out companies (perhaps with the identification tool) whose market value is particularly high or low compared to their sales, without being distracted by the extreme cases.

> TIP
>
> If you accidentally "lose" the points in a plot, by shifting them off the screen, choose **Home** to restore the points.

The {Modify ▶ Scale} **Scale to Selected Points** command performs a similar refocusing of the plot. It redraws the plot with a scale determined as if only the selected points were to be plotted. This is somewhat different than the tool, which can select large amounts of empty space in a plot and include it in the refocused plot. When a plot's scaling has been altered with any of the tools or commands, you can return to the original scaling by selecting all points (⌘-A is a good shortcut for Select All) and choosing Scale to Selected Points, or with the {Modify ▶ Scale} **Home** command.

9.10 Resize

Data Desk's plots ordinarily expand or shrink to fit in their windows. Many plots resize automatically when you resize their windows. You may want to inflate any plot to see details or to concentrate on a part of the plot, even though other parts will no longer fit in the window. Or you may want to shrink the plot to see all of it after the refocus tool has dropped points off the edge of the plot window.

RESIZE

to shrink

to expand

To resize the contents of a plot, select the Resize tool and move the mouse cursor inside the plot's window. Near the center of the plot, the cursor looks like this: . Click the mouse to halve the size of the plot contents. Near the edge of the plot, the cursor looks like this: . Click the mouse to double the size of the plot contents. The plot window remains the same size, so expanding a plot may push points beyond the edge of the window. These points are still part of the plot; you can see them by sliding the plot with the Grabber .

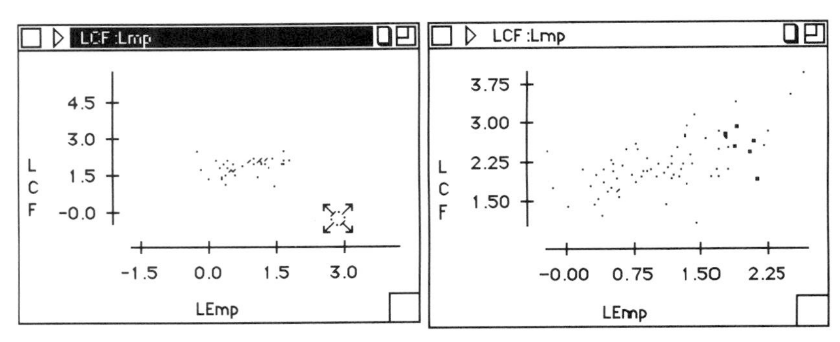

Figure 9-13. Using the Resize tool doubles (or halves) the scale of the plotted points.

The tool is available immediately in most plots by pressing Shift-Option-⌘.

9.11 Substitute

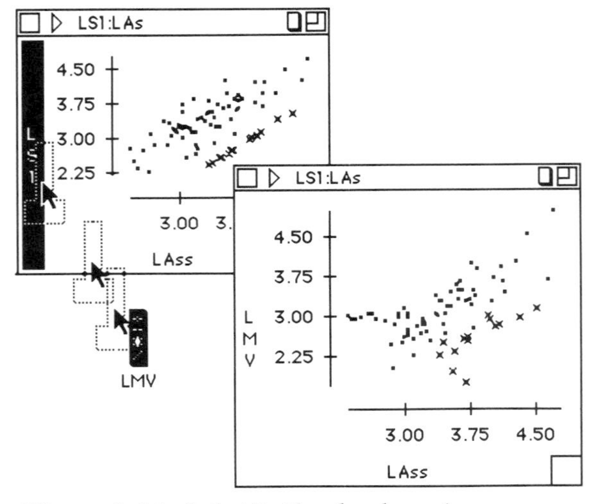

Figure 9-14. Substitution by dragging an icon over the axes label of the old plot.

We learn more by looking at our data from many different perspectives than by seeing only a single view. Any plot can suggest another that is just slightly different. Many of the actions discussed thus far let you modify a plot slightly to improve the view or clarify an aspect of the display. Data Desk also lets you use some plots as templates for other similar plots. In effect, you can say "I now want a plot just like this one except for the following change."

A typical example of this is the ability to substitute one variable for another in a plot. To do this, simply drag the icon of the new variable over the axis label of the old plot. Figure 9-14 shows an example. Because the plot will change in place, you may wish to first duplicate the plot. (Select its icon alias and choose {Data}

Duplicate. The duplicate plot opens on top of the original.)

Ordinarily, substituting a new variable on a plot axis rescales the plot to suit the new variable. However, if you first choose **Manual Scale** from the plot's HyperView menu, the axis scales will remain unchanged. You can easily compare two plots on identical axes by duplicating the first, selecting **Manual Scale**, and dragging a new variable onto one of the axes. Section 9.15 discusses this and other ways to specify plot scales.

Aspects

Data Desk lets you modify almost any aspect of a plot. Most of the commands that help you do this are in the **Modify** menu, organized into submenus according to the aspect of the plot they modify. Data Desk's plots are data analysis displays rather than presentation displays, so the list of aspects is quite different from the list you might see in a presentation graphics program. The aspects of data analysis plots are chosen to help you understand more about your data.

Unless otherwise specified, each section below discusses a submenu of the **Modify** menu, all commands discussed in a section are in that submenu, and all of them operate on the frontmost window and are active if that window is a plot. Some commands have effects that change other linked windows, but they still operate on the front window. Some commands operate on all selected windows if the Shift key is depressed.

The **Visibility** and **Selection** submenus relate closely to actions discussed in earlier sections of this chapter.

The basic aspects covered in this chapter are:

- *SYMBOLS*
  Data Desk can plot points with one of 8 symbols. Plot symbols identify groups and mark special points.
- *COLORS*
  Data Desk can plot points and lines with one of 64 colors. Plot colors can identify groups, mark special points, or represent a continuous dimension.
- *AXES*
  The axes of a plot describe the relationship between the data values and their graphical representation. However, you can often see the data more clearly without them.
- *SCALE*
  Some plots offer control over the scaling of the plot or over the default scale methods.
- *VISIBILITY*
  You can hide some of the points in a plot to focus attention on the others or to show patterns for different groups.
- *LINES*
  You can draw lines between points, add lines from appropriate variables, and save line information in variables.

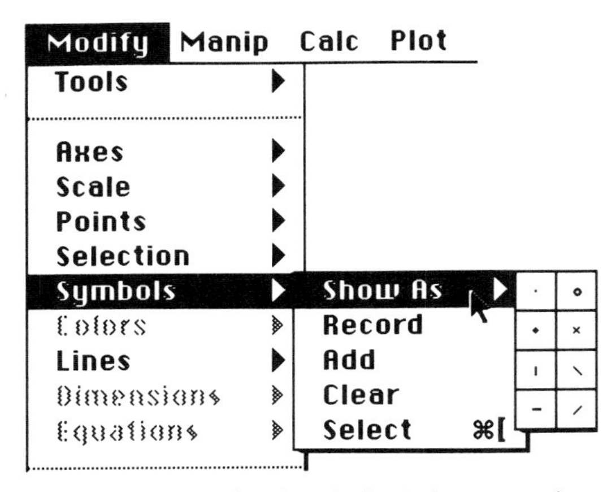

Figure 9-15. The Symbols Palette can be torn off the Symbols submenu and left on the desktop.

9.12 Symbols

The **Symbols** submenu of the **Modify** menu holds commands for working with plot symbols. To assign plot symbols to points in a dotplot, scatterplot, lineplot, or probability plot, select the points and choose a symbol from the Plot Symbols palette found in the {Modify ▶ Symbols} **Show As** submenu. Symbols are linked immediately to all plot windows; a point assigned a new symbol is immediately plotted with that symbol in all plots that use symbols. You can also assign plot symbols to cases in a variable editing window, although you can only see the effect of the assignment in an appropriate plot.

Plot symbols let you mark groups or particular cases. For example, in the Iris dataset introduced by R. A. Fisher, and subsequently used by many writers, measurements are reported for three species of Iris flower. We can plot Petal Length *vs* Sepal Width with each species plotted with a different symbol.

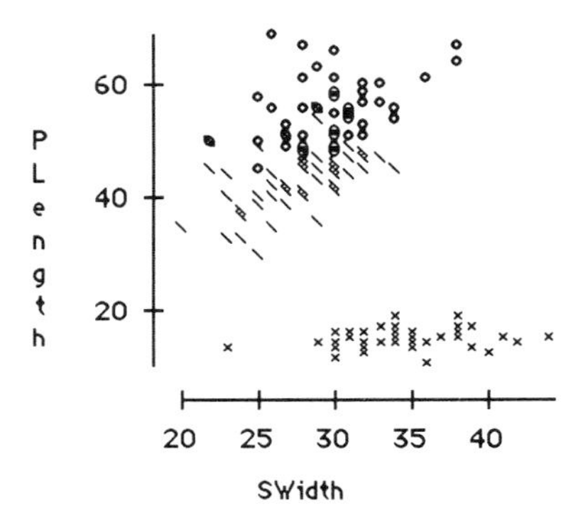

Figure 9-16. The Iris data. Each species is plotted with a different plot symbol.

Data Desk's plot symbols have been chosen to look as different from each other as possible and to overlap each other in ways that make it possible to identify the individual symbols. They are not the standard symbols often used in presentation displays. Even if the o / \ | and – all overlap in a plot, it is relatively easy to see each symbol and know that all five points are plotted there.

The **Record** command creates a new variable named *Group*, records in it codes corresponding to the symbols of all cases in the front plot, and adds it to the data Relation. To assign symbols from the group identities, select the group variable, click on a plot, and choose **Add ▶ By Indices**.

Set Group creates a new variable recording the groups, places it in the Groups button (See Section 12.6), and selects that button for the next command.

To assign symbols so that each symbol represents a group, select the variable that holds group names, click on a plot, and choose **Add ▶ By Group**.

Clear sets all plot symbols to the dot (·).

To select all cases in a group, select at least one point in the group and choose **Select**. All cases whose plot symbols match the selected cases are selected.

9.13 Colors

Like plot symbols, color is a plot attribute that can convey additional information. All commands dealing with color are in the {Modify ▶ Colors} submenu, which parallels the {Modify ▶Symbols} submenu almost exactly. The Color palette shows the 64 colors available in Data Desk. It tears off the **Show As...** submenu. To use color you must have a color monitor and a Macintosh equipped to display 8-bit color and set for the full 256 colors. The Control Panel desk accessory lets you set the number of colors.

Color is an attribute of those plots that show individual cases such as scatterplots and dotplots. Any plot that can display symbols can display color. When a case is displayed in color it has the same color in all plots in the same relation that can display color. Color is also used in pie charts, but there it works in a special way.

You can assign colors in the same way as you assign symbols; select cases and click on one of the color squares in the Color palette.

Alternatively, you can **Add ▸ By Group** from a discrete variable, just as you can assign symbols by group. However, colors are not limited to eight groups as symbols are. Data Desk automatically counts the number of categories and selects colors that are as different from each other as possible to display the categories. Although there are 64 possible colors, it is hard for most people to differentiate more than 10 colors or so.

Alternatively, color can represent a continuous dimension. Adding color from a continuous variable uses a range of colors to represent the range of values in the variable. The range of colors available in Data Desk is circular; the first and last color in the Colors palette are as close to each other as any two adjacent colors in the palette. Data Desk therefore uses only half the color range to represent a continuous variable. The color range represents low values with a pale blue and high values with a deep red, running through green, yellow, and orange for intermediate values.

Data Desk offers several ways to represent values as colors. Often it is best to represent the ranks of the values rather than the values themselves. Many variables follow a humped distribution that groups most values together in the middle, so most points are assigned the same color. Ranking spreads out the values and thus makes better use of the available color range. The **Add** submenu offers to add colors **By Ranks** or by **Linear** mapping of variable value to the integers 0 to 31 (and then to the first 32 colors).

The **Record** and **Set Group** commands record colors (as integers between 0 and 63) just as the same commands do for Symbols. To restore recorded colors, select the recorded variable and **Add ▸ By Indices.**

The **Clear** command sets all colors to the default, light blue color. The **Select** command selects all cases that have the same color as the selected case. (It is hard to tell apart two adjacent colors in the palette.)

TIP

Turn off color when you are not using it to make displays more responsive. Dynamic displays work up to four times faster in black-and-white than in color.

Dynamic plots are slower in color than in black and white because eight times as much information must be copied to the screen. While color plotting does not take eight times as long, it does take longer. You may find that plot brushing slows down somewhat when many windows are open. If you are not using color in your displays we recommend that you turn it off, setting your screen depth to 2 with the Control Panel. Data Desk's dynamic displays will work about four times faster. (Color has no effect on calculation speed. A floating point unit will speed up calculations.)

9.14 Axes

Plot axes tell us what we are looking at and how the plot relates to the data values. The Data Desk commands to modify axes appear in the **Axes** submenu of the **Modify** menu. Axes commands operate on the front window if it is a plot. Alternatively, they work on all selected plots (open or closed) when the Shift key is held down.

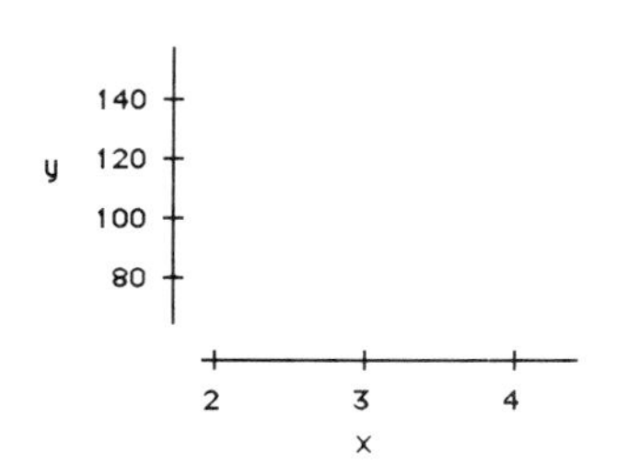

Most displays have a vertical *y*-axis with tick marks, value labels, and the name of the variable plotted in that direction. Many also have a horizontal *x*-axis with the same features. Boxplots, bar charts, and dotplots have no horizontal axis, but instead provide the name of the variable or group displayed in each box, or stripe.

Plot axes serve a different purpose in data analysis displays than they do in presentation displays. Often the numbers underlying a plot are of secondary importance when our goal is insight. We are likely to be more interested in the overall pattern of the data than in the particular values. Many statisticians recommend that data analysis displays have sparse axes. Usually, it is best to show axes in a newly drawn display. However, you may wish to hide them later to be able to concentrate more easily on patterns in the data.

The **Hide Axes** command makes axes invisible. Most plots recover the space previously occupied by the axes and use that space for the body of the plot itself. Hiding axes can thus be especially helpful for plots in small windows. In many plots, patterns are easier to see when the axes are out of the way. The Hide Axes command changes to **Show Axes** for plots in which the axes have been hidden.

You control the axes and their names separately. Hiding the axes leaves their names. Alternatively, **Hide Axis Names** leaves the lines and tick marks, but hides the text of the axis names. The command changes to **Show Axis Names** for plots in which axis names have been hidden.

When you hold down the Shift key, the Axis commands apply to all windows whose icons are selected.

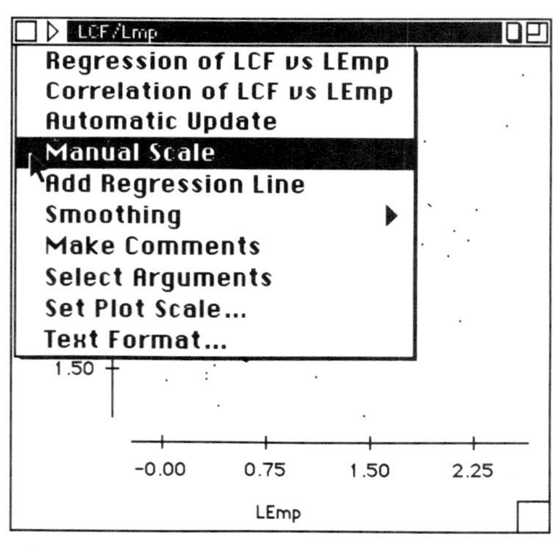

Figure 9-17. *Choose* **Manual Scale** *from the plot window's HyperView menu to freeze the current scale during Updates.*

9.15 Scale

The scale of a plot determines the relationship between the data values and the position of points, bars, lines, and wedges in the plot. The and tools alter the scale of many plots. Alternatively, {Modify ▶ Scale} **Scale to Selected Points** rescales the plot in the front window as if it consisted only of the selected points (but plots all of the visible points). When a plot's scaling has been altered, you can return to the original scaling by selecting all points (**⌘–A** is a good shortcut for **Select All**) and choosing **Scale to Selected Points**, or by selecting {Modify ▶ Scale} **Home**.

All plots automatically scale to all points when they are made. Plots rescale whenever they update in response to changes in underlying data or expressions. (See Chapter 4.) To freeze the scale, choose **Manual Scale** from the plot window's HyperView menu.

Histograms

Stretching and shrinking a histogram window while holding down the Option key rescales the histogram. For more precise control, choose **Set Plot Scale** from the histogram's global HyperView or make the histogram window frontmost and choose {Modify ▶Scale} **Set Plot Scale**. This dialogue gives you control of the bar width and starting point for the first bar.

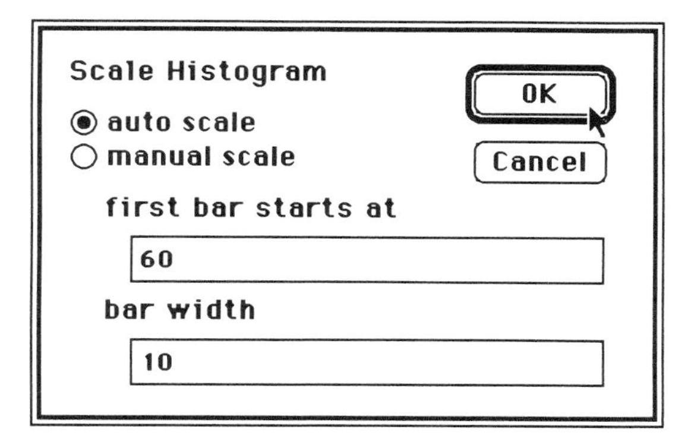

Figure 9-18*. The histogram's scale dialog lets you specify how the histogram in the front window should be drawn.*

Manual scale control lets you draw two histograms with the same scale to facilitate comparisons. Note that unless they are also plotted in windows of the same size, the histograms will not be identical.

Scatterplots

Any scatterplots (in fact all plots) created in Data Desk are scaled to fit all points in the window. You can customize the scale of these plots by choosing **Set Plot Scale** from the plot's global HyperView or by making the plot frontmost and choosing {Modify ▶Scale} **Set Plot Scale**. The options in this dialog allow you to manipulate the plot scale for each axis and change the size of the window.

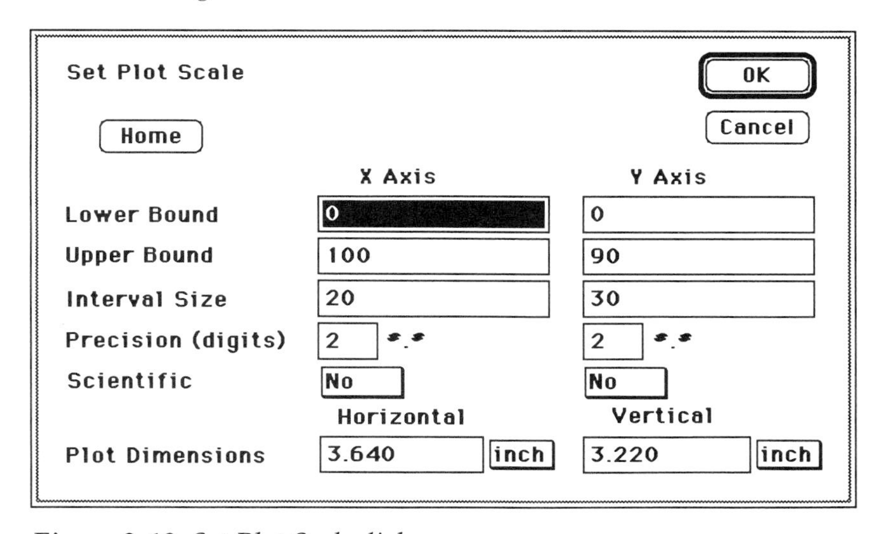

Figure 9-19. *Set Plot Scale dialog.*

The scale of each axis is determined by three values: "Lower Bound", "Upper Bound" and "Interval Size". Data Desk initially sets the lower and upper bounds to the minimum and maximum values for the variable plotted on that axis. The initial interval size is one-fifth of the variable's range or the next largest "nice" number.

You can change the starting and ending points for either or both axes by entering new values in the boxes for "Lower Bound" and "Upper Bound". You can change the amount of space between each tick mark by entering a new value for "Interval Size". If you want the value of the "Upper Bound" to be the last tick mark on your scale, make sure the values of the upper bound and lower bound are whole-number multiples of the interval size. For example if you wanted a scale that went from 0 to 100, entering a lower bound of 0, an upper bound of 100 and an interval of 20 will produce a plot scale with five tick marks, with the last tick labeled 100 {(100/20) = 5 and (0/20) = 0; both whole numbers}. If you enter an interval size of 30 instead of 20, the axis will have 3 tick marks with the last visible mark labeled 90 {(100)/30 = 3.33, not a whole number}.

Plot scaling works best when the lower and upper bounds are whole number multiples of the interval size. If you adjust the interval scale without changing the lower or upper bounds, Data Desk will automatically adjust the values for the lower bound and upper bound to values that are whole number multiples of the the new interval size.

The Precision specifies the number of significant digits in the tick labels. The pop-up menu specifies scientific notation for tick mark labels.

To bring back the original scale, click the "Home" button at the top of the plot scale dialog window or select **Automatic Scale** from the plot's HyperView.

HOW-TO

To asign the scaling parameters of one plot to another, drag the icon alias of the scatterplot whose scale you want replicated into the scatterplot whose scale you want to change.

The **Set Plot Scale** dialog allows you to set the size of the window. Type the values for the Horizontal and Vertical dimensions in the Plot Dimensions fields. The dimensions can be specified in either inches or centimeters.

To manipulate scales so that two plots have the same scaling and size, drag the icon alias of the scatterplot whose scale you want replicated into the scatterplot whose scale you want changed. The scatterplot you dragged into will be scaled and sized exactly like the one you dragged from.

Other Plots

You can change the scale of all the other plots in the Plot Menu by selecting **Set Plot Scale** from the plot's global HyperView. The dialogue presented is the same as for scatterplots with each option providing the same functionality. The only difference is that the x-axis can not be changed for dotplots and boxplots. The global HyperView of a slider provides a Set Plot Scale dialog which lets you scale only the x-axis since there is no y-axis on a slider.

Overlaying Plots

There are times when it is helpful to compare two different plots by overlaying one plot on top of the other. These may be plots of different years, different schools or different species. In Data Desk you can overlay plots using the Layout Window (see Chapter 6 for more information on Layout Windows).

Open a Layout Window by choosing {Data ▸ New} **Layout**. Drag the icon alias of a plot into the Layout window. The picture of the plot in the Layout window is static and will not update like other Data Desk plots. Select all the points in the second plot by making that plot frontmost and typing command-A. Change the symbol or color of the points by clicking on a different symbol or color in the palettes. Drag the icon alias of the first plot into the center of the second plot in order to synchronize the scales. The two plots now have identical scales and dimensions but different colors or symbols.

Drag the icon alias of the second plot into the Layout window. Drag the two plots in the Layout Window until the axes overlay one another.

9.16 *Visibility*

Several kinds of plots represent each case as a point or symbol on the plot. These include scatterplots, dotplots, lineplots, and normal probability plots. In these plots you can select and work with individual cases by selecting and working with the points that represent them in the plot.

The **Visibility** submenu holds commands that help to focus attention on some of the cases by hiding others. These commands thus relate to the ideas of isolation discussed in Section 9.9. In particular, you can make some points invisible with the **Show Only Selected Points** command. **Show All Points** returns any hidden points to visibility. **Toggle Hidden Points** hides the visible points and makes the hidden points visible. This is a particularly effective way to compare two groups. Because it has a command-key equivalent, **⌘-H**, you can toggle hidden points quickly.

9.17 *Lines*

Another way to indicate grouping in scatterplots is to connect points with lines. This method is best suited to small groups, but works well even if you have many small groups. For example, you can group together the before and after values of each case in an experiment by drawing a line between each matched pair of datapoints.

Lines can also show trends and sequences, or even depict simple shapes. Unlike symbols, lines are local to a plot. Connecting two points with a line in one plot does not connect them with a line in other plots.

LINES

The Line tool draws lines one-by-one between pairs of points. To draw a line click on a point and drag the line to another point. To remove a line draw over it again. The line tool is most useful for adding or deleting a few lines from a plot. By contrast, the **Add** submenu in the **Lines** submenu holds commands to add many lines to a scatterplot according to variables that specify the assignment. The **by From/To** command requires that you select two variables; the first holds the case numbers of one end of each line, the second holds the case numbers for the other end of each line. Thus two variables holding the values:

1 5
3 8
5 3

specify lines from the point representing case 1 to the point representing case 5, from case 3 to case 8, and from case 5 to case 3. Values less than one or greater than the number of points in the plot are ignored. The **Record** command records the lines on the frontmost plot as two such variables.

You can also provide a single variable containing the case numbers of points that should be connected with lines and choose the **Add ▶ by Series** command. This is particularly useful for following a time trend through a plot. For example, you can generate a variable that counts from 1 up to the number of cases with the {Manip}**Generate Patterned Data** command, and use it to add lines connecting points in case sequence order.

Finally, you can add separate lines to connect the points in each group. Select a variable containing group identities and choose **Add ▶by Group** from the **Lines** submenu. Data Desk adds a separate line for each group, connecting the points in case order.

For example, the Graduation dataset reports the percentage of entering Freshmen graduating on time from each of several colleges at a major university for each of several years. A plot of %Grad on Time *vs* Year shows a general trend of better graduation performance in the 70's than in the 60's, but we cannot see the trends for individual colleges.

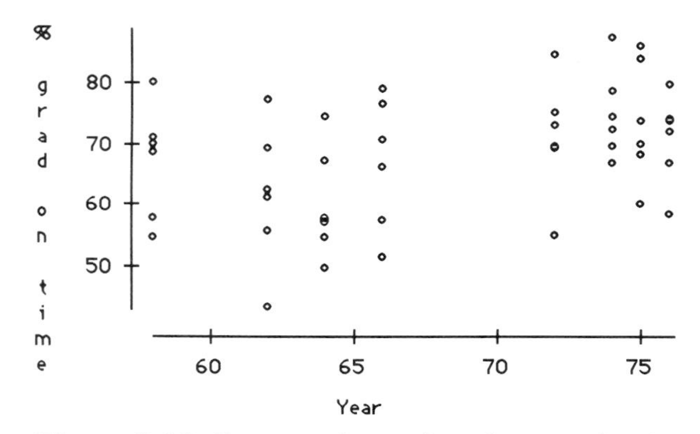

Figure 9-20. *Percent of entering class graduating on time by class Year.*

However, the *School* variable holds the names of the various colleges in the university. Select *School*, bring the plot to the front, and choose {Modify ▶ Lines ▶ Add Lines} **by Group**. Figure 9-21 shows the result.

Clear Lines removes all lines from the front plot. **Hide Lines** makes lines invisible, but remembers them. **Show Lines** replaces **Hide Lines** when lines have been hidden to make all lines visible again.

Drawing a line with the Line tool automatically shows all lines if they have been hidden. You can toggle lines on and off with the ⌘-hyphen keyboard equivalent command.

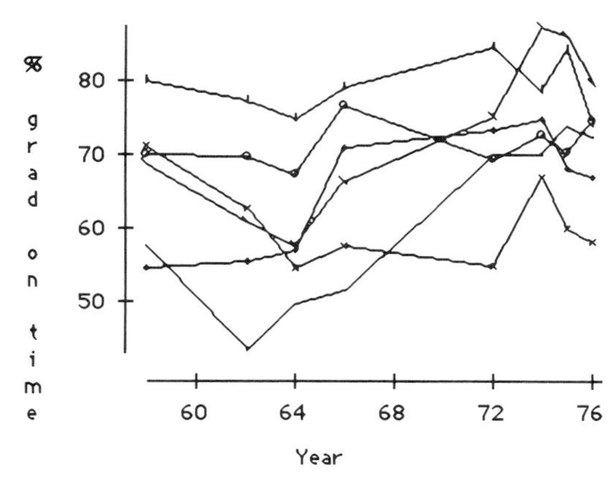

Figure 9-21. *Figure 9-20 with lines added by Group for each School.*

Lineplots plot a sequence variable in order against its case numbers and connect the points in order with lines. Lineplots have the advantage of being generated by the simple {Plot} **Lineplots** command, and the disadvantage that they only generate an equally spaced *x*-axis. The lines in a lineplot are special and cannot be hidden or recorded with the commands of the **Lines** submenu.

The {Plot} **Multiple Lineplots** command plots several variables against a common sequence axis and connects the points in each variable individually. It does this by creating a new relation holding three variables; a variable with copies of the selected variables appended to each other, a variable with a suitable *x*-axis sequence (repeated once for each variable), and a variable holding the names of the original variables as group names. The Multiple line plot is a scatterplot with lines and either symbols or colors added by group. You can work with the variables in the new relation, but because the plot is in its own relation, it does not link to the original relation.

9.18 *Lines and Color*

When lines are added to a plot that is drawn in color, the lines are colored as well. Data Desk assigns to each line a color whose index is the average of the color indices of its endpoints. This means that if you add colors by group to a plot and then add lines by group using the same group variable, the lines for each group have a distinct color. Data Desk does this automatically when making Multiple Line Plots on a color screen.

Plots that are colored according to the values in a variable may show overall trends in color. Lines drawn on these plots will conform to the trends.

9.19 *Working with Displays*

Data Desk offers an extraordinarily rich collection of graphic data analysis tools. It is unlikely that you will ever use all of them. (There are over 1,000 combinations of plots, commands, and tools.) Nevertheless, you may well find each new data analysis leading to a few new methods or plots.

You may also find that there are several ways to accomplish the same goal. For example, you might define an indicator variable by selecting cases in a pie chart or bar chart or by lassoing cases in a scatterplot and saving the selection from the **Modify** menu, by saving the indicator directly with a HyperView from a table (see Chapter 14), or by defining a derived variable (see Section 11.2).

While you can almost certainly *read* Data Desk plots with no practice, you will find that practice improves your facility for *working with* them. Only by doing so will you become more comfortable with basic operations more adept at wielding plot tools, and more knowledgeable about advanced techniques. Nevertheless, statistical graphics rely heavily on intuition and understanding of the data rather than on mathematics. Many people who might be reluctant to learn about multivariate statistics in depth find sophisticated statistical graphics natural and easy to learn.

APPENDIX 9A

Plot Tool Shortcuts

For any display, four tools are at your fingertips regardless of the current plot tool selected from the palette. Holding down the Option key overrides the plot tool selection and offers the tool generally most useful for the current plot. Option-⌘, Option-Shift, and Option-Shift-⌘ offer three more useful tools. These are called *spring-loaded tools* because they override the current plot tool selection only while the keys are held down.

The spring-loaded tools that are available depend upon the kind of display you are manipulating.

Display	**Opt**	**Opt-⌘**	**Opt-Shift**	**Opt-Shift-⌘**
All plots			?	
EXCEPT:				
Histogram				
Bar Chart				
Pie chart				

CHAPTER 10

Brushing and Slicing

10.1 Brushing and Slicing *141*

10.2 Principles of Brushing and Slicing *141*

DYNAMIC PLOTS PROVIDE an extraordinarily effective way to see patterns and structure among several variables. Plots that change over time can use those changes to reveal features of the data. It is usually easier to see changes than to discern patterns in static displays. Change over time thus facilitates recognition of relationships among separate displays. Brushing and slicing provide ways to work with many variables without having to imagine more than three dimensions at once.

This chapter discusses brushing, slicing, and ways in which these techniques can be used. It is not a complete exposition of the uses and usefulness of dynamic graphics methods, but rather is an introduction to basic principles and techniques. Nor is this chapter essential for all Data Desk users. You may find it worthwhile to just glance over the chapter, selecting the sections you find helpful and skipping over others.

10.1 *Brushing and Slicing*

Plot brushing was developed initially by statisticians at AT&T Bell Labs. Data Desk generalizes brushing beyond its initial specialized framework, making the plot brush a tool that works in any appropriate display.

Brushing focuses attention on a selected subset of points while showing them against the background of the rest of the points. A greater variety of plots offers more ways to define the selected subset. Thus, for example, by selecting points in a dotplot you focus on a subrange of the plotted variable to see where those points reside in other displays.

There are a few basic ways to brush plots:

- Brushing with the standard square or large rectangular brush
- Brushing with a tall and thin or short and wide brush to take "conditional" slices in one plot while observing how they look in other plots

10.2 *Principles of Brushing and Slicing*

Brushing and slicing reveal several basic kinds of relationships.

identity

- *IDENTITY*
  When you brush the upper-left corner of a plot, points in the upper-left corner of another plot highlight. This pattern continues throughout the plot. Even though the plots do not match point-for-point, they show similar underlying structure. If there is no *a priori* reason for suspecting such an identity, it is worth investigating further.

equivalence

- *EQUIVALENCE*
  When you brush the upper-left corner of a plot, points in the upper *right* corner of another plot highlight. This pattern continues throughout the plots, with the two plots being nearly identical except for a 90° rotation. The plots show similar underlying structure, provided that the change in orientation is not meaningful. Most kinds of equivalence among variables concentrate on relationships among variables and are thus not concerned with orientations. Sometimes

we may even see plots that are equivalent but for a rotation through an arbitrary angle, or combination of that and a reflection.

conditional distribution

- *CONDITIONAL DISTRIBUTIONS*
  Some relationships change according to values on a third variable. This is sometimes thought of as an *interaction* among three variables. One good way to view this is to consider the interaction *conditional* on the value of the third variable. That is, we restrict attention to a small range of the third variable and consider the distribution or relationship of the cases that fall in that region.

  Brush or slice along one variable (for example, on a dot plot) while watching the histograms of others to see conditional distributions. The histograms highlight to show the *conditional distribution* of the selected points. In this way it is easy to answer questions such as "Is the distribution of the stock price of companies with high research and development expenses about the same as that for companies with low R&D expenses?"

  Similarly, brush one variable and watch scatterplots of pairs of other variables. The highlighted points show relationships among the variables conditional on the selected region of the brushed variable. By seeing the selected points against the background of the entire dataset you can track changes in the relationship due to the conditioning.

simple trend

- *SIMPLE TRENDS*
  Brush along the stripe of a dotplot or along one dimension of a scatterplot and look for trends in other plots. Often slicing up or down the stripe is more effective. Trend in this context is not limited to linear patterns or even to ordered patterns. For example, you might find that as you slice along one variable, a scatterplot of two others grows a ball of highlighted points, starting from its center and spreading outward.

complex trend

- *COMPLEX TRENDS*
  Brush along one dimension of a plot or stripe of a dotplot and look for changes in the *relationship* of two or more variables in their scatterplot. For example, you can brush along a time variable while watching a scatterplot to answer such questions as "Has the relationship between education and expected earnings changed over time?"

CHAPTER 11

Derived Variables

11.1	The Transform Submenus	*145*
11.2	Typing Derived Variable Expressions	*146*
11.3	Updating	*148*
11.4	Expression Types	*148*
11.5	Expression Conventions	*149*
11.6	Dependencies	*149*
11.7	Working with Open Derived Variables	*150*
11.8	Calculating in ScratchPads	*150*
11.9	Boolean Expressions	*151*
11.10	IF/THEN/ELSE	*152*
11.11	Subscripting	*153*
11.12	Identifying Variables by Name	*154*
11.13	Dynamic Parameters	*155*
11.14	Re-expressing Data to Improve Analyses	*157*
11.15	An Example	*159*
11.16	Rules for Re-expression	*161*
11.17	Efficiency	*162*
11.18	Indicator Variables and Logical Expressions	*162*
11.19	Working with Several Relations	*163*
11.20	Subtle Points	*163*
11.21	Common Errors and How to Avoid Them	*164*

APPENDIX

11A	Derived Variable Expressions	*166*
11B	Casewise Functions	*169*
11C	Collapsing Functions	*175*

DATA DO NOT ALWAYS come ready for analysis. Sometimes we need to compute an entirely new variable as a function of other variables. A transformed version of a variable may be easier to analyze than the original data. New variables defined as transformations or algebraic combinations of other variables are called *derived variables* in Data Desk. The derived variable icon looks like an ordinary variable icon with arithmetic symbols.

Derived

You can use a derived variable in plots and calculations just like any other variable. Data Desk automatically computes its values when they are needed, so there is no need to explicitly compute the transformed values.

11.1 *The Transform Submenus*

The **Transform** submenus in the **Manip** menu provide the easiest way to make simple derived variables. The Transform submenus provide simple functions of one or two variables. Select a y-variable or an x- and a y-variable (with x-selection and y-selection, respectively) and choose the appropriate expression from the Transform submenus. Data Desk creates a new derived variable containing the requested expression and places it in the same relation as the selected variables.

For example, if we select *Assets* in the Companies data and choose "log(•)" from the Transform submenu, Data Desk creates a variable named *LAss* holding the expression log('*Assets*') as shown in Figure 11-1.

Figure 11-1. *A derived variable editing window.*

Data Desk selects the new derived variable icons, so you can immediately use them in a plot or calculation.

The Transform submenus organize functions into related groups. The most common data re-expressions[1] — square root, logarithm, reciprocal root, and reciprocal — appear first. Other, less common, functions are grouped into a second level of submenus. Section 11.14 has comments on uses of re-expression in data analysis.

Simple one-variable functions such as the log or square root operate on each selected variable whether it has x-selection or y-selection. (This is why they appear in the menu with a • rather than an "x" or "y".)

Some functions, such as the *mean*(•), or the *max*(•) function, summarize the values in a variable. Unlike other derived variable functions, which transform each case in a variable and thus result in a new variable with as many cases as the original, these functions result in a single value.

Many expressions combine two variables. Such expressions compute a value for each case in the variables. For example, the expression $x + y$ specifies a variable in which each case is the sum of that case's values in x and in y. For two variables to be combined case-by-case like this they must be in the same relation. This is a more stringent restriction than requiring both variables to have the same number of cases, but it preserves the sense of what relations mean. Variables in different relations record data about different individuals, even when they have the *same number of cases*. It makes no sense to combine "apples and oranges" in an expression. (If you defined a new "fruit" relation and dragged both variables into it, you could then combine them.)

[1]Transformations include any function that alters the data values. We use the term *re-expression* for those transformations that stretch or alter the *shape* of the data's distribution. Transformations such as adding a constant or dividing by a constant simply shift location and scale, but do not change the way a histogram or scatterplot of the data look. By contrast, re-expressions, such as taking logarithms or square roots, treat large numbers differently from smaller numbers and thus alter the shape of the distribution. For example, the logarithms of 1, 10, 100, and 1000 are equally spaced, although the original numbers are not.

Functions of two or more variables treat x-selected and y-selected variables differently. If you select one y- and several x-variables (or one x- and several y-variables), the **Transform** commands generate a transformation for each x-y pair. If you select identical numbers of x and y variables, the Transform commands treat them as paired, matching the first x with the first y, the second x with the second y, and so on.

A Note on Expressions:

If you are used to working with a spreadsheet, you will probably find derived variable expressions quite natural. It is common to combine two columns of numbers in a spreadsheet to yield a third column.

If you write computer programs, you have probably worked with arrays of numbers in which a subscript counter directs the calculations to work for each value in turn.

Nevertheless, derived variable expressions are slightly different. In a spreadsheet you would place an expression in each cell (for example, by entering the expression at the top of the column and filling down the column) to complete a row-by-row combination of columns. In a traditional programming language, you might write a loop such as

for i = 1 to n z[i] = x[i] + y[i].

Derived variable expressions operate on all cases in the variables automatically without any need to fill the derived variable with formulas or write a loop to operate on each case in turn. This is a very powerful and effective way to work, but you may want to pause at times to think through how expressions work.

11.2 *Typing Derived Variable Expressions*

Derived variables are not limited to the simple expressions in the Transform submenus. You may type any valid algebraic expression into a derived variable's editing window, either by editing an old expression or starting from scratch. Of course, you edit a derived variable expression in just the same way as you edit any text in Data Desk. All of the common operations in the Edit menu are available, so you can **Copy** and **Paste** parts of expressions within a window or from one window to another. Double-click selects entire words. Data Desk even provides a full **Undo** function.

When you type or edit a derived variable expression you may include comments, which will be ignored by Data Desk, but will remain part of the expression. Anything enclosed in curly braces ({ }) or the symbol combinations (* and *) is considered a comment. It is always a good idea to comment derived variable expressions — especially if they are at all complex.

To create an empty derived variable, choose the **New ▶ Derived Variable** command from the **Data** menu. (The **New Derived Variable** command also appears at the top of the **Transform** menu.) Data Desk creates the new derived variable, appends its icon to the right of the icons in the frontmost data relation, and opens it into an empty editing window.

TIP

To enter variables in the derived variable expression, type the variable names directly or drag variable icons into the derived variable editing window.

You may type any valid expression in the window. For a list of available functions and rules for derived variable expressions see the appendices to this chapter. You enter variables into the expression by either typing their names directly or by dragging their icons into the editing window. If you select part of the expression for editing and then drag the icon of a variable into the editing window, the variable's name will replace the selected text.

For example, to standardize the variable named "x" by subtracting its mean and dividing by its standard deviation, type the expression

(x - mean(x))/StDev(x).

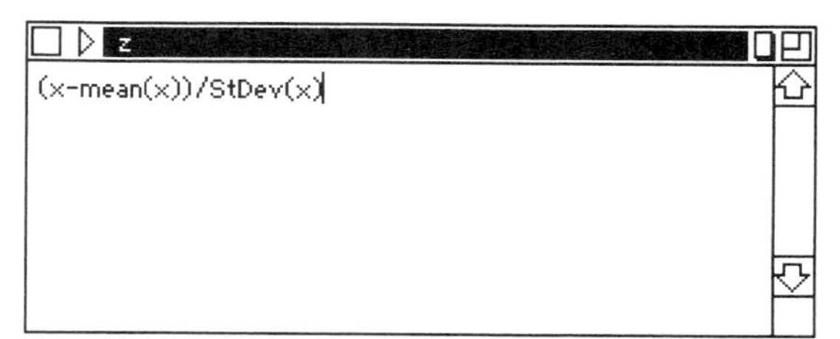

Figure 11-2. *Expressions may contain multiple functions.*

When you close a derived variable window after typing (either to type a new expression or to edit an old one), Data Desk checks the expression to be certain that it recognizes all variable names and function names, and that the expression is proper algebra. It complains if it cannot understand the expression, and offers to re-open the derived variable for editing.

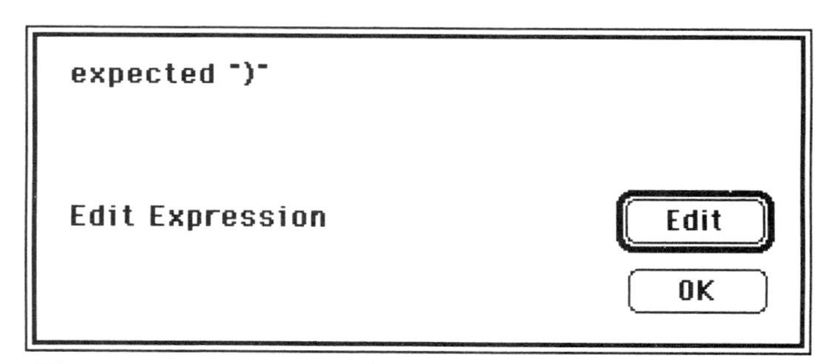

Figure 11-3. *Data Desk notifies you of errors in the expression and offers the chance to edit the expression or close the variable anyway.*

You may leave a bad expression in a derived variable, but you will not be able to use that variable in a plot or calculation until you fix the expression.

Once a derived variable window has closed into its icon, the icon behaves just like any other variable, except that when you open it, you find not numbers but the algebraic expression you typed. Data Desk evaluates the expression to obtain its values when you use the derived variable, but it does not replace the expression with the values.

To see the values of a derived variable, open it and choose **Show Numbers** from its HyperView menu. This is an alternative view of the derived variable; the values cannot be edited and will change whenever the derived variable recomputes. You can take a snapshot of a derived variable as a regular variable containing its current values by selecting it and choosing **Evaluate Derived Variable** from the **Manip** menu.

11.3 *Updating*

Because derived variables are used in other calculations and displays, they always recompute immediately to reflect the current values of their arguments. In this they differ from plots or tables, which offer to update but do not recompute until you tell them to do so. Any changes made to underlying variables are immediately conveyed to any window that uses the derived variable.

Of course, if the window holds a plot or table, it will offer to update, but will not recompute without your permission. When a derived variable uses another derived variable, it recomputes immediately whenever the first derived variable changes, so changes to the base data values may propagate through several levels of derived variables.

Derived variables also update immediately if you edit their expression. The ability to edit a derived variable's expression and have its values recompute (and those changes propagate through all plots, tables, and other derived variables that used the derived variable) offers great power. For example, you can experiment with alternative re-expressions of a variable and immediately see the effect of these changes on a scatterplot.

You can also control parameters in a derived variable with Data Desk's *sliders.* Section 11.13 discusses sliders.

11.4 *Expression Types*

Derived variable expressions can be single numbers or entire variables. Variables are simply lists of individual values thought of as successive cases.

Each value in a derived variable expression can be one of three types: numeric, text, or boolean.

A numeric expression is any number. Numeric expressions include integers and decimal fractions, the special value ∞, and the missing value code, • (typed as Option – 8).

A text expression is any collection of characters. Text constants are enclosed in double quotes. Text values may include numerals. Thus *"20-30"* is a text string (possibly defining an age category for a table), but *20-30* is a numeric expression evaluating to *–10.*

Data Desk assumes that when a variable is named in a derived variable expression, its values are numeric. To specify that you wish to refer to the text of the variable use the *TextOf* function. Thus *TextOf('age group')* may consist of text such as *"10-20"*, *"20-30"*, and so on, but *'age group'* will be evaluated numerically, yielding (for this example) *-10*, *-10*, and so on.

Boolean expressions are expressions that are either true or false. Data Desk provides special functions that work with boolean expressions, so we discuss them at length in Section 11.9.

Advanced Topics

Data Desk's derived variable expressions offer many powerful features. The remaining sections of this chapter discuss more advanced concepts. You may want to read only selected sections at first. The Appendices at the end of the chapter list all the functions Data Desk recognizes in derived variable expressions.

11.5 Expression Conventions

To format a derived variable expression to make it more readable, you can insert any number of spaces within the expression wherever one space could legally occur, except within a variable or function name. For example, extra spaces can go around a + sign or around a variable name. A new line can start wherever a space can appear, so you can split long expressions into several lines to make them more readable.

Derived variables ignore capitalization. You can type variable names and functions with or without capital letters.

single quotes for variable names

Variable names are often enclosed in *single quotation marks*. For example, 'variable name', or log('income'). *You must* use quotation marks if the variable name has a space in it, contains a special symbol or numeral, or is the same as a function name. It is always safest to use the quotation marks; variables dragged into the derived variable editing window, and those included through the Transform menu commands always are enclosed in single quotation marks.

For example, variable names that must be enclosed in single quotation marks include 'two words', '3best' (because it starts with a numeral), '√area', 'Σscores' (because they contain function names), and 'assets – liabilities' (because the minus sign is an operator).

double quotes for text constants

Double quotation marks enclose text constants. Braces and the (* and *) combination symbols enclose comments. You may use parentheses freely in derived variable expressions to specify the order of evaluation just as you would in standard algebraic expressions.

Appendix 11A discusses the details of evaluating derived variable expressions.

11.6 Dependencies

Every derived variable depends on the *underlying variables* used in its definition. Any changes in those underlying variables are immediately reflected in the derived variable, and then passed on to the plots, tables, and other derived variables that depend, in turn, on its values.

The chain of dependency can extend through many levels. If derived variable D1 uses derived variable D2, which in turn uses derived variable D3, which uses the ordinary variable x, then changing x changes the corresponding values in all three derived variables. Any plots or analyses that use any of these derived variables will notice the change and offer to

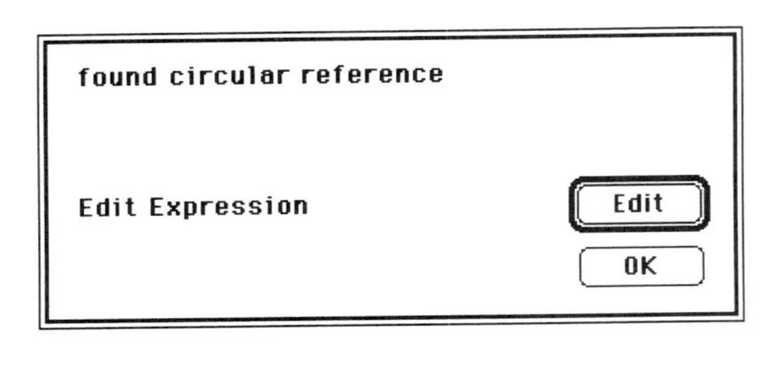

update. The dependency chain cannot be circular. That is, you may not have a chain in which D1 uses D2, which uses D3, which uses D1. Data Desk will notice if you inadvertently define a circular chain, and warn you of the problem.

Variables created with the **Evaluate Derived Variable** command in the **Manip** menu do not change with subsequent changes in the underlying variables. They are frozen snapshots of the derived variable at the time of evaluation. (If you need a variable equal to any other variable that does update, define a derived variable whose expression is simply the name of the variable you wish to copy.)

The {Special ▶ Locate} **Users of** command locates all icons that use any variable (including derived variables that use it) and selects them. Repeating the command (with those icons selected) locates the next level of usage — the icons that use the originally selected icons indirectly.

The {Special ▶ Locate} **Arguments of** command searches in the reverse direction, locating and selecting the icons of the variables that were arguments of the selected icons. If you select a derived variable icon and choose **Locate Arguments of**, Data Desk will locate the icons of the variables used in the derived variable expression.

11.7 *Working with Open Derived Variables*

You need not close a derived variable in order to work with it. You can select its (gray) icon or the icon alias in the title bar and choose a command. If the derived variable is used in a plot or analysis, you can open it and edit its expression and then update or redo the plot or analysis without closing the derived variable. Press the Enter key or click on another window to end expression editing and update dependent windows.

The ability to edit derived variable expressions and update values provides a powerful tool for investigating alternative expressions. For example, you can alter an expression and see immediately how the change affects a plot or analysis.

When you work with an open derived variable, Data Desk parses (translates) and evaluates the expression when it is needed. If it encounters an error or ambiguity, it alerts you at once. This may be disconcerting if your attention is focussed on a plot or analysis. Keep in mind that Data Desk evaluates a derived variable only when it is needed, and parses derived variable expressions either when the window is closed or when the expression must be evaluated.

11.8 *Calculating in Scratchpads*

Any function or expression available in derived variables can be typed in a ScratchPad window and executed immediately in that window. The result of the evaluation appears in the Scratchpad window immediately after the expression. This provides a powerful calculator: You can type any expression you might key into a calculator and evaluate it on the spot.

spot. Even more usefully, you can evaluate expressions that involve variables in the dataset.

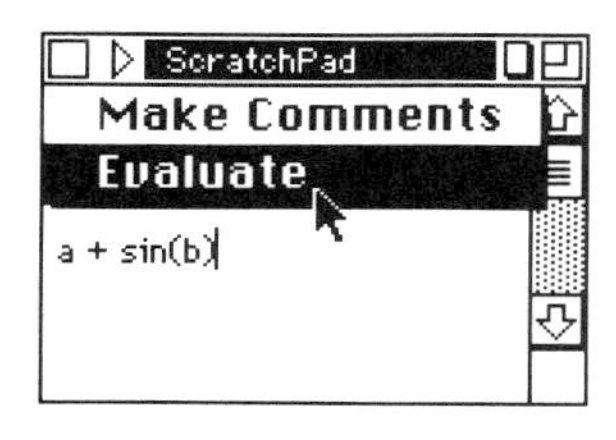

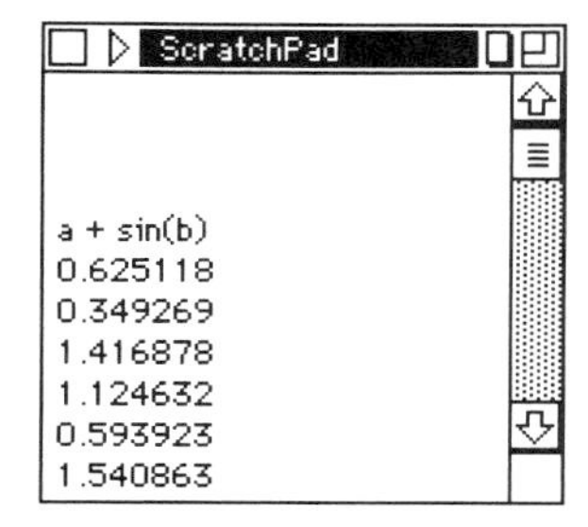

ScratchPads also provide an excellent place to work out complex derived variable expressions. Type the expression into a scratchpad and evaluate it to see whether it behaves as you intended. If the expression is particularly complex or difficult to figure out, try evaluating each part of the expression and then combining the parts into increasingly complex expressions. (This is also an excellent way to experiment with derived variable functions, for example to try out functions as you read about them here.) When the expression is correct, select it with the mouse, Copy, and Paste it into a Derived variable's editing window.

To evaluate an expression in a ScratchPad, place the cursor on the line to be evaluated and type ⌘ = or choose **Evaluate** from the ScratchPad window's HyperView menu. If the expression covers several lines, you can select the expression before choosing Evaluate.

You can, of course, copy a column of results from a scratchpad and paste it back into an open variable or directly into an icon window with the **Paste Variables** command.

11.9 *Boolean Expressions*

A Boolean expression is any expression that checks a condition that is either TRUE or FALSE. Thus, for example, the expression $y < x$ is a Boolean expression. Like most other derived variable expressions, it has a value for each case. It is TRUE for each case whose y-value is less than its x-value, and FALSE for all other cases. A derived variable defined by the Boolean expression $y < x$ (where y and x are variable names), evaluates to a variable holding cases that read *"TRUE"* or *"FALSE"*. Data Desk automatically converts Boolean variables to numbers whenever numeric values are required. When translating from TRUE and FALSE to numbers, Data Desk assigns TRUE the value 1, and FALSE the value 0.

The logical function NOT( ) reverses the truth of any Boolean expression. The Boolean expression NOT($y < x$) is equivalent to $y \geq x$ and evaluates TRUE for each case in which the value of y is not less than the value of x, and FALSE otherwise.

You can combine Boolean expressions into more complex expressions with the operators AND, OR, XOR, and DIFF. The expression

<boolean 1> *AND* <boolean 2>

is TRUE only if both expressions <boolean 1> and <boolean 2> are true, and FALSE otherwise.

The expression

<boolean 1> *OR* <boolean 2>

is TRUE if either <boolean 1> or <boolean 2> is TRUE or if both are true.

The expression

<boolean 1> *XOR* <boolean 2>

is TRUE if either <boolean1> or <boolean2> is true, but not if both are true.

The expression

<boolean 1> *DIFF* <boolean 2>

is TRUE only if <boolean 1> and <boolean 2> have different truth values (one TRUE and the other FALSE).

Most concepts that are commonly represented in statistics with indicator variables are most easily defined by boolean expressions. For example, we can define membership in a group as

'age' = 18

or

texto*f*('religion') = "Protestant" *OR* textof('religion') = "Catholic".

We can define inclusion in a range as

'income' > 20000 *AND* 'income' < 50000

and inclusion in a cell of a two-way table as

textof('year') = "1990" *AND* textof('month') = "April".

11.10 *If/Then/Else*

The logical connectives IF, THEN, and ELSE combine simple expressions to make expressions whose value depends on different conditions. They combine into expressions of the form

IF *<Boolean expression>* THEN *<result1>* ELSE *<result2>*

The expressions labeled *<result1>* and *<result2>* can be any valid derived variable expression. IF/THEN/ELSE expressions evaluate one or the other result expression according to the truth value of the boolean expression. Because the boolean expression typically will have different truth values for each case, IF/THEN/ELSE expressions are a powerful way to select values from each of two variables or select alternative calculations:

REMEMBER

You must always complete an IF/THEN/ELSE combination with an ELSE clause.

IF textof(*'country'*) ≠ *"U.S."* THEN *'km distance'*
ELSE *'km distance'* 1.61*{convert from km to miles}.

You must always complete an IF/THEN combination with an ELSE clause.

Text constants (indicated with double quotation marks) can appear as result expression in an IF/THEN/ELSE expression such as

IF *income < 50000* THEN *"Middle Class"* ELSE *"Rich"*

However, all results of an IF/THEN/ELSE expression must be of the same type, either text or numeric. You cannot mix types. For example, it is **not legal** to write

IF *'age' > 18* THEN *'months since voted'* ELSE *"too young"*

where *'months since voted'* is a variable recording the number of months since the last vote because months since voted is a variable holding numbers but *"too young"* is a text constant.

IF/THEN/ELSE expressions are themselves legal derived variable expressions so they can appear within other IF/THEN/ELSE expres-

example, you can recode ranges of a numeric variable into named categories with an expression like

TIP

The LookUpFirst and LookUpLast functions described in Appendix 11B offer another way to recode data.

IF *'grade'* > *90* THEN *"A"*
ELSE IF *'grade'* > *80* THEN *"B"*
ELSE IF *'grade'* > *70* THEN *"C"*
ELSE IF *'grade'* > *60* THEN *"D"*
ELSE *"F"*

Such nested IF/THEN/ELSE expressions often require careful thought to be sure that you have said exactly what you mean.

The result of the entire IF/THEN/ELSE expression can be a number, so you may write expressions like

StdBonus + (IF *Sales* > *5000* THEN *250* ELSE *100*)

11.11 Subscripting

To refer to individual cases within a variable or derived variable expression that evaluates to a variable, specify the case numbers within square brackets. For example,

age[3]

is the age value recorded for case number 3. The subscript may be an expression that evaluates to a number. Thus,

age[(nummumeric('age') + 1) / 2]

selects the middle case out of age, which will be the median age if *age* is sorted. If the subscript value is not an integer Data Desk truncates the value to its integer part.

More generally, the subscript can be a variable holding integers. When the subscript is a variable, each case in the index variable is used in turn as a subscript, so the resulting subscripted expression evaluates to a variable. For example, given variables with the following values:

case number	data	index
1	1.1	5
2	2.2	3
3	3.3	3
4	4.4	2
5	5.5	
6	6.6	

the expression *data[index]* evaluates to

5.5
3.3
3.3
2.2

The first element is *data[5]*, or the fifth case in the variable *data*. In this example that has the value 5.5. The second element is *data[3]*, which is 3.3.

Unlike the variables in most derived variable expressions, the index variable need not be (and, in fact, usually will not be) in the same relation as the data variable. The expression has a case for each case in the index variable, and is placed in the relation of the index variable.

If a value of the index variable is not an integer, it is truncated to the next smallest integer. If a value of the index variable is missing, negative, or greater than the number of cases in the data variable, the expression evaluates to "•" (missing) for that case.

While the index variable must be numeric (or the result will be missing), the data variable may be a category variable holding text.

The index variable may also be an expression that evaluates to a variable. Thus, if the variable letter grade holds the text values "A", "B", "C" "D", "F", "F", "F", "F", "F", and the variable *test grade* is a score between 10 and 100, the expression

'letter grade' [11 - ('test grade'/10)]

assigns a letter grade to each test grade in the same way as the IF/THEN/ELSE statement in the previous section. For example, a grade of 95 produces an index of 11 - 95/10 = 11 - 9.5 = 1.5, which is truncated to 1. The expression thus returns *'letter grade'*[1], which is "A."

Finally, the variable whose cases are being extracted may itself be an expression that evaluates into a variable, and subscripted expressions may be part of other expressions. Thus,

*('weekly salary' * 52 + 'benefits'['job category']) [3]*

is a legal expression, provided that *job category* and *weekly salary* are in the same relation. It finds the appropriate benefit amount with the first subscript and then returns the sum for the third case. Be sure to enclose any expression in parentheses before subscripting it lest you subscript only the last term in the expression rather than the entire expression.

11.12 *Identifying Variables by Name*

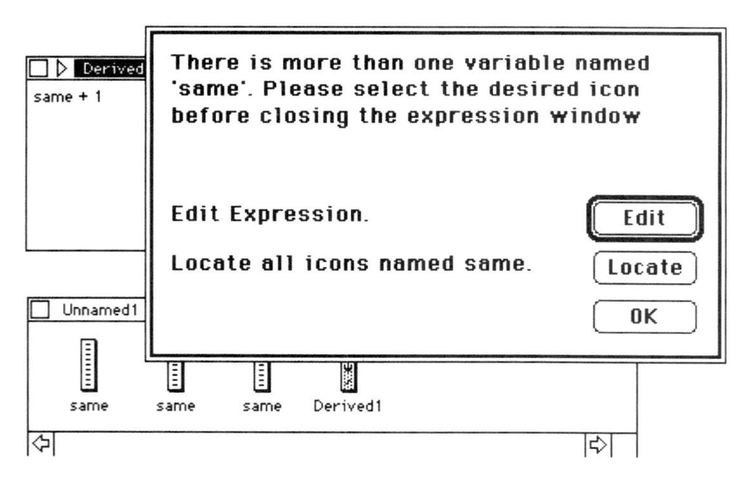

Figure 11-4. *Data Desk asks for help when variable names are ambiguous. The Locate button makes all icons with the name in question visible on the desktop, and selects them.*

Derived variable expressions refer to variables by name. Data Desk thinks variable names are just labels. Two or more variables can have the same name. Ordinarily, this poses no problem; we simply point and click to select the particular variable we want. However, derived variables can get confused when two variables have the same name.

When you generate a derived variable with the Transform menu or if you drag the icon of a variable into the derived variable's editing window, there is no difficulty. Data Desk uses the selected variable icons even if others have the same name. However, if you type a derived variable expression, you may need to tell Data Desk which of several variables with the same name is the one you want to use. Data Desk checks typed (or edited) derived variable expressions when you close the derived variable window or when you use the derived variable in a display or calculation.

To identify a particular variable as the one Data Desk should use, select its icon before closing the derived variable window. Data Desk checks the names of selected icons first when searching for variables to match the names in derived variable expressions. In this way you can specify which variable to use in an expression even if its name is not unique.

Data Desk searches for the variable associated with a particular name in the following order:

- first, any variable whose icon is selected
- second, for expressions that have been edited, any variable used in this expression when it was last evaluated
- third, variables in the same folder as the derived variable's icon.
- finally, any other variable on the desktop

If there are two or more variables by the same name on the desktop (and no variables selected or already in use by the expression), Data Desk asks you to identify the icon of the variable you intend, or to rename it to avoid the conflict.

You may leave a derived variable's window open while it is in use. Indeed, this might be useful if you want to edit the expression and then update plots or analyses. The expression is evaluated when it is needed.

11.13 *Dynamic Parameters*

A derived variable expression consists of functions, arguments (usually variable names), and parameters. Usually the parameters are simply numbers. But, as we have seen with rotating plots, brushing, and slicing, dynamic control often reveals more than static displays and analyses can show. Sliders are graphical controls that offer dynamic control of parameter values in derived variable expression. Thus, you can use them to create your own dynamic graphics or analyses.

The {Data ▶ New} **Slider** command creates a slider. Each slider window shows a plot x-axis and a hairline. You can slide the axis side-to-side with the ☝ tool and rescale it using the **Set Plot Scale...** command in the slider's Global HyperView.

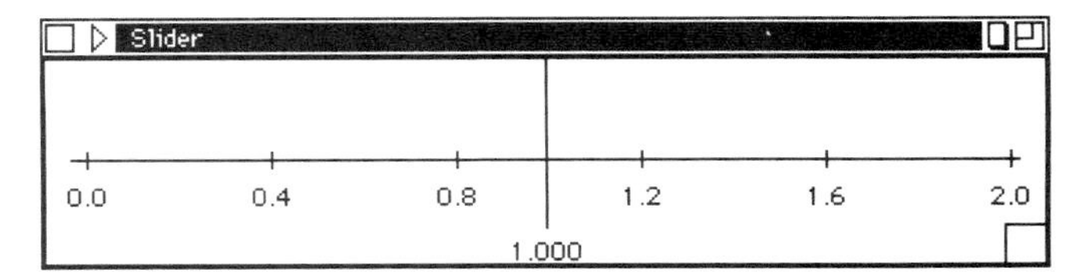

Figure 11-5*. The value of a slider is the number where the hairline crosses the axis.*

A slider always has a value; specifically the value where the hairline intersects the axis. That value is shown below the hairline and is available to any derived variable expression that uses the name of the slider as a parameter. Moving the slider's axis side-to-side changes its value, and (as with any change you make to a part of a derived variable expression) changes the value of the derived variable. Derived variables always recompute immediately and notify all users of their change.

If you hold the Shift key and push the axis, the axis continues to slide in the direction of the push. Grab it again or press the Space bar to stop the slide.

Data Desk plot and table windows typically respond to changes in their underlying variables by offering to update, posting the ! symbol in their global HyperView. However, you can set any plot or table to update immediately as soon as it is aware of an underlying change. The global HyperView of every plot and table offers an **Automatic Update** command. When Automatic Update is selected, the window updates immediately whenever any underlying variable changes. The triangle marking the global Hyperview turns gray to indicate Automatic Update. (Select **Manual Update** from the global HyperView to return to ordinary update-with-approval behavior.)

When a display is set for automatic update and a slider specifies parameters in a derived variable whose values are displayed, then sliding the slider will change the display dynamically. Sliders thus provide a way to build your own dynamic displays.

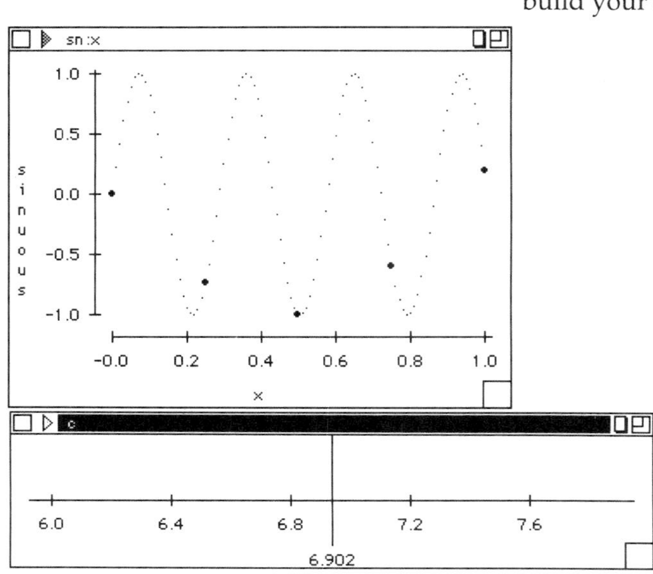

Figure 11-6. *A slider controlling the frequency of a sinusoid.*

To illustrate sliders let's use a simple function. Generate Patterned data values from 0 to 1 in steps of 0.01 and rename the generated variable *x*. (It will be in the Results folder.) Create a new slider named *c* and a derived variable named *sinuous* whose expression is

$$sin(\pi*c*x).$$

(Option-p types π, which Data Desk reads as pi.) Now make a scatterplot of *sinuous* vs *x*. Using the plot's HyperView, choose **Automatic Update** and **Manual Scale**. The first tells the plot to update whenever the slider is moved. The second fixes the plot scale so that it won't change (and speeds response by avoiding plot scaling calculations.)

Grab the slider with the hand tool and move it side-to-side. The slider controls the frequency of the sine wave. For added interest, slide the slider to around 4.0, select (with the selection rectangle) the points where the sine crosses zero, and choose a different plot symbol (and color, if you wish) for them. Now continue sliding and note the patterns of the selected points. Hold the Shift key, give the slider a push, sit back, and watch the show.

Expressions can use more than one slider. For example, define a second slider named *d* and edit the expression in sinuous to read

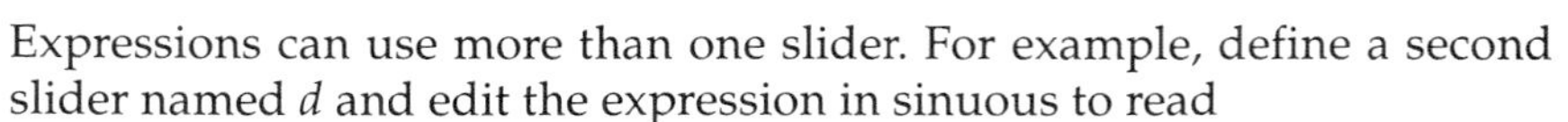

$$sin(\pi*c*x)+sin(\pi*d*x)$$

Now sliding *d* and *c* shows a sum of sinusoids. (You may need to choose **Automatic Scale** to expand the scale.)

Another use of sliders is to determine a parameter value experimentally. Section 11.15 shows an example.

Sliders make it easy to request far more computing power than your Macintosh (or any computer) can deliver. Of course, the more powerful your Macintosh, the more you will be able to do with sliders.

TIP

To reduce the computation load, hold the Option key down while sliding the slider. The values will recompute only when the hairline crosses the axis tick mark.

If a slider seems to stick, then the computations requested by each change in its value are taxing the computing power you have. It is always a good idea to plan derived variable expressions for efficient computing, but this is especially important for expressions that use sliders to create dynamic displays.

You can reduce the computing load by having the slider change only when the hairline crosses an axis tick mark. Hold the Option key down while sliding the slider to place it in this "ratchet" mode. The **Set Plot Scale...** command in the slider's Global HyperView provides control over the location and density of tick marks.

11.14 *Re-expressing Data to Improve Analyses*

The broad value of flexible re-expression is one of the most effective and enduring lessons of exploratory data analysis.

As Mosteller and Tukey point out (1977, p.89)

> Numbers are primarily recorded or reported in a form that reflects habit or convenience rather than suitability for analysis. As a result, we often need to re-express data before analyzing it.

There are many examples in everyday experience — such as the Richter scale for earthquakes, the decibel scale for intensity of sounds, average speed in auto races, and gauges of shotguns — in which the data are already in a transformed scale by the time we hear about them.

We often re-express variables to find a transformation that simplifies the analysis of the data. Re-expressing variables to a scale such as the logarithm, square root, or reciprocal can simplify patterns and relationships in several important ways. Most often we choose transformations to:

- Improve additivity of a response in relation to two or more factors. Additive relationships are appropriate for Analysis of Variance and related analyses.
- Straighten nonlinear relationships. Scatterplots reveal nonlinear relationships. Regression and correlation analyses describe linear relationships formally.
- Promote constant variability across groups or constant variance of measurement across the level of another variable (or over time). Dotplots and boxplots display variability across groups. Scatterplots reveal variability that changes across levels of the x-variable.
- Make univariate distributions more nearly symmetric. Histograms show symmetry well. Probability plots and dotplots are also useful.

Fortunately, these benefits tend to occur together; a re-expression chosen for one reason often helps to improve the data with respect to the other reasons. When we must choose, we usually favor the earlier items on the list over the later ones, but sometimes we can only achieve the latter purposes.

After careful re-expression, analyses and displays are often simpler and more likely to reveal both patterns and unexpected deviations from these patterns. Analyses that were complex and confusing may become simple and straightforward when the data are re-expressed appropriately.

The most common re-expressions are ordered in the sense that the effect of each is slightly more or less than that of its neighbors. Thus you can try a re-expression and, depending on whether it did too little or too much, readjust easily.

The most common data re-expressions are the powers and the logarithm. They are naturally ordered according to the exponent of the power, with the logarithm occupying the "0" position. Thus, we can array the most common re-expressions in order like this:

"ladder of powers"

exponent	function	
2	y2	Square
1	y	Raw data
1/2	$\sqrt{y}$	Square root
"0"	Log(y)	Logarithm
1/2	$-1/\sqrt{y}$	Reciprocal root (the minus sign preserves order)
-1	$-1/y$	Reciprocal or inverse (the minus sign preserves order)

Other powers and roots fit into the order naturally. Thus, for example, a cube root (1/3 power) alters raw data less than a logarithm but more than a square root.

The most common re-expressions are at the top of the **Transform** submenu. Other re-expressions are available in the Transform submenus or by typing the expression. Appendix 3A discusses re-expressions most likely to be useful for various kinds of data.

Because these re-expressions are ordered, we can search effectively for an appropriate re-expression. Data Desk makes this particularly easy by automatically building an appropriate derived variable controlled by a slider. The slider controls the exponent of the re-expression so it is easy to slide up and down the list of powers in search of an appropriate one.

To search for a re-expression in this way, select the variable you wish to re-express and choose {Manip ▶ Transform ▶ Dynamic} **Box – Cox Transformation**. Data Desk creates a new folder containing a slider named *power* and a derived variable containing the expression

> IF *power ≠ 0* THEN *exp(ln('variable')*power)/power)* ELSE *ln('variable')*

When the slider value is not zero, the expression is equivalent to

$$(variable^{power} - 1) / power$$

a form of re-expression discussed by Box and Cox (Box, G.E.P. and Cox, D.R. 1964.) Data Desk uses the equation form with logarithms and exponentiation because it is more efficient to compute, especially on machines with floating point units. The expression reverts to *ln(vari-*

able) when the slider value is zero. This prevents dividing the entire expression by zero.

To select a power, make an appropriate display of the derived variable (depending on the purpose of the re-expression), set the display to Automatic Update (with the command in the display's global HyperView menu), and slide the power slider until the display looks right. Normal probability plots and dotplots work well for individual variables. If you are looking at the relationship between two variables, use a scatterplot. For computing efficiency, you may want to hold down the Option key so that the display recomputes only as the power crosses tick marks. We usually prefer integer or half-integer powers for simplicity.

Once you find a suitable power, you can move the derived variable to the window holding other variables for analysis, and use it in subsequent analyses. Later in your analysis you may wish to return to the slider and investigate how sensitive your analysis is to the choice of power. For example, you may compute residuals, display them, and set that plot to Automatic Update. Now sliding the slider initiates recomputation of the analysis and residuals. The resulting animated display of residuals will give you a good feeling for their stability of the analysis at nearby powers, and thus a feeling for the sensitivity of your conclusions to the choice of power.

11.15 *An Example*

To illustrate both how re-expressing data can improve our ability to work with it and how Data Desk makes it easy to find suitable re-expressions, we consider data from the Companies dataset. We consider two ways to experiment with re-expressions: using sliders and editing typed expressions. If you are working along with the example on your computer, choose the method you find easiest. (If you are using a Mac Plus or Mac SE you may prefer the text editing method. The functions in this example can stress the computing power of the smaller Macs.)

Select the variables *Assets* and *Sales* and choose {Manip ▶ Transpose ▶ Dynamic} **Box – Cox Transformation**. Data Desk creates a folder for each variable containing a slider and a derived variable. The sliders are both named *power*. The derived variables hold expressions of the form

IF *power ≠ 0* THEN *(exp(ln('Assets')*power)– 1) / power)*
ELSE *ln ('Assets')*

This expression is an efficient way to compute the powers discussed in the previous section. The expression

*exp(ln('Assets')*power)*

raises the variable Assets to the power *power*. The natural logarithm (ln) and exponentiation functions compute faster than the "^" operator, especially on machines with a floating point unit. Dividing the result by the power reduces somewhat the tendency of the re-expressed values to change magnitude drastically, and has the additional effect of changing the sign of the expression when the power is negative, thus preserving the order of the data values. The IF/THEN/ELSE structure checks for a zero exponent. Because the logarithm holds the position of the zero

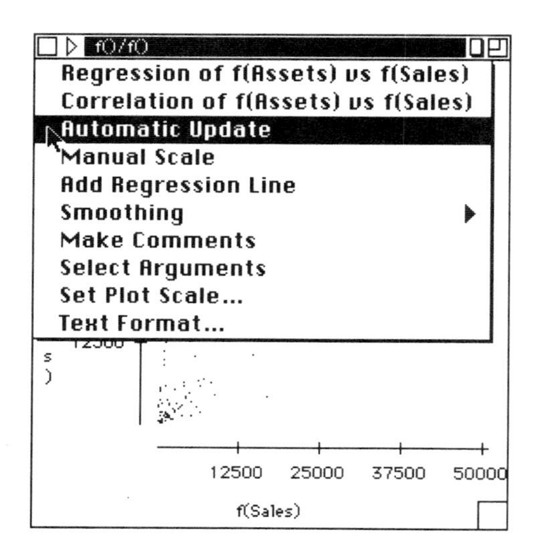

Figure 11-7. *Set Automatic Update in the scatterplot's HyperView.*

power in the ladder of powers (See Section 11-14.), the expression checks for a zero exponent and substitutes the logarithm.

If you are not working with sliders simply define two derived variables with 1.0 in place of the slider names in the expressions above. (You can skip the IF/THEN/ELSE.)

Make a scatterplot of *Assets^p* vs *Sales^p*. Set the scatterplot for Automatic Update using its window HyperView. If you have set the sliders at or near 1.0 (or if you are not using sliders), the plot should show a wedge shape with most points clustered together at the lower left. Such a pattern often indicates that transforming the data might improve the display.

If you are using sliders, try sliding one of them side to side. You can see the plot changing smoothly. If your computer has a hard time keeping the animation smooth, try holding down the Option key as you move the slider. This will limit recomputing to the points where the slider's hairline crosses an axis tick mark.

As the slider approaches zero, the plot will begin to look like Figure 11-8. You can go past zero; the plot continues to distort in the same way. Leave the slider at or near zero, representing a re-expression close to the logarithm.

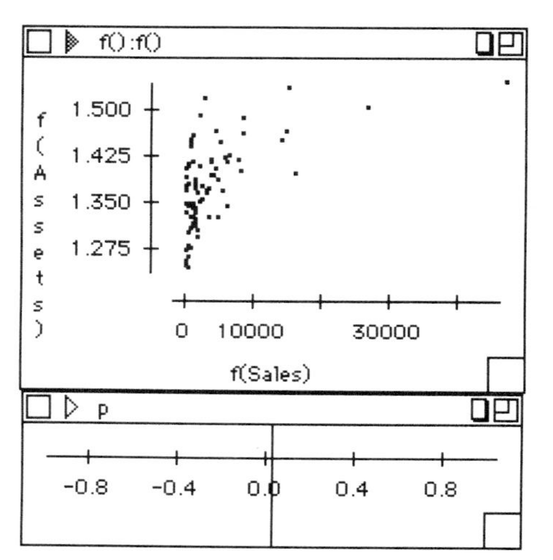

Figure 11-8. *Re-expressing Assets to a power near zero makes the plot look like this.*

If you are not using sliders, edit the expression for *Assets^p*. Substitute 0.5, 0.0, and -0.5 for the 1.0 in the original expression. After changing the expression press the Enter key. The scatterplot will immediately update to show the change. Leave the power set to zero.

Now slide the other slider. As its value gets near zero, the plot begins to look quite reasonable, showing a consistent trend from lower left to upper right.

If you are not using sliders, edit the expression for *Sales^p* in the same way as before, trying different powers. Substitute zero as the final power to get a plot that looks like Figure 11-9.

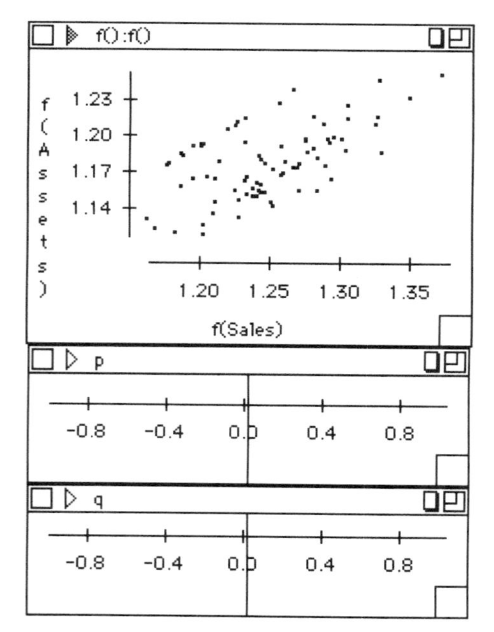

Figure 11-9. *When both re-expressions are near the zero power (log) the plot looks much better.*

This example shows vividly how the choice of expression can alter a plot. The scatterplot of *log(Assets)* vs *log(Sales)* shows much simpler structure and suggests a relationship that might be suitable for further analysis. The placement of the logarithm at the zero position in the ladder of powers is also clear; the shape of the plot for powers very near to zero is virtually the same as the shape for the logarithm.

In general, there is no single correct re-expression for data. Common sense, common practice, and practicality all help to determine the choice of a re-expression. Some re-expressions are standard practice. For example, economists often work with logarithms of values like income, assets, and sales. Some re-expressions are convenient because the measurement units are simple. For example the reciprocal of miles per gallon is gallons per mile. Velocities such as miles per hour or minutes per task often do well in their reciprocal form (hours per mile, tasks per minute).

Data re-expression is a rich and rewarding aspect of data analysis that is dealt with by very few accessible books. We recommend that interested readers consult *Data Analysis and Regression* by F. Mosteller, and J. Tukey (Addison-Wesley 1977) for an excellent and practical discussion.

11.16 *Rules for Re-expression*

Mosteller and Tukey offer general advice on re-expressing data to simplify and improve analyses. They note that the choice of a re-expression function depends upon the nature of the data. The define seven types of data for this purpose. Appendix 3A of Chapter 3 discusses these types.

Mosteller and Tukey suggest that it is more often appropriate to re-express data than to leave values in their original form. Toward this end, they suggest the following re-expressions as a good start, based on the type of data:

Names No re-expression can be suitable

Grades Re-expression is complex and not common.

Ranks are dealt with in Section11.18 below.

Counts May benefit from a square root re-expression.

Amounts cannot be negative. Histograms of amounts often trail off to the high end. Such skewed distributions often benefit from logarithms.

Balances are often differences of two amounts. It may help to transform each of the amounts and then take their difference or ratio.

Bounds

If data values are bounded at one end, we recommend shifting the data so that the bound is at zero, and treating the result as an amount. Thus for $y \geq A$, treat $y - A$ as an amount. For $y \leq A$, work with $A - y$.

Data that are bounded at two ends can be rescaled and treated as a counted fraction. Thus for $A \leq y \leq B$, treat

$$(y - A)/(B - A)$$

as a counted fraction.

11.17 Efficiency

Derived variables generally compute very quickly. Nevertheless, it is wise to think about computing effort when writing a derived variable expression. For example, isolate any repeated computation that can be made separately. The expression

$$sin(2*\pi*c*x/NumNumeric(x))$$

for example, multiplies each value by the constants 2 and π and divides by Numnumeric. It would be more efficient to define a new derived variable named $2\pi x/n$ consisting of the expression

$$2*\pi*x/NumNumeric(x)$$

and then write the above expression as

$$sin(c* \text{'}2\pi x/n\text{'})$$

reducing the repeated calculations to a single multiply and sine evaluation.

Some efficiencies come from restructuring functions. In generating derived variables to explore data re-expressions (as in the example of Section 11.15), Data Desk avoids the exponentiation (^) operation because it is not available on the floating point unit (fpu). When an fpu is present, it is more efficient to use the property of logarithms that

$$x^p = e(p*ln(x))$$

11.18 Indicator Variables and Logical Expressions

Many statistical analyses can be made easier — or possible — by the use of specially constructed indicator variables. Indicator variables typically take on only the values 0 and 1, or the values –1, 0, and 1. Indicator variables are specially constructed to isolate a subgroup or an individual in the data, typically by assigning a 1 to all cases in the subgroup and a 0 to all others.

Derived variables offer a simple and intuitive way to generate indicator variables. When a derived variable containing a logical expression is used as a numeric variable, it generates a 1 for all cases for which the expression is true, and a 0 for all cases for which the expression is false. Thus, for example, the derived variable expression

TextOf ('season') = "spring"

defines an indicator variable that isolates two subgroups. It evaluates TRUE (1) for all cases for which the variable season is the word "spring" and FALSE (0) for all others.

Indicator variables constructed to select ranges can be specified in much the same way. For example, the expression

25000 < income AND *income< 40000*

evaluates as TRUE for those individuals whose incomes fall in the specified range and FALSE otherwise, isolating this range of incomes.

11.19 *Working with Several Relations*

Ordinarily all of the variables in a derived variable expression must be in the same relation. However, some derived variable functions can work across relations. Several functions provide the kind of capability commonly found in relational database programs. With these functions you can construct complex datasets to reflect the true structure of your data.

The *Lookup* and *GetCase* functions accept arguments from two different relations. The Lookup functions (Lookup, LookupFirst, LookupLast) have the form *Lookup(<var>, <value>)* where *<var>* is any expression evaluating to an array of values that could be a variable and *<value>* is any expression evaluating to an array of values to be found in the variable. The Lookup functions return an array of indices of the cases in *<var>* that match the values in *<value>* (or the values just below or just above the values if LookupFirst or LookupLast function is used). The result is a variable in the relation of the *<value>* array. LookupFirst and LookupLast require that the variable being searched (the first argument) be sorted from low to high.

GetCase takes the form *GetCase(<var>, <indices>)* where *<var>* is as in Lookup, and *<indices>* is an expression evaluating to an array of indices. GetCase extracts the values in *<var>* whose case numbers are the values in *<indices>*. It is a variable in the same relation as *<indices>*. If an index is not an integer it is truncated. If its value is zero or less or is beyond the last case in *<var>* GetCase returns the missing code, "•".

An alternative form that may be more convenient than GetCase is the subscript form using square brackets: *<var>[<indices>].*

11.20 *Subtle Points*

Ordinarily, derived variables behave just as you expect. However, some subtleties of design may be worth a brief explanation. You can skip this section and miss nothing of great importance, but reading it may help you to understand derived variables (and Data Desk) better.

The relation of a derived variable is an ephemeral thing. Surely an expression transforming one variable or combining two variables that are both in the same relation belongs in that relation. But it is easy to write derived variable expressions that change relations. For example, all collapsing functions yield a single number. Subscripted expressions (Section 11.11) are in the relation of the index variable. You can change the relation of a derived variable just by editing its expression, or even by editing one of its arguments.

When you create a derived variable by selecting a command from the **Transform** menus, Data Desk tries to place the new icon in the window of the relation to which it is most likely to belong. Thus, if you select a variable and choose log(•) from the Transform menu, the resulting derived variable is placed in the same window as the icon of the variable. However, if you create a New Derived Variable, Data Desk has no idea what expression you will type, and thus no idea of the relation of the variable. The new icon is placed in the Results folder, but the relation of the variable is determined by the expression.

If you change the relation of a derived variable after it has been used, plots and tables that use it may complain of mismatched relations and refuse to recompute.

While derived variables always update immediately, you may still see the ! alert in derived variable HyperViews. The numbers view of a derived variable shows a snapshot of the values at the time they were computed. If the expression or arguments of a derived variable change, the numbers view will post the ! in its global HyperView rather than updating instantly. This is appropriate since it is just a view of the variable.

Although derived variables refer to variables by name, once a derived variable expression has been parsed, Data Desk knows the identities of the variables in the expression. If you then rename one of the argument variables, Data Desk will not forget which variables were used in the expression. Indeed, the expression window of the derived variable will post a ! and offer to update, substituting the new variable name for the old one in the expression wherever it occurs.

However, the name of the derived variable icon will not change automatically even if that name was generated to reflect the derived variable function. Thus, if you select the variable *Sales* and choose {Manip > Transform} **Log(•)**, Data Desk will create a derived variable named *LSales* containing the expression *log('Sales')*, and place it in the same icon window (and relation) as *Sales*. If you now rename *Sales* to *Proceeds* and open the icon *LSales*, the expression will read *log('Proceeds')*. If you had already opened the derived variable, the expression editing window posts a ! when the window is again frontmost, and offer to update. However, the derived variable's name will still be *LSales*. This is because you might have renamed the icon yourself. Data Desk has no way of knowing whether you want the variable renamed, so it leaves control over variable names to you.

11.21 *Common Errors and How to Avoid Them*

Derived variables are very powerful. However, some subtle errors can be difficult to find. This section lists some hints for ferreting out such bugs.

- Always complete an IF/THEN/ELSE combination. A common error is to omit the ELSE clause. It is **not legal** to write a statement such as

  IF *income > 0* THEN *profits*

  because Data Desk does not know what to do when the condition is not true.

- **Do not** type an "invisible" character in the expression. Some keys (for example, the Enter key) do not print on the screen. Data Desk complains that you typed an illegal character. Try backspacing from the end of the expression until the rightmost character is deleted. If you can't find the invisible character, try deleting everything in the window and retyping the expression.

- Avoid typing an extra space in a variable name. Spaces may be inserted freely in derived variable expressions between functions and around variable names, but variable names must be typed literally. An extra space between two words of a variable name is hard to see, but the name will not match the icon name. Try retyping the variable name. To be certain to get a variable's name right, drag its icon into the derived variable editing window.

- Remember the TextOf function when working with the text version of a variable. Derived variables work with the numeric version of variables by default. This is often a confusing error because the expression seems to say what you meant.

- **Do not** mix two different types of outcomes in an IF/ THEN/ELSE expression. For example, the following expression is **incorrect**:

  IF *contribution* > *1000* THEN *'bigGiver'*
      ELSE IF *contribution* > *100* THEN *2*
          ELSE *"form letter"*

  because in the first two cases it evaluates to a number (if the variable "bigGiver" is numeric) and in others to a text string.

APPENDIX 11 A
Derived Variable Expressions

The following components may be used in constructing the expression for a derived variable:

NUMBERS

Any number that could appear as a data value may be part of a derived variable expression. Numbers may have a leading + or -, followed immediately by the digits of the number. Numbers may be in *scientific notation* in which the number is written as some value times a power of ten. The power of ten is separated from the rest of the number by an "E" (for exponent). Thus, the number 1000 can be written as 1.0E3 (that is, 1.0 times 10^3). There must be no blanks between the E and the numbers on either side of it.

Both the symbol ∞ (typed as Option–5) and the "word" *INF* represent infinity, and may be preceded by a minus sign to indicate negative infinity. The symbol π (typed as Option–p) and the "words" *PI* or pi are recognized as the value pi (= 3.1415926...). Note that you must type 2 * π rather than 2π to mean "two times pi". The symbol • (typed as Option–8) represents a missing value.

TEXT

Text must be enclosed in double quotes, " ".

NAMES OF VARIABLES

Data Desk ignores capitalization, so 'fun', 'Fun', and 'FUN', all refer to the same variable.

All variables in a derived variable expression must be in the same relation except for the arguments of the relational functions Lookup and GetCase, and subscripted references of the form *variable[index]*. Derived variables may use other derived variables (provided they are in the same relation).

Variable names can be typed without any extra punctuation unless they are potentially ambiguous. A variable name is ambiguous if it contains a space or punctuation mark, if it contains another variable name or a function name, or if it is or contains a number (such as '1968'). Ambiguous variable names must be enclosed in single quotation marks ('). It is generally a good idea to enclose variable names in single quotation marks to reduce confusion.

Data Desk searches for variables to match the variable names in expressions starting with the icon window containing the derived variable icon. If it cannot find a match, it searches all icon windows that share the same parent with the home window (in effect, all the "first cousins" of the derived variable's icon). The search widens until it includes all variables in the datafile. If, at any level, Data Desk finds two variables with the same the same name, it asks you to declare which one you intended. Click on the variable you choose and try again.

Arithmetic Operators

The operators + - * / (or ÷) and ** (or ^) indicate respectively, addition, subtraction, multiplication, division, and exponentiation (raising to a power). When no parentheses are used, operations are performed in the following order:

1) unary negation (minus sign), parentheses, NOT, IF, other unary operations
2) ^, ** (exponentiation (raising to a power))
3) * (multiplication), / or ÷ (division), MOD, DIV, Lesser, Greater
4) + (addition), - (subtraction)
5) $<, \leq, =, \neq, >, \geq$ (comparisons)
6) AND, XOR (exclusive OR), DIFF
7) OR

Multiple operations at the same level are performed from left to right. Thus 12 - 3 - 2 equals (12 - 3) -2, or 7 rather than 12 - (3 - 2), which would be 11, and 12/3/2 equals (12/3)/2, or 2, not 12/(3/2), which would be 8. Similarly, 2**3**4 is (2**3)**4, which is $(2^3)^4$.

Parentheses and Brackets

Use parentheses ( ) freely to indicate the order in which the expression should be evaluated. Parentheses must be balanced. That is, every left parenthesis must be closed with a corresponding right parenthesis. Square brackets, [ ], denote selected cases in the variable expression they follow. Thus *myvar[3]* is the value of the third case of *myvar.*

Comments

Enclose comments in {braces} or in (* comment parentheses *). Comments may be nested to any depth and may appear any place a space can appear.

Casewise Functions

Casewise functions produce one value for each case in the variables they use. The arguments of a casewise function are variables specified within the function's parentheses. Arguments can be variables or expressions that evaluate to variables. If a case has a nonnumeric value or is missing, casewise functions produce a missing value called a NaN (for Not a Number) for that case. Capitalization does not matter in function names. Some functions have two or three synonyms. Casewise functions are listed in Appendix 11B.

Collapsing Functions

Collapsing functions use a variable as an argument, but produce a single number, which can then be used for further calculation or reported by itself. Nonnumeric values in the argument variable are treated as missing and are ignored both in arithmetic and in counting numeric cases. Infinities are treated as missing values. Collapsing functions are listed in Appendix 11C.

Relational Functions

Relational functions refer to variables in different relations and perform basic relational operations.

HotResults™

Data Desk uses a number of internal derived variable functions to provide computation *HotResults* that update automatically when the analysis on which they are based updates. For example, the residuals, predicted values, and diagnostic statistics from a regression analysis (see Chapter 22) are HotResults that refer to that analysis. Any change in the analysis causes the HotResult to update.

HotResults are not editable, although you can open them and see the function name. They are not available for you to type as part of an expression. To build a derived variable expression using, for example, the predicted values of a regression, refer to the HotResult holding the predicted values rather than to the *predicted values* function itself.

APPENDIX 11B
Casewise Functions

The following casewise functions are available either from the **Transform** submenus or by typing them as part of a derived variable expression. Capitalization is optional.

COMMON RE-EXPRESSIONS

The re-expressions on the ladder of powers (discussed in Section 11.8) are commonly used to make patterns and relationships clearer and easier to describe.

Sqr(·)	square, x2
√, Sqrt(·)	square root
Log(·)	base 10 logarithm
inv(·), 1/	reciprocal
ln(·)	natural log
ln1(·)	ln(x+1)
exp(·)	e^x, e = 2.7182818..., the base of the natural logarithm
exp1()	exp(x)-1
^, **	general exponentiation (raising to a power)

ARITHMETIC FUNCTIONS

y + x	addition
y - x	subtraction
y * x	multiplication
y / x, y ÷ x	division (type ÷ as Option -/)
y DIV x	integer divide; truncates result to an integer
y MOD x	remainder of y/x
Lesser(y, x)	pairwise minimum
Greater(y, x)	pairwise maximum
neg(·)	negation. neg(y) = -y.

LOGICAL FUNCTIONS

Data Desk maintains logical values internally as Boolean (that is, TRUE or FALSE) values. Data Desk translates Boolean values to numbers according to the rule FALSE = 0, TRUE = 1, producing indicator variables suitable for use as selectors.

y < x   Returns TRUE (1) for cases in which y < x.

y <= x Returns TRUE (1) for cases in which y ≤ x.

Alternatively, use "≤", typed as Option–comma.

y = x Returns TRUE (1) for cases in which y = x.

y >= x Returns TRUE (1) for cases in which y ≥ x

Alternatively, use "≥", typed as Option–period.

y > x Returns TRUE (1) for cases in which y > x.

y <> x Returns TRUE (1) for cases in which y ≠ x.
Alternatively, use "≠", typed as Option – =

NOT(·) Negates the truth value of its argument. When applied to variables rather than expressions, it first converts the variable to a logical expression using the rule that 0 means FALSE and anything else means TRUE. In this form NOT is equivalent to *(1 - ABS(SGN(y)))*. Alternatively, use the unary negation sign, ¬ typed as Option-l.

y AND x Logical AND of two Boolean arguments. When applied to variables rather than expressions, it first converts the variables to logical expressions using the rule that 0 means FALSE and anything else means TRUE. In this form AND is equivalent to *(ABS(SGN(y * x)))*.

y OR x Logical OR of two Boolean arguments. When applied to variables rather than expressions, it first converts the variables to logical expressions using the rule that 0 means FALSE and anything else means TRUE. In this form OR is equivalent to *(ABS(SGN(y + x)))*.

y XOR x Logical exclusive OR of two Boolean arguments. When applied to variables rather than expressions, it first converts the variables to logical expressions using the rule that 0 means FALSE and anything else means TRUE. In this form XOR is equivalent to *(ABS(SGN(y)) ≠ ABS(SGN(x)))*.

y DIFF x Logical difference of two Boolean arguments; y DIFF x is the same as y AND (NOT x).

Rounding Functions

ABS(·) Absolute value; | y |.

INT(·) Integer part, sometimes denoted [y]. The whole number nearer to zero or equal to the argument value. Int(-2.5) = -2.0. Int(2.5) = 2.

Floor(·) The whole number less than or equal to the argument value. Floor(-2.5) = -3.0.

Ceiling(·) The whole number greater than or equal to the argument value. Ceiling(-2.5) = -2.0.

Sign(·), SGN(·) The sign of its argument. Returns –1, 0, or 1 according to whether its argument is negative, zero, or positive, respectively.

RoundEven(·)	The value rounded to the nearest whole number. The fraction .5 rounds to the nearest *even* whole number: RoundEven(2.5) = 2.0. RoundEven(-2.5) = -2.0.
RoundUp(·)	The value rounded to the nearest whole number. The fraction .5 rounds up to the *next largest* whole number: RoundUp(2.5) = 3.0. RoundUp(-2.5) = -2.0.
RoundDown(·)	The value rounded to the nearest whole number. The fraction .5 rounds down to the *next smallest* whole number: RoundDown (2.5) = 2.0. RoundDown(-2.5) = -3.0.

Trigonometric Functions

All trigonometric functions work in radians.

sin(·)	sine (argument in radians)
cos(·)	cosine
tan(·)	tangent
arcsin(·), asin(·)	arcsine
arccos(·), acos(·)	arccosine
arctan(·), atan(·)	arctangent
sinh(·)	hyperbolic sine
cosh(·)	hyperbolic cosine
tanh(·)	hyperbolic tangent
arcsinh(·), asinh(·)	archyperbolic sine
arccosh(·), acosh(·)	archyperbolic cosine
arctanh(·), atanh(·)	archyperbolic tangent

Miscellaneous

Rank(·)	For each case, returns the rank of that case. The rank is ascending and ranks either numerically or alphabetically according to the type of its argument.
NScores(·)	NScores are the values computed for a Normal probability plot. The ith NScore is the median of the sampling distribution of the ith order statistic based on a sample of size n drawn from a standard Normal distribution. See Appendix 8B for details of how NScores are computed.
CumSum(·)	Cumulative sum. The first case is the same as the first case of the argument variable. The second case is the sum of the first and second cases. The third case is the sum of the first three, and so on.

Dynamic Functions

Box – Cox Transformation

Creates a slider and derived variable that can be used to find a good re-expression for the selected variable(s). See Sections 11.14 and 11.15 for details and an example.

Mix X and Y — Creates a slider and derived variable that blends two variables together. The blend can be anywhere from 100% of variable X and 0% of variable Y through 50% of each to 0% of variable X to 100% of variable Y. The percentage of the blend is determined by the value of the slider.

Tukey's Lambda — Creates a slider and derived variable that can be used to re-express counted fractions. The counted fractions must be expressed as values between 0 and 1.0.

Control Functions

TextOf(·) — The text of the named variable. Ordinarily derived variable functions operate on the numeric values of variables. The TextOf function returns the text values. It is most often used to identify specific cases within a category variable.

IF/ THEN/ ELSE

IF *<logical expression>* THEN *<expression1>* ELSE *<expression2>* evaluates as *expression1* if the logical expression evaluates to TRUE, and as *expression2* if the logical expression evaluates to FALSE. For nested expressions, the ELSE works with the closest IF. Either result expression can be a number. Hence, IF/THEN/ELSE can be used to recode variables:

IF 0 < PAY AND PAY < 10000 THEN 0
ELSE IF 10000 <= PAY AND PAY < 25000
THEN 1
ELSE IF (etc.)

Alternatively, the expressions can be text, so the above recoding could take the form

IF 0 < PAY AND PAY < 7000
THEN "Poor"
ELSE IF 7000 < PAY AND PAY < 25000
THEN "Middle"
ELSE IF (etc.)

The logical expressions themselves can use the TextOf function to select groups from a categorical variable:

IF (TextOf('gender') = "female") THEN 0.57
ELSE 1.00

Similarly, the result expressions can simply name other variables:

IF TextOf('Country') = "US" THEN HtInch

ELSE HtCm

The result of the entire IF/THEN/ELSE expression can be a number, so you may write expressions like

StdBonus + (IF Sales > 5000 THEN 250
ELSE 100)

The following functions are advanced capabilities:

Manipulation Functions

Lag(y, k) — Shifts the cases in variable *y* down k cases, insertıng missing cases as the first k cases and dropping excess cases off the end to preserve the length of the variable. If *k* is negative, then Lag shifts the variable up *k* cases and inserts missing cases at the end.

Relational Functions

Data Desk's relational functions provide facilities for looking up values across relations. They thus provide the basic operations on which to build a Relational Data Analysis™.

GetCase(y, x) — Each value of *x* is taken to specify a case number in *y*. The corresponding case value of *y* is returned. A constant or an expression evaluating to a constant may take the place of *x*.

Thus GetCase('income', 5) returns the income value in the fifth case. Non-integral case numbers are truncated. Case numbers of zero or less or case numbers greater than the number of cases in y return the missing value code, •.

To obtain the text value of y at case number *x*, specify TextOf:

GetCase(TextOf('y'), x)

The two arguments need not be in the same relation. The result is in the same relation as *x*.

An alternative form of GetCase is to use the square bracket notation. Case numbers enclosed in square brackets following a variable name or expression select the cases with those case numbers. Thus *y[3]* is the third case of *y*. However, the argument in the brackets may itself be a variable holding case numbers. Thus, *GetCase(y, x)* is equivalent to *y[x]*.

LookUp(y, x) — If *k* is a constant, and *y* is a variable sorted in ascending order, *LookUp(y, k)* is the case number of a case of *y* for which $y = k$.

If *"text"* is a quoted string then *LookUp(y,"text")* is the case number of a case of *y* for which *TextOf(y)* = *"text"*.

If no match is found, LookUp returns missing.

If x is a variable, then *LookUp(y, x)* returns for each element of *x*, the case number of a case of *y* for which *y* equals the corresponding value in *x*.

To search on the text values of a variable, use TextOf: *LookUp(y, TextOf(x))* returns for each case in *x*, a case in *y* whose text matches the text of *x*.

The two arguments need not be in the same relation. The result is in the relation of *x* because a lookup is performed for each element of *x*.

LookUpLast(y, x) If *k* is a constant, and *y* is a variable sorted in ascending order, *LookUpLast(y, k)* is the case number of the the last case of *y* for which $y \leq k$.

If *x* is a variable, then *LookUpLast(y, x)* returns a value for each element of *x*.

If *y* is not sorted in ascending order, LookUpLast finds a case satisfying the condition, but it may not be the last one.

The two arguments need not be in the same relation. The result is in the relation of *x* because a lookup is performed for each element of *x*.

LookUpFirst(y, x) If *k* is a constant, and *y* is a variable sorted in ascending order, *LookUpFirst(y, k)* is the case number of the the first case of *y* for which $y \geq k$.

If *x* is a variable, then *LookUpFirst(y, x)* returns a value for each element of *x*.

If *y* is not sorted in ascending order, LookUpFirst finds a case satisfying the condition, but it may not be the first one.

The two arguments need not be in the same relation. The result is in the relation of *x* because a lookup is performed for each element of *x*.

The LookUp functions can be used to group numeric variables into categories. For example, we can define the following variables: If

letterGrade contains the text values "F", "D","C", "B", "A".

gradeBounds contains the "cut" values 0, 60,70, 80, 90.

numberGrade contains students' numeric averages.

Then *GetCase(TextOf('letterGrade'), LookupLast('gradeBounds', 'numberGrade'))* assigns the appropriate letter grade to each student.

APPENDIX 11C
Collapsing Functions

Collapsing functions take a variable as an argument, but produce a single number. Nonnumeric values in the argument variable are treated as missing and are ignored both in arithmetic and in counting numeric cases. Infinities are treated as missing values except where noted.

Min(·)	The minimum value of the argument. Min( ) returns *-INF* if the argument contains a negative infinity. Min( ) returns *NaN* if the argument contains no numeric cases.
Max(·)	The maximum value of the argument. Max( ) returns *INF* if the argument contains an infinity. Max( ) returns *NaN* if the argument contains no numeric cases.
Σ(·), Sum(·)	Sum of cases in the variable. Type Σ as Option–w.
SSQ(·)	Sum of squares of cases in the variable.
Mean(·)	Sample average or mean of the variable.
StDev(·), SDev(·)	Standard deviation.
StdError(·)	Standard error of the mean. Equivalent to *StDev(y)/sqrt(numnumeric(y))*
Variance(·)	Variance of the variable.
NumNumeric(·), NumNum(·)	The number of numeric values in the argument. This is the denominator for the mean.
NumNonNumeric(·), NumNonNum(·)	The number of nonnumeric values.
NumCases(·)	The total number of cases. Note that NumCases = NumNumeric + NumNonNumeric.

CHAPTER 12

Manipulating Variables

12.1	Sorting	*179*
12.2	Ranking	*181*
12.3	Generating Patterned Variables	*181*
12.4	Appending and Splitting Variables	*183*
12.5	Transpose	*184*
12.6	Selectors	*185*
12.7	Performing Analyses Group by Group	*186*
12.8	Samples	*188*
12.9	Duplicating Icons	*189*
12.10	Data Tables	*189*
12.11	Copying and Printing Results	*190*

APPENDIX

12A	Missing Values	*191*

THIS CHAPTER DISCUSSES DATA manipulations that are different from data editing or transformation. These manipulations act on all the cases of one or more variables to create an entirely new variable.

The commands discussed in the first part of this chapter are in the **Manip** menu. The remaining Manip commands are **Transform** (discussed in Chapter 11) and **Generate Random Numbers...**(discussed in Chapter 15). Manipulations found in other menus are discussed in the final sections of this chapter.

12.1 *Sorting*

The **Sort on Y, Carry X's** command in the **Manip** menu reorders the cases in a variable. To sort a variable, select its icon and choose **Sort on Y, Carry X's**. Data Desk creates a new variable holding the same data values reordered with the lowest (most negative or smallest positive) value in the first case and the highest (most positive or least negative) value in the last case and places it's icon in a new relation.

The new relation also holds a variable of *Unsort Indices*. These record the case number of each case before sorting.

sort key

If you select a y-variable and one or more x-variables, Data Desk makes copies of all of the variables, reordering them in the same order as the corresponding cases of y. The original variables are not changed. When you sort several variables, the y-variable is called the *sort key* because it determines the new order of the cases in all of the sorted variables.

For example, the Nuclear Plants dataset contains 32 cases that are not ordered on any of the three variables. While it might be interesting to sort the data on any of the variables, we will sort the cases chronologically. Select *Date*, extend the selection to the other two variables, and choose **Sort on Y, Carry X's**. The sorted variables are shown in Figure 12-1.

stable sorting

When cases are equal, Data Desk preserves their original order. For example, in the Nuclear Plants data, we can see by checking the unsort indices that whenever two cases have the same Date, the one originally first in the data is still first. Sorting methods with this property are said to be *stable*. Stable sorting allows you to sort on several sort keys by

Date	Cost	MWatts	Unsort In...
67.17	288.48	821	28
67.25	207.51	745	27
67.25	217.38	745	31
67.33	452.99	1065	3
67.33	443.22	1065	4
67.83	284.88	886	29
67.83	280.36	886	30
67.83	270.71	886	32
67.92	345.39	514	1
68.00	652.32	1065	5

Figure 12-1. *Nuclear Plants data sorted according to Date. Note that cases are re-ordered consistently across all three variables.*

sorting on them one after the other, last key first. Thus to sort on year and on month within year, sort first on month and then sort the resulting values on year.

If the sort key has any infinities, they are sorted to the ends of the variable (according to whether they are +∞ or -∞). If the variable has any missing values, they sink to the bottom. Alternatively, missing cases can

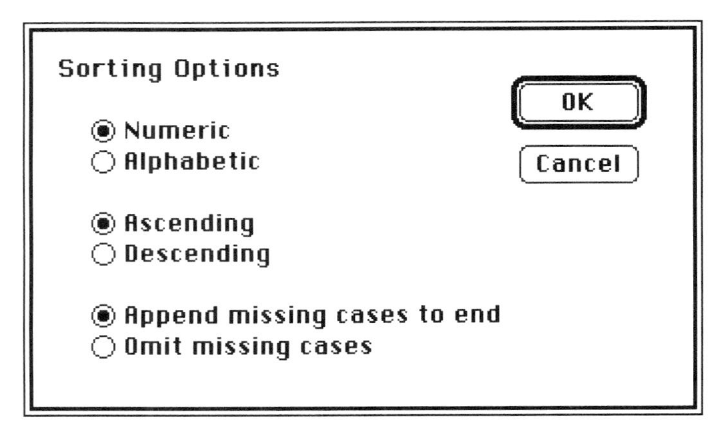

Figure 12-2. *The Sorting Options Dialog.*

be omitted entirely from the sorted variables.

The Sorting Options dialog offers three options to control sorting. It is shown in Figure 12-2 with the default choices selected. Choose **Set Sorting Options** from the **Manip Options** submenu to show the dialog. Changed settings affect all future Sort commands.

Choose Alphabetic sorting to order cases according to the alphabetic ordering of the text of the sort key variable. You can control the alphabetizing conventions and determine such things as the correct ordering of non-English characters and accents with the Macintosh System International resources. Consult your Macintosh documentation for details.

Sorting a numeric variable alphabetically reorders the numbers according to the sequence of their *numerals* rather than according to their *values*. This can be confusing because the two orders are similar but not identical. For example, the numerals from 1 to 100 sort alphabetically as

1, 10, 100, 11, 12,… ,19, 2 , 20, 21, … ,29, 30, …, 99

Choose descending ordering to place the highest (largest, most positive) values of the sort key first rather than last. Descending alphabetic ordering places "z" before "y", and "b" before "a".

Although the sorted versions of variables deal with the same individuals and have the same number of cases, those cases are in a different order. The sorted variables are thus not in the same relation as the original variables. Data Desk creates a new relation for the sorted data. Although you can drag the variables back to the original relation (Data Desk would check only that they have the same number of cases), this would create an anomalous relationship in which cases were linked incorrectly. You should only place sorted and unsorted versions of your data in the same relation if you are very careful about the operations you then perform.

12.2 *Ranking*

ranks

Many statistics methods (especially nonparametric methods) work with the *ranks* of the cases in a variable rather than with their numeric values. The rank of a case is the case number of its position in a sorted version of the variable. The lowest valued case has rank 1, the next largest has rank 2, and so on.

HOW-TO

To compute the ranks of variables:
- Select one or more variables.
- Choose **Rank** from the **Manip** menu.

Each selected variable generates a variable of ranks.

To find the ranks of some variables, select them and choose **Rank** in the **Manip** menu. A new variable is created for each variable selected, and named *Rank:<varname>*.

When two or more values are identical, they are usually given the same rank. The rank assigned is the average rank of all cases with that value. For example, the ranks of 1, 2, 2, and 3 are 1, 2.5, 2.5, and 4, respectively. Except for the fractions that can result from this averaging, rank variables contain integers.

One advantage of working with ranks is that the most common data transformations preserve order and thus do not change the ranks. Transformations that preserve order include the logarithm, square root, and negative reciprocal. Statistics procedures based on ranks therefore yield the same results when such transformations are applied to the data. Rank assigns a missing value to any cases that are missing in the variable being ranked.

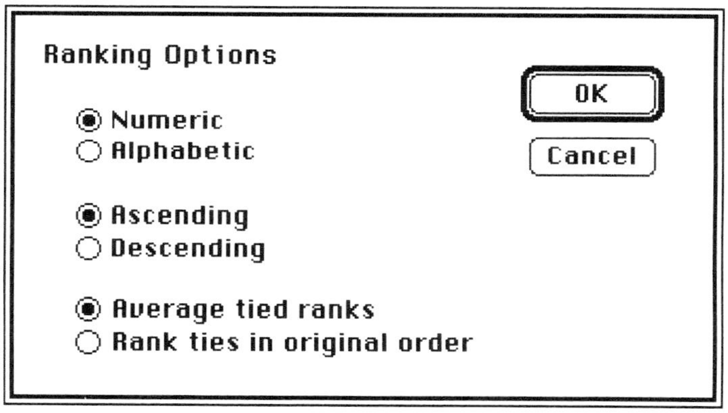

Ranking Options

OK

Cancel

◉ Numeric
○ Alphabetic

◉ Ascending
○ Descending

◉ Average tied ranks
○ Rank ties in original order

Figure 12-3. *The Ranking Options Dialog.*

The {Manip ▶ Manip Options} **Set Ranking Options** command, whose dialog is shown in Figure 12-3, lets you specify how ranking should be performed. Ordinarily, tied values are assigned the average of their ranks, but one option allows them to be ranked in their original order. You can also specify that cases are to be ranked alphabetically rather than numerically.

12.3 *Generating Patterned Variables*

Variables whose case values follow a pattern are useful in many ways. For example, the numbers from 1 to 100 might be useful as the *x*-axis of a hundred-point scatterplot where case order was of interest. Patterned variables often can label categories for an Analysis of Variance. **Generate Patterned Data...** generates new variables named *Pattern1*, *Pattern2*, and so on.

Generate Patterned Data... presents a dialog to specify a sequence of numbers from any number to any number, in steps of a specified size. For example, a sequence from -3 to 9 in steps of 3 is -3, 0, 3, 6, 9. By default it offers the sequence that counts from one up to the number of cases in the frontmost relation.

Generate Patterned Data OK Cancel

Generate numbers from -3 to 9
in steps of 3

Replicate each number in the sequence 3 times.
Replicate the entire sequence 1 times.

Figure 12-4. The Generate Patterned Data Dialog.

The dialog also offers to repeat each value any number of times and to replicate the sequence any number of times. For our example, repeating each value in the original sequence 3 times generates -3, -3, -3, 0, 0, 0, 3, 3, 3, 6, 6, 6, 9, 9, 9. Repeating the example sequence twice generates -3, 0, 3, 6, 9, -3, 0, 3, 6, 9. Both kinds of replication can be used in the same data generation. (It doesn't matter which kind of replication is performed first, the resulting sequence is the same.)

Data Desk places newly generated patterned variables in the frontmost relation that has the same number of cases. If it can find no appropriate relation it creates a new one, places its icon on the desktop, and opens it to show the new variables. If you generate patterned data to use with existing data, but the patterned variable will not go into the data's relation, check that you are generating data with the correct number of cases.

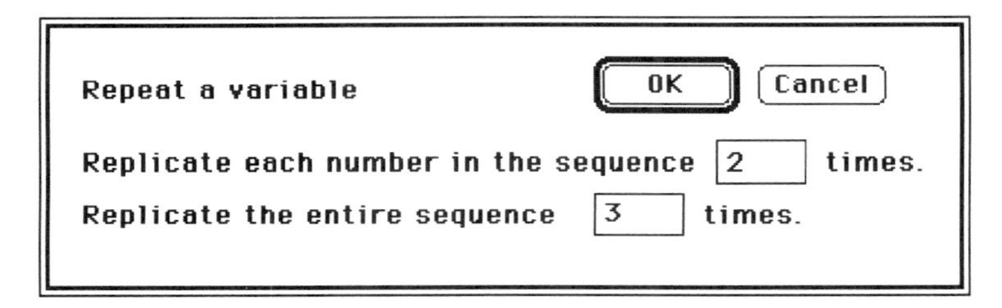

Figure 12-5. The Repeat Variables... dialog.

The **Repeat variables...** command generates a patterned variable by replicating the cases of an existing variable. This might be valuable, for example to create a factor variable for an experimental design model. Select one or more variables you wish to repeat and choose **Repeat variables...** from the **Manip** menu. Specify how you want the new variable to be constructed in the dialog as shown in Figure 12.5.

You can replicate each case, the entire variable, or both any number of times. The selected variable may hold numeric values, text values, or both. For example, if the original variable sequence is Hi, Lo replicating each number in a sequence twice and replicating the entire sequence three times generates a variable holding Hi Hi Lo Lo Hi Hi Lo Lo Hi Hi Lo Lo. Data Desk names the generated variable with the same name as the original variable, places it in a relation that has the appropriate number of cases. If Data Desk cannot find a relation with the same number of cases, it creates a new one and names it *Data*. Repeated variables are often useful along with variables constructed by appending different variables together, as discussed in the following section.

12.4 *Appending and Splitting Variables*

Data in which each case belongs to one of several groups can be represented in two different ways. Data for the groups can reside in separate variables in separate relations. Alternatively, all of the data can reside in a single variable, with a second variable supplying a category name for each case.

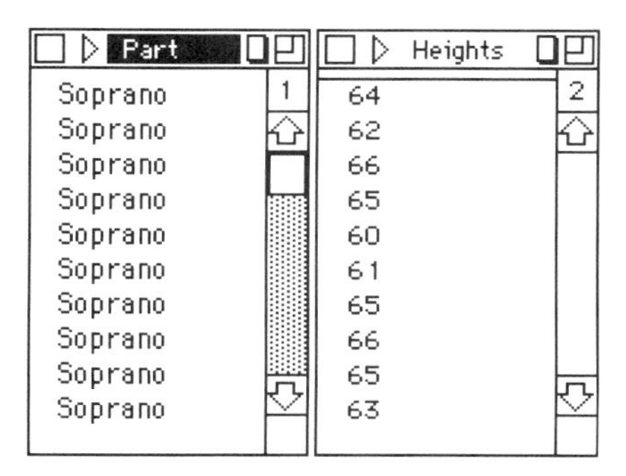

Figure 12-6. *Part and Heights of singers as two variables.*

The Singers dataset has data arranged each way. There are four relations; sopranos, altos, tenors, and basses. The variables in these relations are appropriate for producing histograms to compare the distributions of heights according to vocal part. The dataset also has a variable called *Heights,* that holds the heights for *all* of the singers. These are the same heights as were found in the individual variables, but stacked one after the other. Another variable, named *Part,* holds the vocal part associated with each case in the *Heights* variable. Thus the first height is recorded for a soprano, and the group variable *Part* says "Soprano" for the first case.

(In the datafile, the variable icons have been removed from their relations and placed in a common folder. Even though they appear together in this folder, they still belong to separate relations.)

The **Manip** menu provides two commands for moving between these two forms. The **Append & Make Group Variable** command works when two or more variables are selected. It appends the case values of the second selected variable to the end of the first. It then appends the third variable's cases to that, and so on. The result is a single variable containing all of the data values of the selected variables named *Data*. In addition, the command creates a group variable containing, for each case, the name of the variable from which it originally came, and names the variable *Group*. It then creates a new relation to hold these variables and opens it to show them.

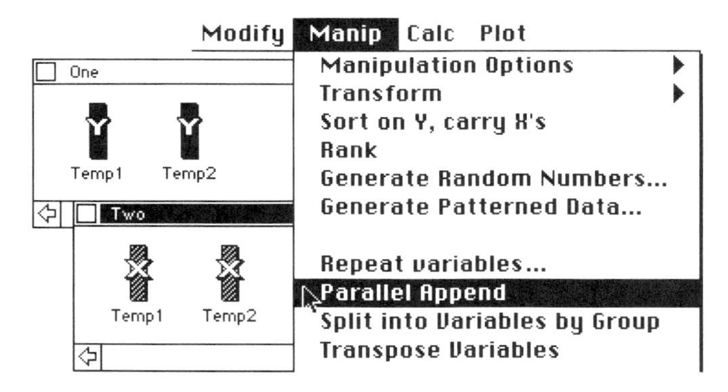

Figure 12-7. *When you select an equal number of x and y variables the Append & Make Group command changes to Parallel Append.*

If you select several y-variables and an *equal* number of *x*-variables, the command changes to **Parallel Append** as in Figure 12-7. Parallel Append appends the first *x*-variable selected to the end of the first *y*-variable selected, the second *x*-variable to the end of the second *y*-variable, and so on. It creates a Group variable with the group names *y* and *x*.

Parallel Append is an ideal tool for appending new data to an existing set of variables. You may have a dataset that has been split into two files or that has come to you as separate files. If the files have different cases but report the same variables, then parallel append

easy to append the cases from one file onto the end of the cases in the other file.

One file may already be a Data Desk datafile. If both files are text files or in some other form (a spreadsheet or database, for example), Import or Paste the data from one file into Data Desk. (Chapter 6 discusses importing data.) Now Import or Paste the data from the second file into the first. Data Desk will place the new data in its own relation.

Select as y each of the icons in the relation's window from left to right. (Option-⌘-A selects all icons in the frontmost window as y. Alternatively, if the first and second relations have the same variables in the same order, just select the first relation's icon as y.) Select as x all of the icons in that relation's window in the same order as their corresponding variables in the first relation. (Shift-⌘-A selects all icons in the frontmost window as x. Alternatively, if the first and second relations have the same variables in the same order, just select the second relation's icon as x.) Now, if equal number of icons have been selected as y and as x, the **Append & Make Group Variable** command changes to **Parallel Append.**

Parallel Append creates a new relation named *Parallel Append* and places in it variables named to correspond with the y-selected variables. Each holds the cases from the first relation followed by the cases from the second relation. Each variable in the second relation is appended to the end of the corresponding variable in the first relation. Parallel Append also creates a Groups variable as did Append & Make Group Variable command.

12.5 *Transpose*

The **Transpose** command exchanges rows and columns for all selected variables. The cases that were in each selected variable become a row (and thus a case) in these newly created variables. There is one new variable for each case in the selected variables. The effect is the same as transposing a matrix.

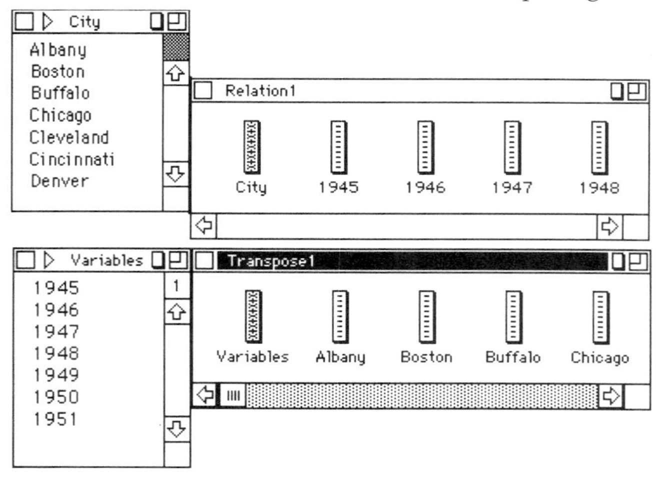

Figure 12-8. *Transposing variables creates a new relation holding variables one for each case in the variable City and each holding population values for 30 years in order.*

To transpose, select one or more variables as x-variables (hold the Shift key while selecting them) and choose **Transpose** from the **Manip** menu. Data Desk creates new variables, one for each case in the selected variables, named *case1, case2,* and so on. Data Desk also creates a variable named *Variables* that holds the names of the original variables corresponding to each row of the new transposed variables.

If you have a variable that names cases, you can select that identifying variable as the y-variable and the other variables as x-variables. **Transpose** will now name the newly created variables with the case identifiers found in the y-variable.

Thus, for example, data reporting the population of each of 40 cities for each of 30 years might be recorded in 30 variables (one for each year), each holding 40 cases (one for each city), along with a variable that named the cities. Selecting the city name variable as y, the other variables as x, and choosing **Transpose** will create 40 variables, each named with a city name and each holding 30 population values in order. An additional variable will hold the names of the original variables, in this case year.

Transpose may stretch the memory constraints of your Macintosh. Before transposing variables with over 100 cases consider allocating more memory to Data Desk. Chapter 4 discusses how to allocate extra memory to Data Desk.

12.6 Selectors

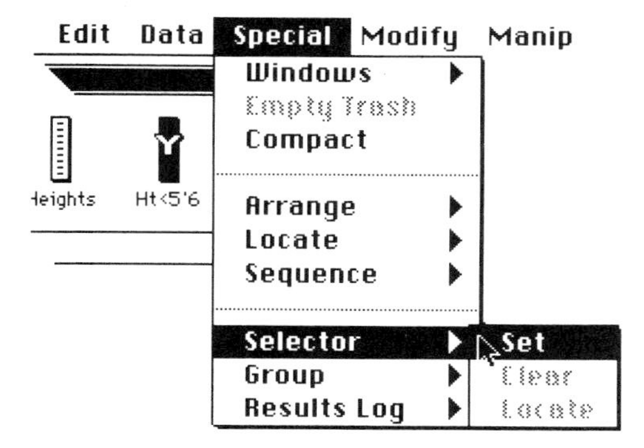

Figure 12-9. Assigning a selection criterion to the Selector Button.

It is often useful to select a subset of cases for analysis without having to construct new variables and copy over the cases. Data Desk provides a powerful tool for case selection.

Perhaps the best way to define a selection criterion is with the logical operators in a derived variable. For example, you could select all the men in a sample with the selector expression *TextOf(gender) = "male"*, or the richest members of the sample with the expression *'income' > 100000*.

Once you have defined the selection criterion in a derived variable, select that variable's icon and choose {Special ▸ Selector} **Set**. The selection criterion is assigned to the *Selector button*, which appears in the lower left of the Data Desk Desktop just below the identifying label. Initially it is turned on (highlighted).

Clicking the Selector button toggles it off and on. When the Selector button is highlighted, Data Desk commands operate *only on those cases for which the value of the selector expression is TRUE or nonzero*. After a command has used selection, the Selector button turns off. It must be clicked again to invoke selection for another command.

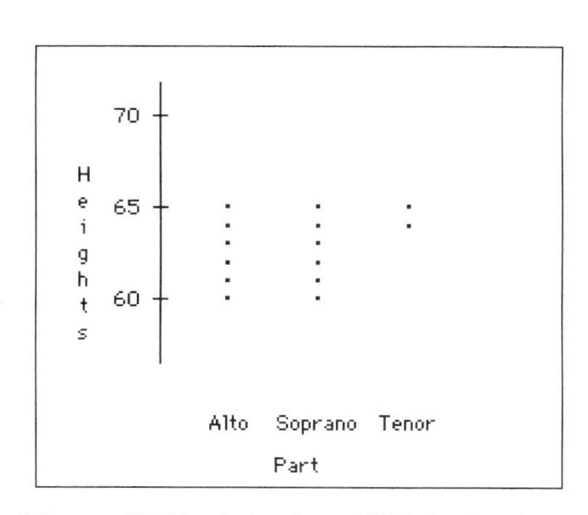

Figure 12-10. A dotplot of Heights by Part with cases selected for Height<66. (No Bass singers are represented because all of them are at least 66 inches tall.)

For example, the derived variable named *Ht < 5'6"* is defined by the expression *'Heights' < 66*. Figure 12-10 shows the Dotplot of Heights by Part, with *Ht < 5'6"* set as the selector. The effect of the selection is clear in the plot.

You can assign an ordinary variable or a derived variable that evaluates to numeric values to be a selector. In this case, the selection is done as if the selection expression were *variable* $\neq 0$. That is, all nonzero cases are selected and all zero-valued cases are not.

If the Selector button is on when you make a plot or do an analysis, the resulting window remembers its dependency on the selector and is thus a user of the selector variable. If you change the selector variable, the dependent plot or analysis offers to update to reflect the change. This happens even if that variable is no longer set as the selector in the Selector button.

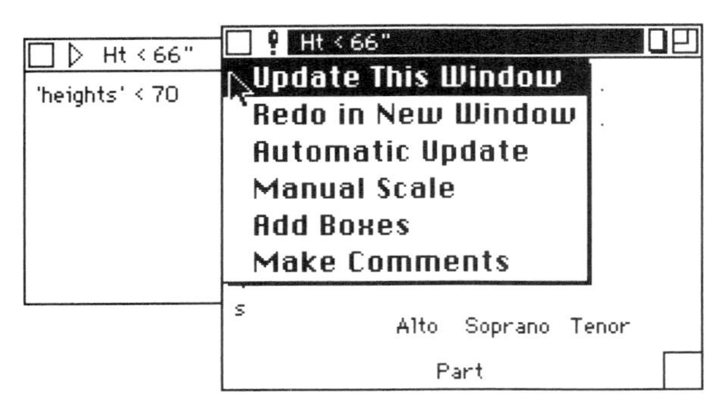

Figure 12-11. *Updating the dotplot after changing the expression in the selector variable.*

For example, now that it has been a selector variable, the derived variable *Ht < 5'6"* is permanently identified as the selector associated with the dotplot created in Figure 12-10. We can edit it without making it the selector variable, for example, by opening the derived variable and changing the 66 to a 70. The dotplot immediately offers to update. Choosing the Hyper-View menu under the ! and updating the window as shown in Figure 12-11, generates the new plot shown in Figure 12-12. Notice that three basses are now represented.

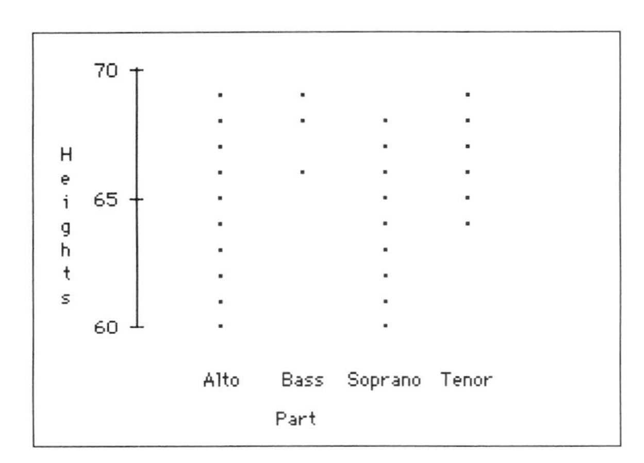

Figure 12-12. *The dotplot of Figure 12-10 updated.*

The selector variable must be in the same relation as the variables for the command. It would make no sense to select cases in one relation but apply the selection to another. Data Desk alerts you of mismatches between selectors and variables.

Selection is a fundamental property of all plots and results windows. HyperViews pass along the selector. Thus, for example, if you make a scatterplot of a subset of cases selected with the Selector button and then use the scatterplot's HyperView to request a histogram of one of the variables, the histogram will display only the cases shown in the scatterplot. This guarantees that HyperView analyses properly describe the same data as their parent windows. Chapter 13 discusses HyperViews in greater depth.

You can replace the selector variable by selecting another variable and choosing {Special ▶ Selector} **Set**. To clear the selector and remove the selector button from the desktop choose {Special ▶ Selector} **Clear**. Discarding the selector variable also clears the selector. The {Special ▶ Selector} **Locate** command locates the selector variable so you can examine or edit it.

12.7 *Performing Analyses Group by Group*

As Section 12.4 noted, data in which each case belongs to one of several groups can be represented in two different ways. Data for each group can reside in separate relations in separate variables, or all of the data can reside in a single variable, with a second variable supplying a category name for each case. The latter arrangement is almost always more

each of the groups. If the groups are in separate relations, you must select the appropriate variables and perform the analysis in each relation. If the groups are specified with a single grouping variable, Data Desk offers a more convenient alternative.

Group Variable : Cylinders

Select the variable that holds the group names and choose {Special ▶ Group} **Set**. A *Group* button similar to the selector button is placed on the lower left of the desktop and highlighted. Whenever the Group button is highlighted, any command in the **Calc** or **Plot** menus generates a folder of results, repeating its analysis or plot for each group specified by the grouping variable.

Let's use the Cars dataset as an example. We can examine the relationship between Drive Ratio and Displacement for cars with different numbers of cylinders. The Cylinders variable names four different categories: 4, 5, 6, and 8 cylinder cars, so setting it as the Group button makes four different plots or tables for any display or analysis.

Select Cylinders and choose {Special ▶ Group} **Set**. A button labeled *Group Variable: Cylinders* appears on the bottom of the Data Desk desktop and is highlighted. Select Drive Ratio as y and shift click to select Displacement as x. Choose **Scatterplots** from the **Plot** menu. Data Desk opens four different scatterplots, one for 4-cylinder cars, one for 6-cylinder cars, and so on. Figure 12-13 shows three of them (excluding the plot for the single 5-cylinder car.)

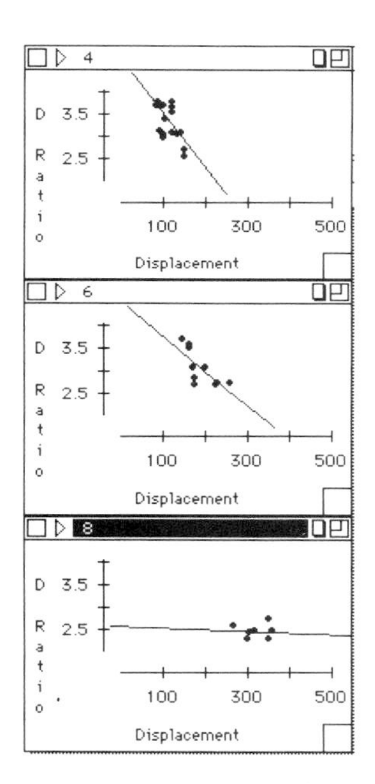

Figure 12-13. *Scatterplots of Drive Ratio vs Displacement. Plot scales have been matched and regression lines added to show differencesin the slope by group..*

To compare these plots it helps to set them to the same scale. Choose **Set Plot Scale...** from the HyperView of any of the plots, and set the scale as shown in Figure 12-14. Then drag the icon of the re-scaled plot into the other plots to have them match scales. We can now see that the slope of the relationship between these variables does change according to number of cylinders.

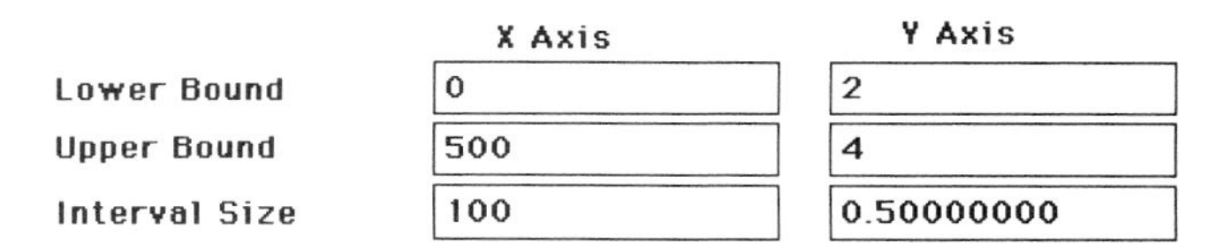

	X Axis	Y Axis
Lower Bound	0	2
Upper Bound	500	4
Interval Size	100	0.50000000

Figure 12-14. *Set the plot scale as shown here.*

Commands chosen from HyperView menus in any of these plots operate only on the group shown in the plot, so you can pursue a pattern without having to re-specify the group. For example, the lines added to the plots in Figure 12-13 are computed only for the points in each plot. You can even drag the icon of another variable onto an axis of one of these plots; the plot continues to focus only on the points in its group.

Although the windows holding the results for each group are in the same relation, you cannot brush or slice to relate windows because any one case can be in only one group and thus can appear in only one window. Data Desk places the icons of the plots for each group in the same folder so you can close all of them by choosing {Data ▶ Close All} **Siblings** when any one of the plots is the frontmost window.

If the Group and Selector buttons are active simultaneously Data Desk still produces a window for each group, but only those cases that are included *both* by the Selector and by the Group are used for each plot or table.

12.8 Samples

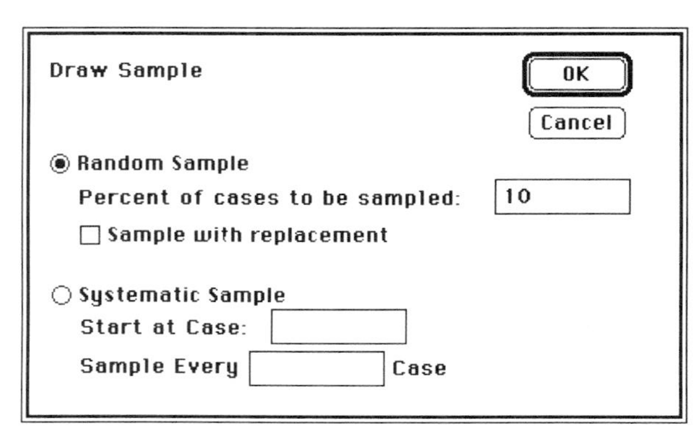

Figure 12-15. *The Sampling dialog offers several ways to extract cases from the selected variables.*

The {Manip} **Sample...** command extracts subsets of cases from the selected variables and places them in a new relation. Random samples are representative subsets of the full dataset.

Select the variables from which to sample and choose **Sample...** from the Manip menu. The Sample... dialog (Figure 12-15) offers several ways to draw samples. Random samples give each case an equal and independent chance of selection. They thus draw what is formally called a *Simple Random Sample.*

Simple Random Sample

Data Desk draws a sample of cases from the selected variables so that the specified percentage of cases is sampled. The sampled cases are placed in new variables (but with the same variable names as the original variables) in a new relation. In addition, Data Desk creates a variable named *Sample Indices,* which holds the case numbers of the sampled cases. The sampled cases remain in the same order as they were in the original variables.

Ordinarily each case can appear only once in a sample. However, clicking the *Sample with replacement* option, gives each case the chance to be selected more than once. When sampling with replacement, each selection is made from the entire set of possible cases (with the specified chance of being selected applied to all cases). Since each draw samples from the full population of cases, any particular case may be selected more than once.

Systematic Sample

An alternative way to select a subset of cases is with a *Systematic Sample.* Systematic samples select cases in a regular pattern. You must specify which case to select first, and how many cases to skip between selected cases. Systematic samples have the advantage that they select cases equally from each subrange of cases. If cases are in an important order (for example, all the Freshmen first, followed by Sophomores, Juniors, and Seniors in that order), then Systematic sampling *stratifies* the sample so that an equal fraction of cases is selected from each subrange.

However, a systematic sample may impose a spurious pattern or hide a real one if the sampling interval interacts with a pattern in the data. Thus if variables hold data for husbands, each followed immediately by data for wives, a systematic sample of every other case will include only men or only women.

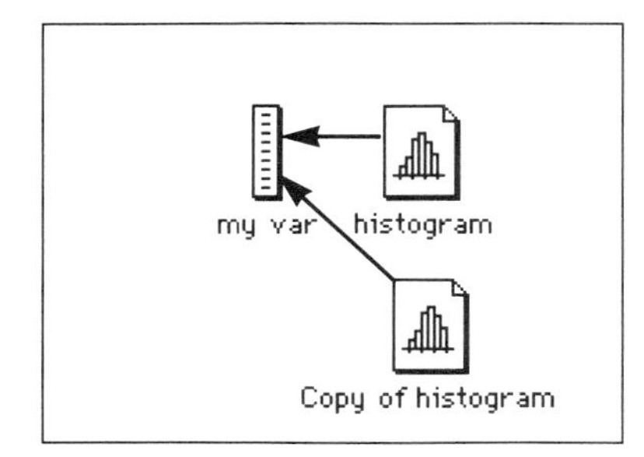

Figure 12-16. *Duplicating an icon by itself makes a copy that uses the same variables as the original.*

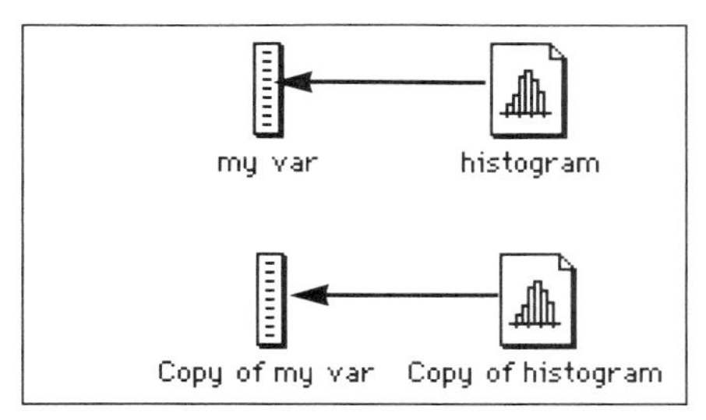

Figure 12-17*. Duplicating both the "user" and "used" icons simultaneously copies the entire structure. Here the duplicate histogram uses the duplicate variable.*

12.9 *Duplicating Icons*

The **Duplicate** command in the **Data** menu duplicates each selected icon. The copies have the same name as the original icons.

Duplicate is an extraordinary powerful command because it performs a *smart duplication*. Any result icon, such as a derived variable, plot, or table, may use other icons. When an icon that uses others is duplicated, its duplicate uses the same icons as the original did. However, if an icon and the icons it uses are duplicated together, the entire structure is duplicated so that the duplicated result icon uses the corresponding duplicated variable icons.

For example, the histogram in Figure 12-16 uses the variable *my var*. Duplicating *histogram* produces a copy also named *histogram* that uses the original variable *my var*. Duplicating *both my var* and *histogram* in Figure 12-17 produces a copy of *histogram* that uses the *copy* of *my var*.

Smart duplication makes it possible to reproduce entire data structures. For example, reproducing a folder that contains both a relation and any combination of derived variables, plots, and results icons that use the variables in the relation, makes a full copy of both the icons and their interrelationships, preserving all relation structure and the dependencies of icons upon each other. You can then experiment with the copy. Changes you make to the copy do not affect the original plots and analyses.

12.10 *Data Tables*

Data Tables display the contents of variables side-by-side in a spreadsheet-like table. They are a convenient way to view the contents of variables, especially when you want to read across several columns. The Selector and Group buttons restrict a Data Table to a subset of cases. (See Section 12.6 and 12.7 for more information on Selector and Group buttons.)

To make a data table, select the icons of the variables you want to put in the table and choose {Manip} **Make Data Table**. The resulting table shows the variables in left-to-right order in selection order. Drag the icons of additional variables into a Data Table to add their contents to the table.

Table1

Car	MPG	Weight	Drive Ratio
Buick Estate Wagon	16.9	4.360	2.73
Ford Country Squire Wagon	15.5	4.054	2.26
Chevy Malibu Wagon	19.2	3.605	2.56
Chrysler LeBaron Wagon	18.5	3.940	2.45
Chevette	30.0	2.155	3.70
Toyota Corona	27.5	2.560	3.05
Datsun 510	27.2	2.300	3.54
Dodge Omni	30.9	2.230	3.37
Audi 5000	20.3	2.830	3.90
Volvo 240 GL	17.0	3.140	3.50
Saab 99 GLE	21.6	2.795	3.77
Peugeot 694 SL	16.2	3.410	3.58

Figure 12-18. *A Data Table displays the contents of the selected variables.*

The **Use Colors** command in the Data Table's global HyperView displays each case in the Data Table in the color with which it is plotted. While you can view the colors assigned to each case, you cannot add or change the color of the cases from a Data Table.

The variable names at the top of the table do not scroll with the table, so they are always available to label the data. HyperViews attached to the variable names offer to select the named variable. As with other selection commands in HyperViews, you can hold the Shift or Option keys to select as x or y, respectively. If you wish, you can use this feature to establish an interface similar to spreadsheet-oriented programs, displaying your data in a Data Table, making selections by clicking on the column names, and choosing commands from the menus.

12.11 Copying and Printing Results

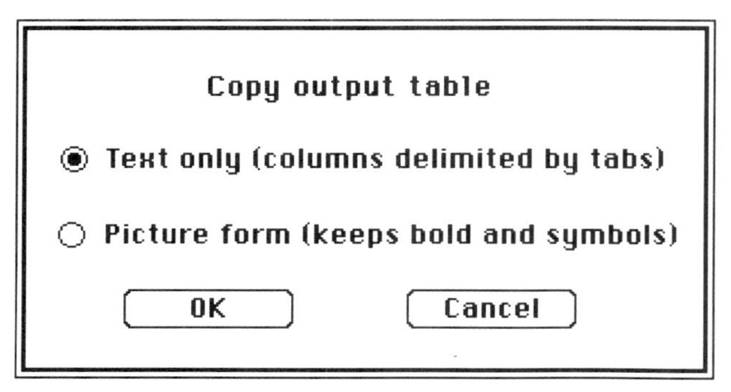

Figure 12-19. *The* ***Copy Window*** *command offers the choice of tab-delimited text or bit-mapped picture form.*

The **Copy Window** command in the **Edit** menu places a copy of the contents of the frontmost window on the Clipboard. For output tables (regression summary tables, correlation tables, ANOVA tables, frequency count tables, etc.) a dialog (See Figure 12-19) offers the choice of saving a picture version of the table or saving a tab-delimited text version of the table which can be edited and reformatted in a word processor or page composition program.

While tab-delimited text form is usually the most useful, Data Desk uses some special characters (typically Greek letters, superscripts, and subscripts) that are not found in some fonts. When you paste the text into another program you may find that a few symbols are missing or display strangely and need to be replaced with substitutes appropriate to the fonts you are using.

The **Print Front Window** command in the **File** menu prints the contents of the frontmost window on the default printer. As with any Macintosh program, select the printer with the Chooser, usually found in the Apple menu.

Both Copy Window and Print Front Window generate a PIC2 form of plots that takes full advantage of the higher resolution of a LaserWriter. Copying a plot in PIC2 form allows you to edit each part of the plot as a graphic object in graphics editors such as MacDraw™ and SuperPaint™.

Data Desk also offers a way to lay out several plots and tables on the same page. The {Data ▶ New} **Layout** command opens a window in which you can position plots and tables for printing. To place a plot or table in a layout window, bring it to the front and **Copy** it. For tables, choose Picture form. Then click on the layout window and **Paste.** Data Desk places a copy of the window in the layout window. You can then drag it with the mouse to position it as you please. When you print the layout window all of the pictures placed in it print as shown. (See Chapter 6 for a description of Layout windows.)

Alternatively, you can drag the icon (or icon alias) of any open window into a Layout window. This is particularly useful for placing the contents of variables and scratchpads. (Copying them would have copied their text rather than a picture of the window, and would thus have been inappropriate to paste into a Layout.)

The pictures in a layout window are just that; pictures of plots and tables at a fixed time during the analysis. Subsequent changes to plots or tables are not reflected in the layout window (although you can always copy them and paste them into the layout window).

To delete a picture from a Layout window, hold the key labeled either *Backspace* or *delete* on your keyboard, and click on the picture.

APPENDIX 12A
Missing Values

Data Desk treats nonnumeric values as missing in numeric computations. The treatment of missing values is a subtle matter. This Appendix documents Data Desk's treatment of missing values. Missing values are ordinarily excluded from calculations. In most cases infinities are treated as missing and also excluded.

Missing values propagate correctly through expressions and calculations. The result of any calculation involving a missing value is missing. Thus, for example, 1 + missing = missing, log(missing) = missing, and so on.

The result of any logical comparison with a missing value is FALSE. Thus, for example, the expressions $x < 0$, $x = 0$, and $x > 0$ are all false for cases in which x is missing. The rule that logical comparisons with missing values are false extends even to expressions that might appear to be true. Thus, for example, *log(-1) = log(-1)* evaluates FALSE. The logarithm of -1 does not exist, so *log(-1)* is missing, and comparing it with anything, even itself, yields FALSE.

To construct a variable that identifies all the non-missing cases in the variable x, use a logical tautology such as $x = x$. This expression is false only for the missing cases, so it can be used in a selector to select non-missing cases.

You can identify all the missing values in a variable with the {Edit} **Find** commands, which offer the option of searching for nonnumeric cases. To recode the missing values, locate them with {Edit ▶ Find} **Find** and then alter them with the {Edit} **Replace** command.

Data Desk commands that operate on categories rather than numbers do not treat ordinary text as missing values. The only cases considered missing in categorical variables are those containing the missing value symbol, •, typed as Option-8, or those that are completely empty.

While it is possible to remove missing cases from a variable, this is usually unwise. Missing values are place-holders that preserve the overall case identity in variables. Values missing in one variable may be present in others. When you remove a missing value, you remove that case from the entire relation, which may not be desirable.

MISSING VALUES AND SELECTORS

You can select a subset of cases by assigning a Selector variable with the {Special ▶ Selector} **Set** command and pressing the Selector button that is placed on the Desktop. Cases omitted with a Selector are different from cases omitted because they hold missing values. The key operational difference is that Selector variables are arguments of the calculations in which they are used. Changes to the Selector are treated as changes to one of the underlying variables of the plot or analysis.

Most Data Desk output tables report when any cases have been omitted from the analysis. This report includes the omission of cases with a Selector variable.

Group variables work by creating temporary Selector variables. These variables are not ordinarily visible, but they are available to Data Desk for internal calculations.

Pairwise and Listwise Deletion

The question of how to compute meaningful statistics in the presence of missing values has generated many suggestions, much heated debate, and some light. Because Data Desk is so highly integrated, its missing value decisions have been chosen with consideration of consequences that may extend beyond any one particular analysis.

Many advanced statistics calculations (for example, those discussed in Chapters 20-23) begin with the correlations, covariances, or simple cross-products among all pairs of variables. Each of these calculations contains a *cross-product* calculation of the general form $\Sigma_i x_i y_i$. Missing values are ordinarily dealt with when these calculations are performed. The two most common ways of dealing with missing values are commonly known as pairwise deletion and listwise deletion.

In *pairwise deletion*, each cross-product includes all cases present in the two variables involved in *each* sum. In *listwise deletion*, any case with a missing value in *any* of the variables in the analysis is omitted from *every* cross-product calculation.

Pairwise deletion poses a problem for statistics involving several variables. Specifically, different cases can participate in each cross-product calculation, so the matrix of values is not really a proper cross-product matrix. (In particular, it may have negative eigenvalues. Since it is theoretically impossible for a true correlation, covariance, or cross-product matrix to have a negative eigenvalue, such an occurrence can invalidate computations based on those values.) Some research indicates that even when the matrix is not obviously illegal, the structure of the matrix can be profoundly perturbed by pairwise deletion calculation.

In listwise deletion, cases that have a missing value on *any* of the variables in the analysis are excluded. Listwise deletion guarantees that the resulting matrix is mathematically proper, but it can result in many cases being deleted. Data Desk uses listwise deletion in all multivariate calculations.

Data Desk lets you request multiple calculations with a single command, but this does not generally make the resulting calculations multivariate. For example, Data Desk omits cases on an individual or *pairwise* basis for:

- Summary Reports computed for several variables.
- Several scatterplots computed with a single **Scatterplot** command.
- The individual correlation coefficients in a table of correlation coefficients.

Missing Values and Multivariate Calculations

Data Desk insures that HyperView plots and analyses are based upon the same cases as the analysis that hosted the HyperView menu whenever the plot or secondary analysis generated by the HyperView directly extends the analysis. When you perform a multivariate analysis on data with missing values, Data Desk takes special care to treat missing cases appropriately.

Specifically, whenever a Selector or Group variable is specified for a plot or calculation, it operates in all windows created by HyperViews from that plot or calculation *even if the Selector or Group button is no longer active.* Data Desk regards the Selector or Group variable as one of the variables used by the plot or analysis. Changes made to the Selector or Group variable cause Data Desk to notify you of the change by posting a ! in the window of the plot or analysis, offering to update or redo the analysis.

Apparent Ambiguities

Because Data Desk selects cases consistently in all windows, some analyses and plots that appear at first to be equivalent may, in fact, differ in subtle ways. For example, Data Desk computes tables of correlation coefficients generated by the **Correlations** submenu commands (Chapter 21) with *pairwise* deletion of cases. Thus a scatterplot generated by HyperViews to depict the relationship between any pair of variables is the same as a scatterplot of those two variables made independently. However, the correlations (or covariances) may not be the same as those computed with listwise deletion as part of a multiple regression analysis (Chapter 22).

Similarly, if you perform a multiple regression with the Selector button on and compute residuals and predicted values, the generated residuals and predicted values will include values for cases that were omitted from the regression computation by the Selector variable. The regression equation is found using only the selected cases, but predictions and residuals are computed from that equation for all available cases. A scatterplot of the generated residuals *vs* predicted values will show points for all cases in the data, whether they were selected for the regression or not, unless you press the Selector button again before making the scatterplot.

However, if you request a scatterplot of residuals *vs* predicted values from the global HyperView of the regression table, the scatterplot will automatically be restricted to the selected points. This insures that the HyperView analysis corresponds to the regression it is intended to describe. The resulting scatterplot is the same as the one you could make directly from the saved residuals and predicted values *with the Selector button on*, but differs from the plot you would see if you simply plot residuals *vs* predicted values without the selector.

CHAPTER 13

Integrated Analyses

13.1	Context-based Expertise	*197*
13.2	Data Analysis Expertise	*197*
13.3	Global and Context-sensitive HyperViews	*198*
13.4	Updating Window	*199*
13.5	Following the Links	*199*
13.6	Case Identities	*200*
13.7	Consistency	*201*
13.8	Recording the Analysis History	*201*
13.9	Avoiding Window Overload	*201*

DATA ANALYSIS IS A JOURNEY of successive discoveries. Each step reveals a new aspect of the data and suggests new questions or explanations. The path of the analysis is guided both by knowledge about the context of the data ("The South seems different from the North. I wonder if temperature is important here.") and knowledge about the statistics ("The correlation is lower than I expected. I wonder if there is an extreme datapoint.")

13.1 *Context-based Expertise*

data context

The *data context* of your analysis is informed by your knowledge of the world and of your data. If you understand the background of your data so that you can recognize a meaningful pattern, then you are likely to learn a great deal more from the *process* of analyzing the data than from reading a table of statistics or just looking at a plot. Data Desk is designed to help you bring your real-world knowledge to bear on your data analyses, even if you are not expert at advanced statistics.

statistics context

In contrast to the data context, the *statistics context* of a data analysis consists of the properties and relationships among the statistics and graphics methods. Data Desk incorporates this context in HyperView pop-up menus. Most Data Desk results windows suggest additional or alternative analyses or plots. These might be checks on the underlying assumptions of a procedure (for example, a histogram to check how a variable is distributed) or they might be naturally related analyses (for example a frequency breakdown to provide the counts and percentages graphed in a pie chart).

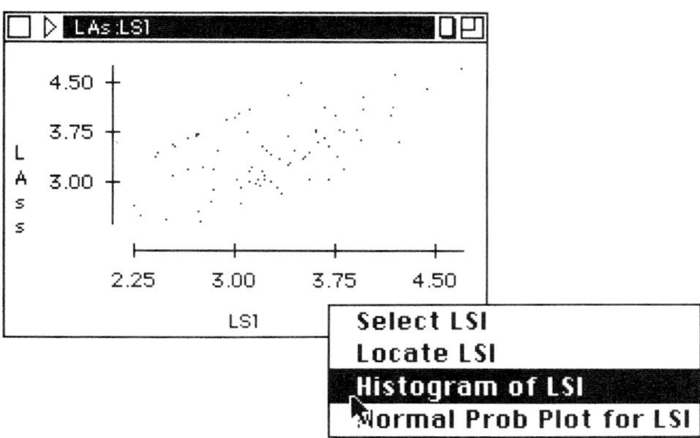

Figure 13-1. *One HyperView in a scatterplot suggests ways to look at the x-axis variable.*

HyperViews neither restrict your choices nor require any action. Most HyperView commands could be given from the desktop in the ordinary way. HyperViews are simply suggested steps placed at your fingertips and set apart from the full array of Data Desk capabilities so you can select among them easily. HyperViews offer expert guidance, but they are designed to assist you rather than to take control of the analysis.

13.2 *Data Analysis Expertise*

It is said that no two data analyses are exactly alike. Each analyst has favorite tools and methods and favorite ways to combine them. The HyperViews in Data Desk are only a selection from all the alternative paths that are possible. This selection embodies the program designer's many years of data analysis and consulting experience and reflects a philosophy of data analysis founded in the tradition of exploratory data analysis.

13.3 Global and Context-sensitive HyperViews

Global HyperViews are attached to the window as a whole. Most results windows have a submenu arrow ▷ located along the bottom or right of the window. The HyperView attached to that arrow suggests general actions related to the analysis or display in the window. Figure 13-2 shows an example.

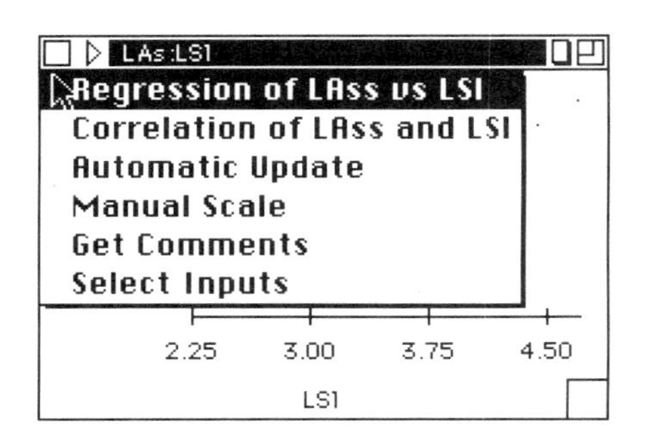

Figure 13-2. Global HyperViews pop-up from the ▷ symbol at the top of a window.

Context-sensitive HyperViews are attached to parts of a plot or table and suggest analyses or plots specific to those parts. For example, the axis labels of a plot usually offer to locate the icon of the displayed variable or to show it in a simple one-variable display, such as a histogram.

Context-sensitive HyperViews attached to parts of a display or table offer particularly effective expert suggestions. They often can be more context-specific than the global HyperViews. For example, a HyperView attached to a test statistic may offer checks of some of the assumptions required for that test to be valid, rather than just another view of the data. While not all HyperView paths are documented in this book, context-sensitive HyperViews that embody statistical expertise are discussed in chapters that discuss the specific analyses.

When a part of a plot or table has a context-sensitive HyperView underneath the mouse cursor changes to ☝ when it passes over that part. Whenever the cursor looks like the "button presser", you can press the mouse button to see a Hyperview menu (see Figure 13-3).

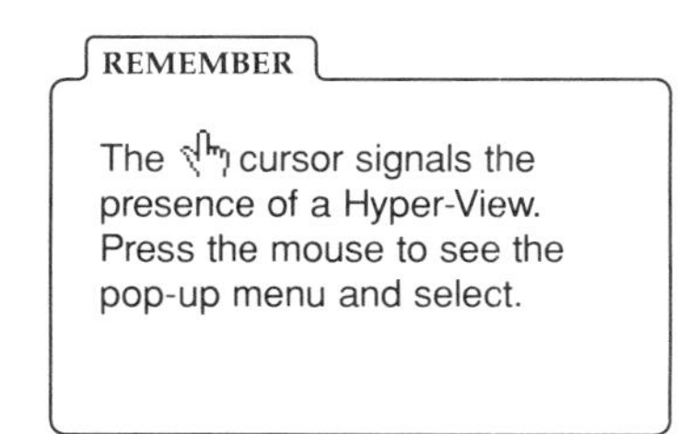

REMEMBER
The cursor signals the presence of a Hyper-View. Press the mouse to see the pop-up menu and select.

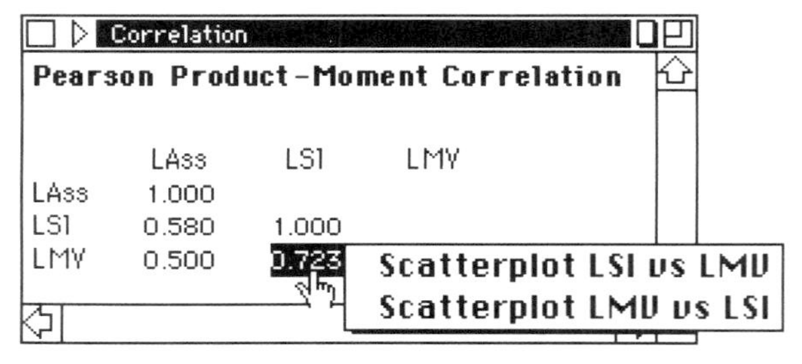

Figure 13-3. Context-sensitive HyperViews are attached to parts of an analysis or plot. In Data Desk , for example, an individual correlation coefficient suggests scatterplots to depict the association measured by the coefficient.

HyperViews maintain consistency by working with any selector that was active when the window contents were first computed. Thus, for example, if you compute an analysis using the Selector button to select a subset of cases, any plot generated by a HyperView from that analysis uses the same selector variable, so it displays the cases that contributed to that analysis and no others.

If a selector was active when the original analysis was performed, the HyperView commands are *not* identical to selecting the named variables and choosing the specified command from the menu, but rather, are identical to setting the Selector button as it was when the original analysis was performed and then selecting variables and choosing a command.

13.4 *Updating Windows*

Data change during an analysis. You might transform the data, correct a value, or temporarily eliminate an extraordinary case. During an analysis you might try out alternatives: "What would the prediction be if we lowered the price?", "What if we isolate the Southwest region and repeat the analysis for those cases?" Many analyses turn up errors in the data that can be corrected on the spot.

For these reasons and others, you may change your data, either permanently (for example, to correct an error), or temporarily. Each change may require that you repeat parts of your analysis to see how they adjust. Data Desk's results windows indicate immediately when the variables they use have been changed. The submenu arrow, ▷, that marked the general HyperView menu changes immediately to an exclamation mark ! to alert you to the change, and a new HyperView menu offers to Redo the analysis in a new window using the updated variables, or to update the window in place.

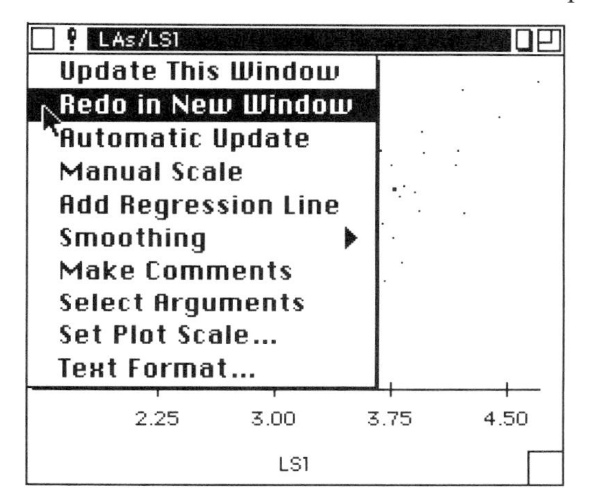

Figure 13-4. *When a variable is changed, plots and analyses that use that variable offer to update or recompute into a new window.*

REMEMBER

A ! indicates that a window is out-of-date. The HyperView menu offers to update.

Window updates guarantee that HyperViews reflect the state of the data shown in the plot or analysis. That is why all other HyperViews are turned off when a window needs updating, and all that is offered is the chance to update the window. It is often better to Redo the plot or computation to a new window than to update in place because it lets you compare "before" and "after" views of the data and keep a history of the analysis.

Many windows also offer the option of automatic updating. Select **Automatic Update** from the window's HyperView menu. The window's HyperView button (▷) will turn gray to indicate the change in status. Windows set to automatically update do just that. Whenever they are notified of a change in an underlying variable they immediately recompute. The most common use for automatic updating is to create custom dynamic plots using sliders. An automatically updating plot of a derived variable that uses a slider changes dynamically as you move the slider. Section 11.13 discusses sliders and gives an example.

It is generally not a good idea to use automatic updating except in special situations. Windows that change when you are not watching them are confusing and may cause you to miss important aspects of your data. We advise you to set only selected windows to automatically update and then to watch them while they change.

13.5 *Following the Links*

Data Desk maintains and follows arbitrarily long links among variables, derived variables, and results windows. For example, changing a variable affects all derived variables that depend upon it, any plots that depend upon those variables, any tables that depend upon those variables, any secondary statistics that depend upon the tables, any plots or tables that depend upon the secondary statistics, and so on. All depen-

The icon "Assets" cannot be discarded because it is still being used. You must discard all icons that use Assets before discarding it.

Locate all icons that use Assets. (They will be selected so you can drag them to the Trash if you wish.)

Locate

Cancel

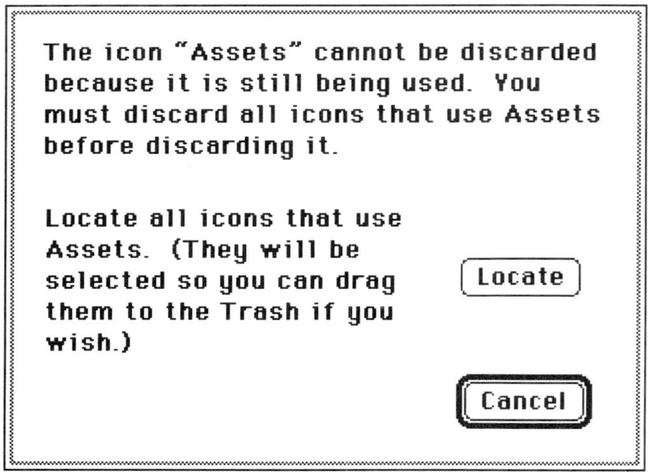

Figure 13-5. *You cannot discard an icon that is still in use. Data Desk offers to locate icons that use the one you wish to discard.*

dent plots and tables can be updated or recomputed after a change. For example, a scatterplot of the *logarithm* of Assets, defined with a derived variable will offer to update when data values in the *original* underlying variable are changed.

The update flag appears as soon as you make a change in a variable's editing window. You can open a variable, change values, and immediately update dependent windows without closing the variable's window. This makes it easy to try a number of "what if?" experiments by varying data values, selection criteria, or functions, and then updating results windows. If a dependent window is set to automatically update, press the Enter key to stop editing the derived variable expression and update the window.

You may not discard any variable or derived variable that is still used by a results window. Data Desk alerts you and offers to locate the icons that still use the one you wish to discard (Figure 13-5).

You may discard a variable along with all of its dependent windows. If Data Desk cannot empty all of the Trash, it displays an alert like Figure 13-5. You can locate any dependent icons by pressing the Locate button in the dialog. You can then pick up any selected icon and drag it into the Trash. Because all the dependent icons are selected by the Locate command, they will all be dragged into the Trash together.

13.6 *Case Identities*

Data Desk preserves case identities across windows. New plots and tables generated by HyperViews deal with the same cases as their parent windows do and are linked appropriately. You can always ask "who is this?" with the **?** tool or select a subset in one view and see it in other views.

Data analyses rarely proceed on a straight path. Rather, they spiral toward new understandings. Thus, you might perform an analysis, look at a related plot, detect an extraordinary case, select it in the plot, edit it to correct or omit it, and repeat the analysis, the plot, and even the editing (of yet another, more subtle anomaly, perhaps).

Case identities are at the heart of Data Desk's use of relation structure. To Data Desk, relations are distinguished primarily by the collection of cases they hold. Variables with measurements on the same individuals in the same order belong in the same relation. Variables that deal with different individuals, a subset of individuals, or the same individuals in a different order belong to a different relation.

Each relation maintains a list of case numbers; you can always find a case by its case number. The case number is the one individual identifying part of a case that is guaranteed to be unique. Other cases may have the same values on measured variables, but they will have different case identities.

13.7 *Consistency*

Data Desk plots and analyses always know who they are and where they came from. They know the identities of the variables that were used to make them and whether a selector or group button was active when they were made. They know if points were hidden. Whenever you derive a new analysis or plot from an existing one — whether by updating or recomputing the window or by following HyperView suggestions to create new results — Data Desk keeps the new analysis path consistent with the original window.

Good data analysis requires many views of the data. Views that are mutually inconsistent (especially if the inconsistency is subtle and hard to see at first glance) are especially dangerous for they can lead to false conclusions and obscure vital relationships.

13.8 *Recording the Analysis History*

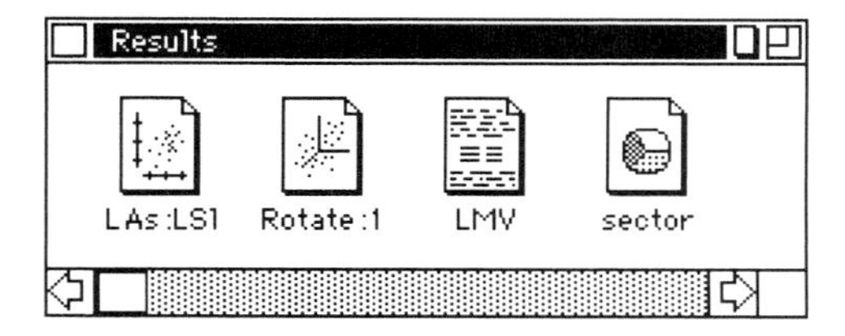

Figure 13-6. *The results folder holds icons of plots and analyses in the order in which you create them.*

Data Desk places the icons of the results of each command in the Results folder one after the other. They thus record the history of your analysis. You can review this history at any time by opening old icons. As you work you can discard plots and analyses that are clearly of no further interest, so they do not clutter the history. (Any easy way to discard an open window is to drag its icon alias into the Trash.)

Of course, you can also record your thoughts and actions in a ScratchPad, in the Comments window called up by the window's HyperView, or on Jot Notes attached to the windows. (They will become part of the history as well.)

While you can change Results folders during an analysis (with the {Special ▶ Results Log} **Set** command), we recommend that you do so only at major breaks in your analysis path to preserve the history of your analysis for future reference.

13.9 *Avoiding Window Overload*

Data Desk is a multi-window environment. The truth of this statement is never quite so clear as when you are following HyperViews or substituting variables and recomputing into new windows. It is easy to generate a deep stack of windows on your desktop, because each has lead naturally to the next with no requirement to close as you go. Unfortunately, windows fill up computer memory. You may find that Data Desk becomes sluggish or even that you eventually run out of space and get a warning message. Simply close the windows to free up the space.

CHAPTER 14

Tables

14.1 Frequency Counts and Percentages *205*
14.2 Two Factors *206*
14.3 Contingency Tables *207*
14.4 Table Contents *207*
14.5 Independence and Chi Square *210*
14.6 HyperViews in Tables *211*
14.7 Copying and Printing Tables *212*

APPENDIX
14A Equations *213*
14B Multi-Way Tables *214*

SOME DATA IDENTIFY GROUPS or categories rather than reporting values. It is not appropriate to perform arithmetic on category identifiers (even if they are numerals), so group data must be analyzed differently than numeric data. An average family income is a reasonable concept when you know incomes in dollars, but you can't average "lower middle class", "middle class", "upper middle class", and "rich".

14.1 Frequency Counts and Percentages

frequency table

The natural thing to do with category data is to count it. A *frequency table* reports how many cases fall into each category. To make a frequency table in Data Desk, select the category variable and choose **Frequency Breakdown** from the **Calc** menu.

For example, a frequency table of companies by market sector in the Companies data appears in Figure 14-1.

Frequency breakdown of **sector**

Group	**Count**	%
Finance	17	22
HiTech	12	16
Oil	3	3.9
Other	14	18
Retail	14	18
Utilities	14	18
Medical	3	3.9
Total	77	

Figure 14-1. *Frequency table by Sector for the Companies data. The default table displays counts and percentages.*

Frequency tables are a good way to see

- Patterns or trends across categories
- Individual categories that are extraordinarily large or small
- The relative allocation of cases to different categories.

Frequency tables can provide more information than counts and percents. The Frequency Options dialog offers a variety of alternatives. You can open this dialog from either the **Calculation Options** submenu or the Frequency Table's Global HyperView. Changes made to the dialog from the Calculation Options submenu affect the frontmost frequency table if it is active window, and any future frequency tables. If there are no frequency tables open, this dialog will only affect future tables. The dialog attached to a table's Global HyperView only affects that specific table.

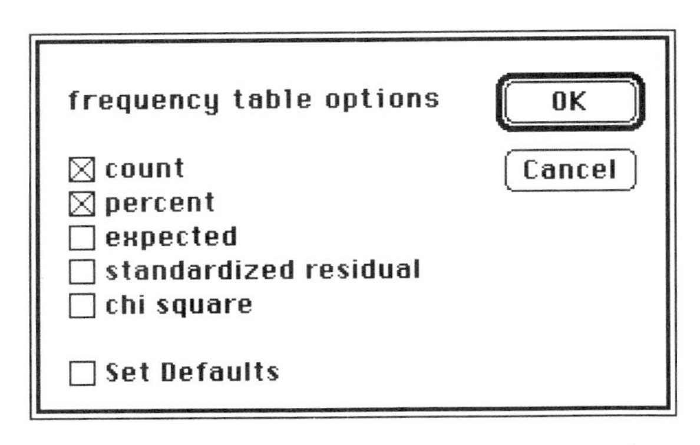

Figure 14-2. The Frequencies Options dialog offers alternative choices for frequency tables.

Percents report the percentage of the total sample falling in each category. The expected values and standardized residuals are components of the Chi-square (χ^2) statistic, which can be used to test whether the division of cases into categories is essentially equal. A large Chi-square value suggests that the allocation of cases into cells is not equal. Appendix 14A presents the mathematics underlying the Chi-square statistic.

Pie charts and bar charts show graphically much of the same information as frequency tables do. You can access these plots from the frequency table's HyperView. Frequency tables also link with other Data Desk windows. The HyperViews connected to the category labels of a row of a frequency table offer to select all of the cases falling into that row in all other open windows. You can extend your selection to more than one row of cases by holding down the shift key and choosing *Select* from the HyperView menu of the category you wish to add to the selection.

REMEMBER

Category variables treat values that are empty or contain the symbol • as missing.

Data Desk uses the text of variables to determine categories. Cases whose values are empty or contain the • symbol (typed as Option-8) are considered to be missing.

14.2 *Two Factors:*

factors

levels

cells

Data that can be categorized in two ways are often arrayed in a two-way table. A typical table has two *factors*, one with categories labeling the rows of the table and another with categories labeling the columns. A category within a factor is called a *level* of the factor. Every combination of a row level and a column level specifies a *cell* of the table.

Columns are levels of **Cracker**
Rows are levels of **Bloat**

	bran	**combo**	**control**	**gum**
high	0	2	0	5
low	4			2
med	1			3
none	7	2	6	2

table contents:
count

The factor levels are named at the top and side of the table.

The intersection of each row and column is a cell in the table.

Figure 14-3. *A two-way table shows the counts of cases falling in each combination of levels on each of two factors.*

Tables are particularly good at showing:

- Patterns or trends across rows (where the row level stays the same while the column level changes) and down columns (where the column level stays the same while the row level changes)

- Individual cells that are extraordinarily large or small
- Indications of whether the factors are statistically independent or whether they are related to each other.

14.3 *Contingency Tables*

Contingency tables are tables of counts used primarily to investigate the dependence (hence the term "contingency") of two categorical factors on each other. Each case in the data falls in one of the levels on each of the two factors. Each cell of the table represents a combination of a level on the row factor and a level on the column factor, so each case falls in one cell. Contingency tables count the number of cases falling in each cell and report the counts and related statistics.

As in frequency tables, the variables used to construct contingency tables are categorical. This means that rather than holding measurements or amounts, each case identifies a category. Categories may be numeric, but they are identified with text labels, so numeric labels are treated as category names made up of numerals. Thus a variable that names an individual's religion is categorical. A variable reporting the number of cylinders in a car engine might also be considered categorical because there are relatively few possible values.

To make a contingency table, select two categorical variables identifying the two factors, and choose **Contingency Tables** from the **Calc** menu. You can change the category variables in the table by dragging the new category variable into it. If the variable is selected as *y*, then the columns factor is replaced, if the variable is selected as *x*, the rows factor is replaced. To change both column and row factors, select *y* and *x* variables and drag them into the summary table.

14.4 *Table Contents*

Data Desk can place a variety of information in each cell of a contingency table. The Table Options dialog from the **Calculation Options** submenu, offers a choice of items.

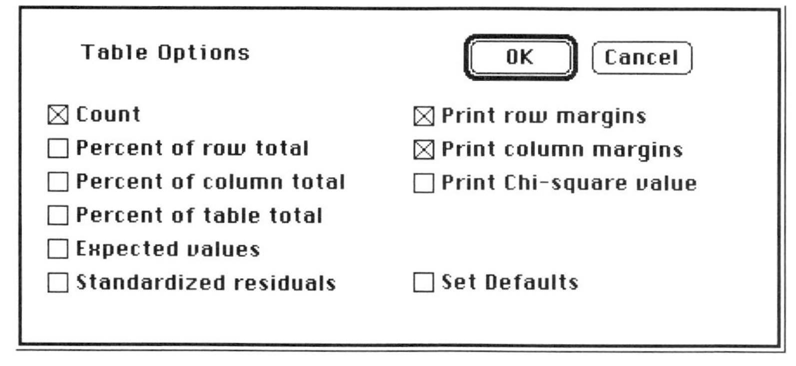

Figure 14-4. *The Table Options dialog.*

One way to describe alternatives for filling contingency tables is by example. The Fiber dataset is from a study in which subjects were fed one of four types of fiber-filled cracker before meals. Bran fiber, gum fiber, and a combination of the two were used, along with a control cracker containing no fiber. One of the observed effects was occasional gastric upset in the form of bloating. The contingency table of Cracker *vs.* Bloat is one way to examine this effect.

count

The *count* is the number of cases falling into each cell of the table. Counts sometimes show cells with unusually many or unusually few cases, or patterns and trends. The counts of Fiber *vs.* Bloat are shown in Figure 14-5.

Columns are levels of Cracker
Rows are levels of Bloat

	bran	combo	control	gum	total
high	0	2	0	5	7
low	4	5	4	2	15
med	1	3	2	3	9
none	7	2	6	2	17
total	12	12	12	12	48

tablecontents:
Count

Figure 14-5. *A contingency table of Cracker vs. Bloat for the Fiber data.*

margin

The category names appear around the edges of the table in boldface to label the rows and columns. This is one good reason for using category identifiers that name the category rather than numeric codes. The *margins* of the table, on the right and bottom, report row and column totals. We can see that there are 12 datapoints for each cracker. This was by design — each of the 12 subjects was fed each of the crackers in the course of the study. We can also see some pattern in the counts. For example, neither the control nor the bran crackers seem to have caused high levels of discomfort.

column percent

Columns are levels of Cracker
Rows are levels of Bloat

	bran	combo	control	gum	total
high	0	16.7	0	41.7	14.6
low	33.3	41.7	33.3	16.7	31.2
med	8.33	25	16.7	25	18.8
none	58.3	16.7	50	16.7	35.4
total	100	100	100	100	100

table contents:
Percent of Column Total

Figure 14-6. *A column percent contingency table for the Fiber data. The percentages in each column sum to 100%.*

The counts can also be converted to percentages. Percent of column total computes, for each column, the percentage of its values in each cell in that column. As a result, the percentages sum to 100% down each column. A column percent table for the Fiber data is in Figure 14-6.

Here we can see that half (50%) of the subjects fed the control cracker reported no bloating, while almost 42% of the subjects fed gum fiber reported a high level of bloating. The column totals show that about 35% of the time subjects reported no bloating and about 14.5% of the time they reported high levels of bloating.

Percent of row total computes the corresponding percentages within each row. Here the percentages sum to 100% across each row.

Columns are levels of Cracker Rows are levels of Bloat					
	bran	combo	control	gum	total
high	0	28.6	0	71.4	100
low	26.7	33.3	26.7	13.3	100
med	11.1	33.3	22.2	33.3	100
none	41.2	11.8	35.3	11.8	100
total	25	25	25	25	100

table contents:
Percent of Row Total

Figure 14-7. A row percent table for the Fiber data. The percentages in each row sum to 100%.

In this table we can see, for example, that of all the instances of high levels of bloating, 71% of them occurred when subjects were eating gum crackers. This was of particular interest to the researchers who were investigating the possible use of the gum fiber as a healthful appetite suppressant, but concluded that the bloating would discourage people.

table percent

Percent of table total reports, for each cell, the percentage of the total count for the table falling in that cell. Now the percentages in all the cells of the table (excluding the row and column totals) sum to 100%.

Two of the remaining options specify whether row and column totals should appear. While some tables might be displayed more conveniently without the extra row or column, most tables benefit from including them. The remaining statistics are related to the chi-square test for independence discussed in the next section.

Columns are levels of Cracker Rows are levels of Bloat					
	bran	combo	control	gum	total
high	0	4.17	0	10.4	14.6
low	8.33	10.4	8.33	4.17	31.2
med	2.08	6.25	4.17	6.25	18.8
none	14.6	4.17	12.5	4.17	35.4
total	25	25	25	25	100

table contents:
Percent of Table Total

Figure 14-8. A table percent contingency table for the Fiber data. The percentages over all cells sum to 100%.

14.5 Independence and Chi Square

It is possible to construct a hypothesis test to investigate whether the two factors in a contingency table are related or independent. Chapter 16 of this book discusses hypothesis testing in detail; this section gives an introduction to the chi-square test, and Appendix 14A presents the underlying formulas.

chi-square test

The null hypothesis associated with the *chi-square test* for independence states that the two factors are statistically independent. Formally, this means that the probability that a randomly selected case falls in a specified cell depends only on the probability that the case falls in the specified column and the probability that it falls in the specified row.

For the Fiber data, we hypothesize that bloating is independent of the type of fiber eaten. That is, if we knew what fiber a subject ate we would have no additional information about the likelihood of bloating, and if we knew the level of bloating reported by the subject, we would have no additional information about the kind of fiber eaten.

expected values

For each cell in the table, we calculate the number of cases we expect there to be were the null hypothesis true. These are called the *expected values*. If the null hypothesis *is* true, then the observed cell counts will tend to approximate the expected cell counts. If the null hypothesis is false, then the observed cell counts will tend to differ from the expected cell counts in some way.

standardized residual

For each cell, we calculate a *standardized residual* to describe the extent to which the observed count differs from the expected count. Data Desk can display the standardized residuals, but you need not do so to compute χ^2. The χ^2 value is the sum of the squared standardized residuals across all the cells in the table. If the value of χ^2 is relatively large, then we reject the null hypothesis of independence. Otherwise, we fail to reject the null hypothesis. The prob value (listed below the χ^2 value) reports the probability that we would observe a χ^2 value at least as large as the one we have observed if the null hypothesis were true.

degrees of freedom

Prob values of χ^2 depend upon the *degrees of freedom*. A χ^2 statistic associated with a contingency table of r rows and c columns has $(r - 1) \times (c - 1)$ degrees of freedom. If the prob value is relatively small, we can reject the null hypothesis.

For example, Figure 14-9 shows the counts, expected values and standardized residuals for the Fiber data. Examining the standardized residuals, we can pick out the cells where the violation of the null hypothesis is greatest.

Some authors recommend that the chi-square statistic only be computed for tables in which the expected value of each cell is at least 5. However, Data Desk does not enforce this restriction.

alternative hypothesis

The *alternative hypothesis* for the chi-square test is that the two factors are *not* independent. There are many ways in which a lack of independence could appear. There might be a trend across rows or down columns, one row or column could be unusual, or even a single cell could make the test significant. It is therefore a good idea to examine the standardized

Columns are levels of	Cracker				
Rows are levels of	Bloat				
	bran	combo	control	gum	total
high	0	2	0	5	7
	1.75000	1.75000	1.75000	1.75000	7
	-1.32288	0.188982	-1.32288	2.45677	0
low	4	5	4	2	15
	3.75000	3.75000	3.75000	3.75000	15
	0.129099	0.645497	0.129099	-0.903696	0
med	1	3	2	3	9
	2.25000	2.25000	2.25000	2.25000	9
	-0.833333	0.500000	-0.166667	0.500000	0
none	7	2	6	2	17
	4.25000	4.25000	4.25000	4.25000	17
	1.33395	-1.09141	0.848875	-1.09141	0
total	12	12	12	12	48
	12	12	12	12	48
	0	0	0	0	0

Chi-square = 16.94 with 9 df
p ≤ 0.049621

table contents:
Count
Expected Values
StandardizedResiduals

For the Fiber data, χ^2 = 16.94. The table has (4 - 1) x (4 - 1) = 9 degrees of freedom. The probability of obtaining a value of χ^2 at least this large is less than 5%, so we reject the null hypothesis of independence.

Figure 14-9. *Counts and chi-square components for the Fiber data.*

residuals for hints about how the data might violate the null hypothesis — whether a few cells are making the value of χ^2 large, or whether the entire table shows a lack of independence between the factors.

Fisher's Exact

Data Desk computes *Fisher's Exact* test for any 2x2 contingency table. Fisher's exact test considers all possible 2x2 tables with the same marginal frequencies. From among these tables, it classifies tables according to the strength of the association between the categorizations. The exact test computes the fraction of tables with the same marginal frequencies for which the association would be as strong or stronger than the association in the observed table. This is a value between zero and one, where a value of zero indicates the virtual absence of association and a value of one indicates the strongest possible association.

14.6 *HyperViews in Tables*

The HyperView menus connected to each cell of a table offer to select the cases corresponding to that cell. The selection highlights cases in all windows that display data from the same relations. Alternatively, you can record the selection as a 0/1 indicator variable, or record it and place it

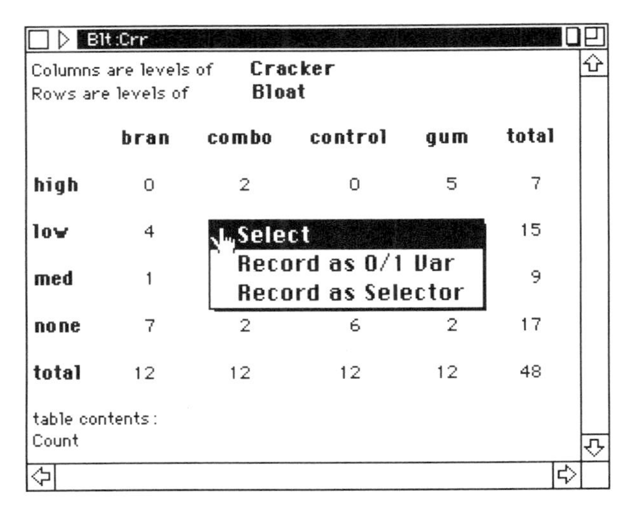

Figure 14-10. *The HyperView on a cell of a table offers to select the cases counted in that cell.*

in the Selector button immediately. To extend the selection to cases in more than one cell in the contingency table, press Shift key and choose **Select** form the HyperView of the cell to be added.

The HyperViews connected to each row or column title offer to select the cases corresponding to the named levels of their respective factors.

By recording a selection as the selector and then making a new table of other factors, you can generate any of the parts of a three- or four-way contingency table. Multi-way contingency tables categorize data on more than two factors. They can be visualized as having a two-way contingency table within each cell of another one- or two-way table. Appendix 14B shows how the By Group button can be used to make a three-way contingency table.

14.7 *Copying and Printing Tables*

The **Copy Window** command, which replaces the **Copy** command in the **Edit** menu when an output window is frontmost, offers a choice of copying the contents of the frontmost window as a picture (as it appears on the screen) or as text.

If you copy in Picture form, the Clipboard will hold a PICT image of the table. If you copy in Text form, the text of the window is copied. The columns of the table are separated by Tab characters, so you need only set tab stops in a word processor to make the table look as you would like. The **Text Format...** command in the table's Global HyperView offers control over the font, size, and style of both the body and titles of the table.

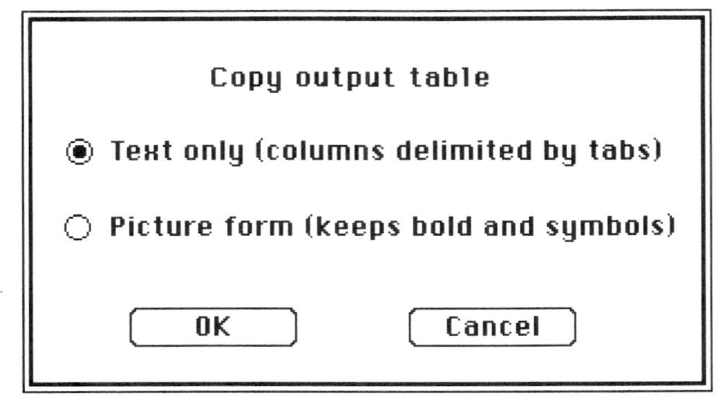

Figure 14-11. *The Copy Front Window dialog offers a choice of tab-delimited text and bit-mapped picture.*

When a table is the frontmost window, the **Print Front Window** command in the **File** menu prints a copy of the table. The table is printed using an intermediate PICT form, so it appears at the full precision of the printer.

APPENDIX 14A
Equations

If we denote the probability that a randomly selected case is in the ith row by π_i, the probability that it is in the jth column by π_j, and the probability that it is both in the ith row and the jth column by π_{ij}, then the null hypothesis of statistical independence is

$$H_0: \pi_{ij} = \pi_i \pi_j$$

This is the hypothesis usually tested with the chi-square (χ^2) test of independence. The χ^2 statistic is computed as

$$\chi^2 = \sum_{i,j} \frac{(O_{ij} - E_{ij})^2}{E_{ij}}$$

where O_{ij} is the observed count for the cell in the ith row and j^{th} column, and E_{ij} is the expected count for that cell, found as

$$E_{ij} = \frac{O_i O_j}{n}$$

The *expected values* are the E_{ij} terms in the formula for χ^2 above. Comparing them with the observed counts gives some idea of the nature of the differences.

The *standardized residuals* are the individual contributions to the χ^2 statistic. The *i, j* cell contains

$$\frac{(O_{ij} - E_{ij})}{\sqrt{E_{ij}}}$$

The sum of the squared standardized residuals across all cells of the table is equal to χ^2, so each standardized residual reports the contribution to that sum due to its cell.

The degrees of freedom is *(#rows – 1)* x *(#cols – 1)*, which depend on the table size rather than on the number of cases, but the value of χ^2 is likely to increase as the sample size increases if the null hypothesis is not strictly true. In most situations the null hypothesis can be at best only approximately true, so the χ^2 test is likely to reject the null hypothesis for very large samples. It is always a good idea to examine the standardized residuals to get some idea of what is really going on in the data. If you see no consistent patterns or significantly deviant cells and have a large sample size, you should be suspicious of marginally significant χ^2 values.

APPENDIX 14B
Multi-Way Tables

While the basic Contingency Tables command makes a two-way table, you can construct tables with three or even more category variables by using the Group button. For example, a three-way table is often thought of as simply a collection of two-way tables, one for each level of a third variable.

To make a three-way table, select one of the category variables and choose {Special ▶ Group Variable} **Set**. Then select the other two variables and choose the Contingency Tables command as before. Data Desk will make a separate table for each level of the Group variable. The resulting tables together form a three-way table.

If there is no particular reason to favor one variable over the others, you will have a better chance to see more of the results by making the variable with the fewest number of categories the group variable. This will produce fewer windows, so you can see more of them together on the screen.

One advantage of this kind of three-way table is that you can rearrange the component tables in any order to make it easier to look for patterns across tables. A disadvantage is that the χ^2 statistic is computed for each two-way table and not adjusted for a three-way table computation.

CHAPTER 15

Random Numbers and Simulation

15.1 Randomness 217
15.2 Creating Random Samples 218
15.3 Distributions 219
15.4 Generating Random Samples 220
15.5 Bernoulli Trials 221
15.6 Binomial Distribution 221
15.7 Poisson Distribution 221
15.8 Uniform Distribution 222
15.9 Normal Distribution 222
15.10 The Law of Large Numbers 223
15.11 Sampling Distribution 224
15.12 Central Limit Theorem 226

APPENDIX
15A Details and Formulas 228

THUS FAR WE HAVE considered ways to depict and explore data. This chapter and those that follow discuss ways to draw firm conclusions from data. These two approaches are sometimes called *exploratory* and *confirmatory* analyses, but the imaginary line between the two is really quite fuzzy. Many so-called confirmatory techniques are excellent tools for exploring data and many explorations lead to firm conclusions.

15.1 Randomness

The techniques in this book pertain to formal methods for drawing conclusions from data. The chief concern of these methods is that the inferences we draw from the data reflect genuine relationships and not random patterns that might appear structured just by chance.

Human vision is a marvelous tool for finding patterns. Unfortunately, sometimes our mind is so good at seeing patterns that it sees them where there are none. The best way to get a feeling for what truly unstructured random values look like is to look at some examples. Data Desk provides ways to generate random values from a variety of distributions.

Of course, random values have other important uses. Chief among them is their use in selecting a sample of cases from a larger population. In general, to draw conclusions about data or make inferences about the population from which the data were drawn, we must take into account how the data were collected. In particular, most statistics methods require that the data we work with be sampled *at random* from the population of interest.

Randomness is a concept that everybody has some general feeling for. Many card tricks, for example, play upon our expectations of what ought to happen when a card is selected "at random" from a shuffled deck. But when it comes to discussing randomness precisely, most of us need some help. Computers provide particularly good ways to experiment with randomness to get a better feeling for the consequences of random sampling.

simple random sample

A *simple random sample* is one in which each member of the population has an equal and independent chance of being selected. In this context, being independent just means that the chance of selecting a particular individual is not affected by the selection of any other individuals.

In much of statistics theory, we assume that the population of interest is infinitely large. It would seem to be impractical to experiment with simple random samples from infinite populations, but Data Desk can get very close to this ideal model. The secret is that computers can generate numbers that appear to be selected randomly from practically infinite populations. Although the numbers are not truly random and the populations are only very, very large, the differences between what we can simulate and the ideal model are negligible.

pseudo-random numbers

To be precise, the random numbers computers generate are called *pseudo-random numbers*, and are designed to be almost indistinguishable from truly random numbers. (See Appendix 15A.)

15.2 *Creating Random Subsamples*

There are many situations when we would want to work with random samples of our datasets. We might have a dataset that is too large to analyze practically on our machine. Or we might want to work with several small samples and use the full dataset for confirmatory analyses. In Data Desk there are several ways to create random samples.

This section discusses how to create a random sample from an existing dataset. The methods discussed in Sections 15.5 and 15.8, use 0/1 indicator variables and a Selector button to simulate random samples within the existing relation. Section 15.5 deals with samples that include a specified percentage of the original cases. Section 15.8 demonstrates how to generate random samples with an exact number of cases.

The major advantage of having a random sample placed in its own separate relation is that less memory is required for processing the sampled data in subsequent analyses. Random samples generated using 0/1 indicator variables hold all the cases in memory for your analysis.

To draw a random sample select the variables from which to sample and choose **Sample...** from the **Manip** menu. A dialog shown in Figure 15-1 appears. Click the button labeled Random Sample, specify a percentage of cases you wish to include in your sample and click the OK button.

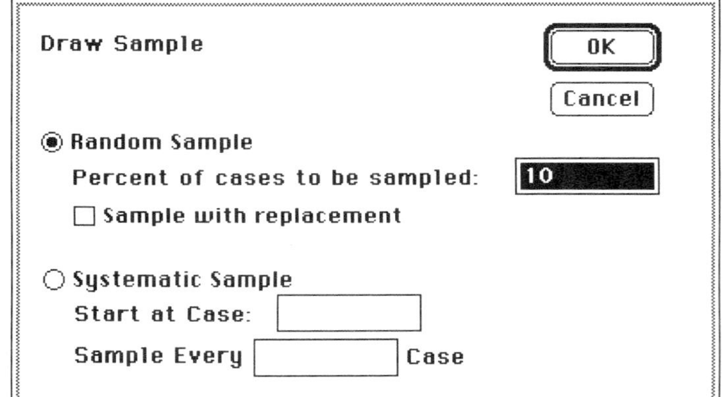

Figure 15-1. *To specify random sample, type the percentage of cases you want to include in the sample. To specify systematic sample, type the starting case and the number of cases to skip.*

For example, to sample 20% of the original cases type 20 and click OK. Data Desk generates a new relation, called *Random1,* holding 20% of the cases drawn from the selected variables. The relation is placed in the Results Folder. It includes a new variable, named *Indices* that holds the original case number for each case.

Data Desk samples without replacement unless you click the *Sample with replacement* box. Sampling with replacement gives every case the same probability of being chosen each time a case is drawn.

Data Desk also draws Systematic Samples. Systematic Samples draw a sample in a specified pattern. Click the **Systematic Samples** button, type the starting case number in the *Start at Case* box and the number of cases to skip between sampled cases in the *Sample Every* box then click the OK button. A new relation labeled *Systematic1* is placed in the *Results* Folder.

15.3 Distributions

Histograms display the distribution of values in a variable. We can consider the distribution of values in a population in a similar way. With an infinite population we must use a mathematical description of the distribution shape rather than counts of observations falling between fixed bounds. Nevertheless, the ideas are the same: a distribution describes the fraction of cases falling within any specified part of the range of possible values.

parameters

statistics

Quantities that characterize a population's distribution, such as its mean and variance, are called *parameters*. Parameters are typically denoted with Greek letters to distinguish them from their sample-based counterparts, called *statistics*.

Most useful distributions are completely determined by one or two simple parameters. For example, a Normal distribution is completely determined by its mean, μ (mu) and standard deviation, σ (sigma). A Poisson distribution is completely determined by its mean, λ (lambda).

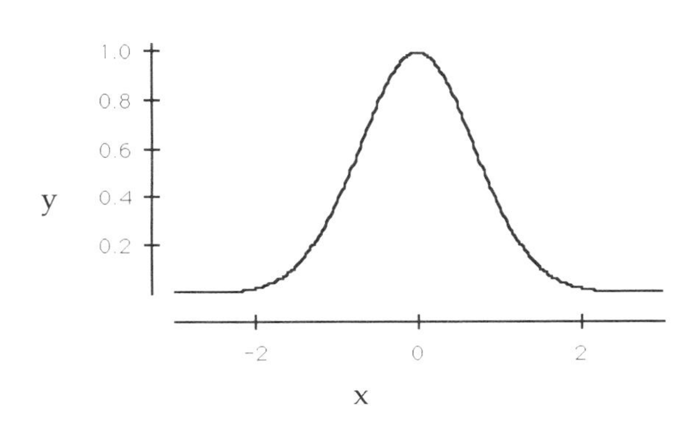

Figure 15-2. *The mathematically described Normal distribution is a smooth curve. The parameters of this particular Normal distribution are* $\mu = 0$ *and* $\sigma = 1$.

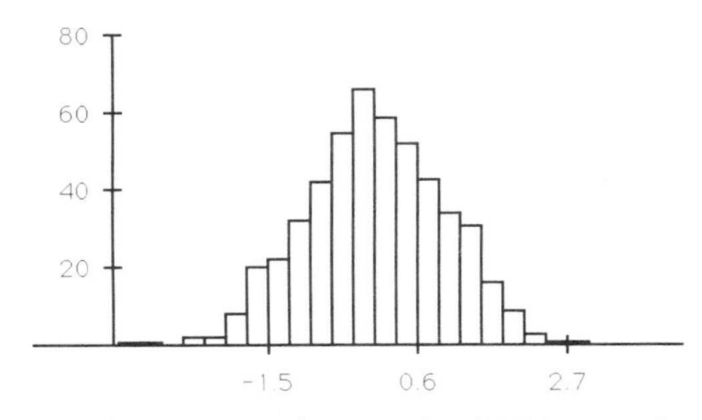

Figure 15-3. *A histogram of a sample of 500 cases from a Normal distribution follows the general shape of the mathematical description but shows that samples do not exactly follow the distribution of the population.*

When we use statistics to draw inferences about the sample's underlying population, we usually must assume that the population's parameters are known quantities. Often we are asked to assume that observations or errors made in the observations follow a particular distribution shape or even that they come from a distribution with a particular mean, standard deviation, or some other parameter. Of course, when dealing with real data we can't be certain of the underlying distribution because we usually can't observe the entire population. However, computers can simulate random samples drawn from specified distributions with specified parameters. The control that this provides is particularly valuable for understanding randomness.

15.4 *Generating Random Samples*

To generate random samples in Data Desk, select the **Generate Random Numbers...** command from the **Manip** menu. Data Desk presents a dialog that looks, in part, like Figure 15-4 The value in the *cases/variable* box is the number of cases in the frontmost relation. This makes it easy to generate samples that will fit into the relation holding the variables you are currently working with. If there are no relations in the datafile, this value defaults to 100. To change this value type the sample size in the box. Data Desk places samples that do not fit into the frontmost relation in their own relation.

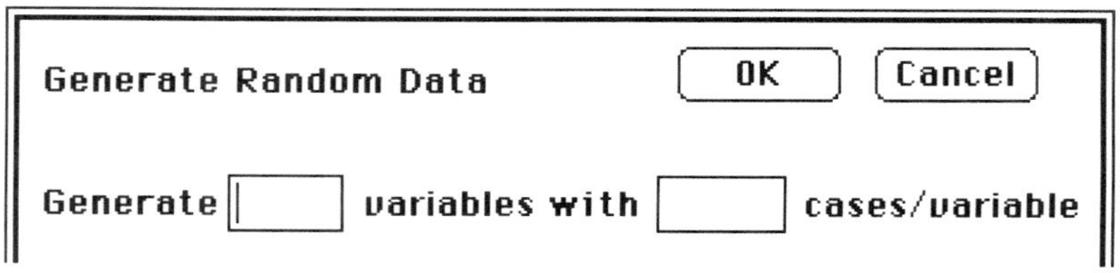

Generate Random Data   OK   Cancel

Generate [ ] variables with [ ] cases/variable

Figure 15-4. Generate Random Numbers... *offers to create any number of randomly generated variables with any number of random values in each one.*

TIP

To move easily among parts of a dialog, press the Tab key. Tabbing to a parameter box automatically selects the corresponding distribution.

The second part of the dialog offers a choice of distribution. All distributions but the Uniform require values for the parameters defining the distribution. When you click the **OK** button, Data Desk generates numbers that behave as if they were sampled from the specified distribution and groups them into variables of the requested sample size, one variable for each sample requested.

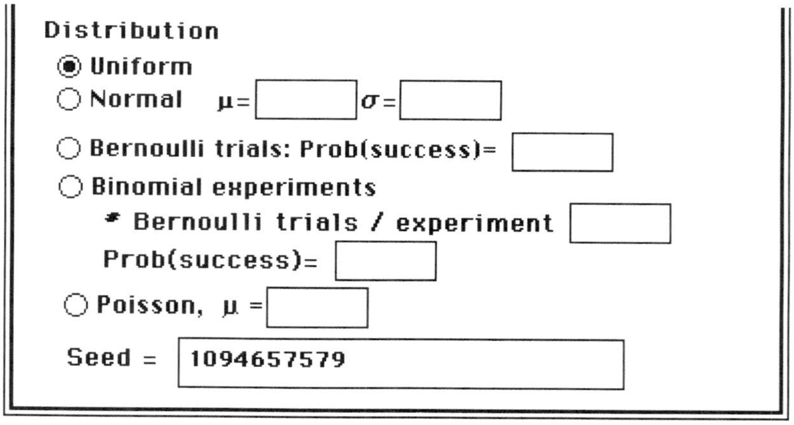

Distribution
- (•) Uniform
- ( ) Normal μ= [ ] σ= [ ]
- ( ) Bernoulli trials: Prob(success)= [ ]
- ( ) Binomial experiments
  - Bernoulli trials / experiment [ ]
  - Prob(success)= [ ]
- ( ) Poisson, μ = [ ]

Seed = 1094657579

Figure 15-5. *You have a choice of distributions for* ***Generate Random Numbers...****; specify parameters in the corresponding boxes.*

seed

Appendix 15A discusses the way in which the Macintosh generates random numbers. The random number generation works with an initial number called the *seed*.

15.5 *Bernoulli Trials*

Bernoulli trials (named for James Bernoulli, 1654-1705) are experiments with two possible outcomes, usually labeled "success" and "failure". The most common Bernoulli trial is flipping a coin. To generate sample Bernoulli trials, specify the probability of a success in the underlying population.

Sometimes we express probabilities in percentage form ("a 56% chance of a success") and sometimes as a fraction between zero and one. Data Desk expects probabilities to be between zero and one, and will ask you to correct any that are not in this range. Data Desk performs the requested number of trials and records either a 0 or a 1 for each, where 1 marks a "success". The result is a variable containing zeros and ones randomly.

One special application of the Bernoulli distribution is to draw a simple random sample from your data. To do this in Data Desk, generate a Bernoulli variable with as many cases as there are cases in the variables to be sampled, and with probability of success equal to k/n, where k is the number of cases you hope to have in the sample and n is the total number of cases. Declare the Bernoulli variable to be a selector variable with the **Set Selector** command from the **Special** menu.

Now all analyses for which the Selector button is active will be performed on a random sample of the data; specifically, those cases with a "1" in the Bernoulli variable will be selected. Note that you are not guaranteed to get exactly k cases in your sample. Instead, you are selecting cases with a probability of k/n, so you can expect to have approximately k cases in the sample. Section 15.4 discusses how to draw a simple random that guarantees the sample size.

15.6 *Binomial Distribution*

The Binomial distribution counts the number of "successes" in some number of Bernoulli trials. The sum of a Bernoulli variable is a Binomial quantity. In the Binomial experiments portion of the **Generate Random Numbers...** dialog, specify the number of Bernoulli trials in each Binomial experiment, and the probability of a success. A Binomial variable consists of integer values between zero and the number of Bernoulli trials specified.

The number of Bernoulli trials in a Binomial variable is a parameter of the distribution and thus is specified in the Generate Random Numbers... dialog.

15.7 *Poisson Distribution*

The Poisson distribution, like the Binomial distribution, describes probabilities of discrete events. It is appropriate in situations where the probability of an event is very small but there are many trials, so that there is a measurable probability of 0, 1, 2,... events.

For example, the probability of a car accident on any short stretch of road on any one day is very small. But if we consider a long stretch of road (for example, all of Interstate 80), the probabilities that on any particular day there will be 0, 1, 2, ... accidents is moderate. If we know the true average number of accidents in a day on the entire road, we can use a Poisson distribution to describe the distribution of the number of accidents.

The Poisson distribution is described by a single parameter, λ, which is defined to be the number of events that are expected to occur over a period of time. The mean and variance of a Poisson distribution both equal λ. A Poisson variable consists of integer values 0, 1, 2, ..., with values near λ more likely than values far from λ.

15.8 *Uniform Distribution*

The Uniform distribution is defined so that every value between zero and one is equally likely. Histograms of samples from the Uniform distribution tend to be relatively flat with several small modes. For example, a histogram of 250 numbers drawn from a Uniform distribution might look like Figure 15-6.

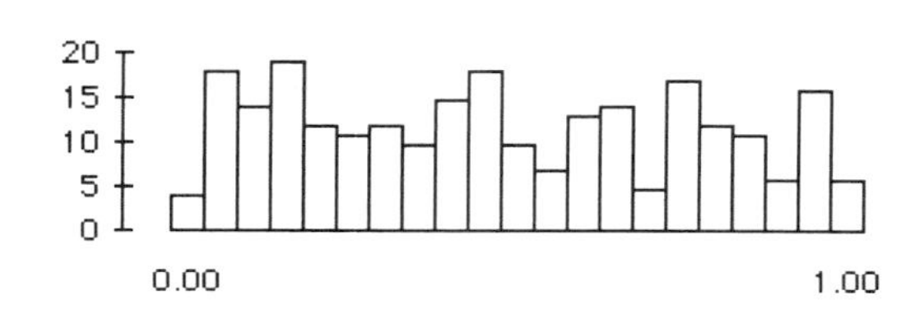

Figure 15-6. *A histogram of 250 random Uniform values is relatively flat.*

The Uniform distribution is a convenient source of randomly distributed values.

To generate a random permutation of cases, generate a variable with the same number of uniform random values using the **Generate Random Data...** command in the **Manip** menu. Select the random Uniform variable as y and the variables to permeate as x and choose **Sort on Y Carrying X's** in the **Manip** menu. This sorts the Uniform variable and reorders the other variables randomly.

15.9 *Normal Distribution*

The Normal distribution is also called the Gaussian distribution because it was discussed extensively by the famous mathematician, C. F. Gauss (1777-1855). The Normal distribution is one of the simplest distributions to deal with mathematically, in part because it is described entirely by its mean, μ, and its standard deviation, σ. Indeed, the most common way to describe the relative locations of values in a Normal distribution is as numbers of standard deviation units above or below the mean.

standard Normal distribution

Data Desk can simulate samples from a Normal population with any mean and standard deviation. A Normal distribution with a mean of zero and a standard deviation of one is called a *standard Normal* distribution because it is the simplest choice of values. Many statistics books include tables of probabilities for standard Normal distributions, but Data Desk has these tables built-in.

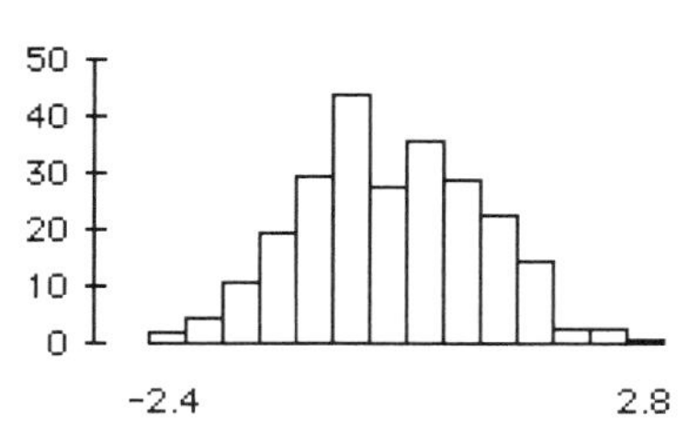

Figure 15-7. *A histogram of 250 random Normal values.*

When we draw a random sample from a standard Normal population, as in Figure 15-7, a histogram of the sample is not a perfect normal distribution shape. If we were to draw another sample of 250 from the standard Normal population, or if you perform the experiment yourself, the histogram would look somewhat different, but would still tend to have a central peak and an overall bell shape shown in Figure 15-7.

15.10 *The Law of Large Numbers*

We all have an intuitive feeling that "in the long run" things average out about right. A particular flip of a coin or roll of the dice may not be predictable, but in the long run, we are pretty confident that about half of our coin flips will be heads and that all six numbers on each die will come up about an equal number of times.

More generally, we can be confident that as we draw larger and larger samples, the average of the sampled values will approach the true population mean. The Law of Large Numbers tells us formally that our intuition is right. Averages of larger samples tend to be closer to the true population mean.

We can illustrate the Law of Large Numbers with a simple simulation. Generate 1000 numbers from a Normal distribution with any mean and variance you please — but remember the mean that you select. Rename the resulting variable y. Generate Patterned Data with the default settings: numbers counting from 1 to 1000. Rename that variable n.

Now, make a New Derived Variable. Name it *means*. Type the following expression into the derived variable window:

cumsum(y)/n

The cumsum function computes the *cumulative sum* of *y*. The first case of the cumulative sum is just the first case of *y*. The second case is the sum of the first and second cases of *y*: *y[1] + y[2]*. The third case sums the first three cases: *y[1] + y[2] + y[3]*, and so on. The variable n gives the number of values summed (1, 2, 3, ...). Thus, the derived variable expression computes first the first case, *y[1]*, then the average of the first two cases, *(y[1] + y[2])/2*, then the average of the first three cases, and so on. As the averages involve more and more cases, the Law of Large Numbers says that they will approach the true mean. (You do recall the mean you specified when you generated the random numbers, don't you?)

Now, a lineplot of *means* shows how these means start out oscillating quite wildly, but settle down close to the true mean. For a slightly better plot, re-express n by taking its logarithm and drag the re-expressed variable onto the x-axis of the lineplot.The log scale stretches out the early means and compresses the later ones, emphasizing the convergence. Figure 15-8 shows an example.

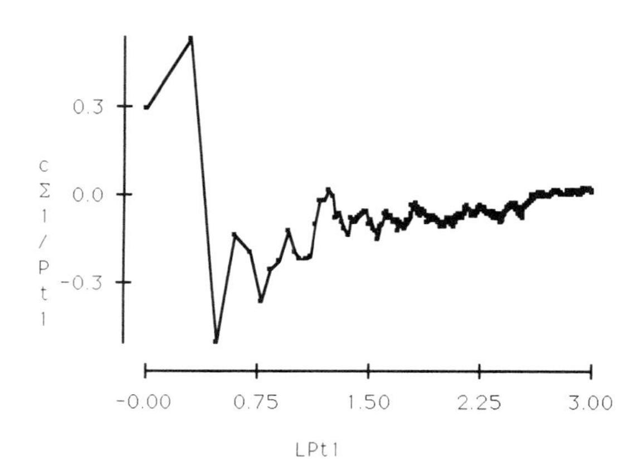

Figure 15-8. *The Law of Large Numbers states that as the sample size increases the average should approach the population mean. This lineplot shows averages for between one and 1,000 random numbers drawn from a standard Normal distribution. The horizontal axis is a log scale to stretch out the early fluctuations. It is easy to see the convergence toward the true mean of zero.*

If you are comfortable with Data Desk try the following sequence of operations to get the same result. **Generate Random Numbers...** . Specify Normal distribution and pick a mean and standard deviation. Data Desk leaves the random variable selected, so you can just **Transform ▶ Summary ▶ CumSum**. The resulting variable is named *cumΣNr1*. **Generate Patterned Data...** and hit Enter — the defaults are right. Select *cumΣNr1* as *y* and *Pattern1* as *x* and choose {Transform ▶ Arithmetic} **y/x** to create *cΣ1/Pt1*. Choose **Plot ▶ Lineplots**. Now select *Pattern1*, choose **Transform ▶ Log(•)**, and drag the resulting variable onto the y-axis of the lineplot.

15.11 *Sampling Distributions*

When we draw a sample from a population and compute a statistic, we know that were we to draw another sample of the same size independently from the same population and compute the statistic on the new sample, the second value would probably differ from the first. We want to take this fact into account when interpreting any statistic we might compute from a specific sample.

Independent samples drawn from the same population differ from each other in the individual numbers observed and consequently, in the values of statistics computed from them. A statistic's *sampling distribution* describes this sample-to-sample variation in its observed values. If we were to draw a multitude of samples of a fixed size from the same population, compute the value of a statistic, and make a histogram of the collected statistic values, the distribution shown by the histogram approximates the sampling distribution of the statistic.

Generate Random Data [OK] [Cancel]

Generate [50] variables with [8] cases

Distribution

(•) Uniform
( ) Normal μ = [ ] sigma = [ ]
( ) Bernoulli trials: Prob(success)= [ ]
( ) Binomial experiments
* Bernoulli trials / experiment [ ]
Prob(success)= [ ]
( ) Poisson, lambda = [ ]

Seed = [909201968]

Figure 15-9. *Generating 50 random Uniform variables.*

Usually, we can't afford to sample repeatedly from a real population. Instead, we draw a single sample as large as we can manage and work with that. Fortunately, the sampling distributions of most common statistics are known mathematically and can be looked up in texts. Nevertheless, we can get a much better idea of how sampling distributions behave by performing the experiment on the computer and seeing the distributions for ourselves.

To simulate a sampling distribution, generate many variables from the same population, and compute a statistic on each of them. For example, we can generate 50 samples of 8 numbers each from a Uniform distribution using the Generate Random Data command. (See Figure 15-9)

Data Desk places the samples together in a new relation named *Ran Data*, and places this relation on the Data Desk desktop (Fig. 15-10).

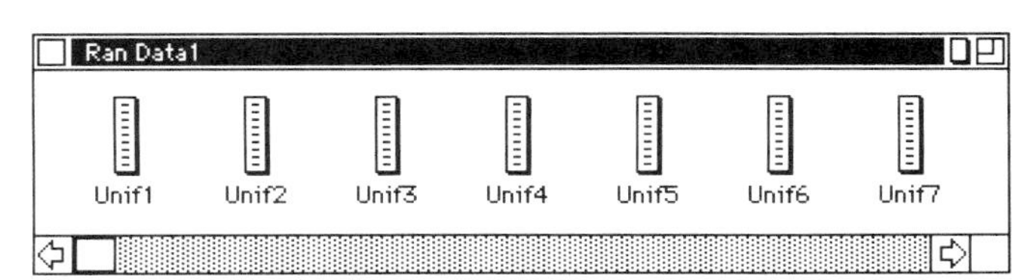

Figure 15-10. *The random Uniform variables are automatically placed in a relation named* Ran Data.

Spreads
- [ ] Standard Deviation
- [ ] Interquartile Range
- [x] Range
- [ ] Variance

Figure 15-11. *Selecting the range.*

We decide to compute the range for each of the 50 samples, selecting it in the dialog presented by the **Select Summary Statistics** command of the **Calculation Options** submenu.

Select the icon of the *Ran Data* relation (whether its window is open or not) and choose {Summaries} **As Variables** from the **Calc** menu. This computes the range for each of the samples in that relation.

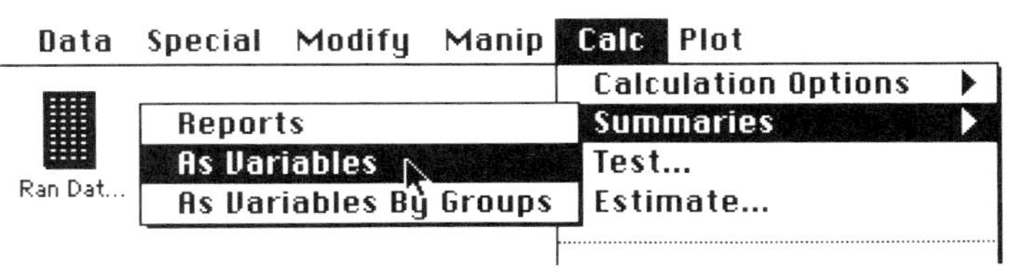

Figure 15-12. *Selecting* {Summaries} **As Variables** *for the entire Ran Data relation.*

The {Summaries} **As Variables** command generates a variable named *Ranges* containing the ranges of the 50 simulated variables.

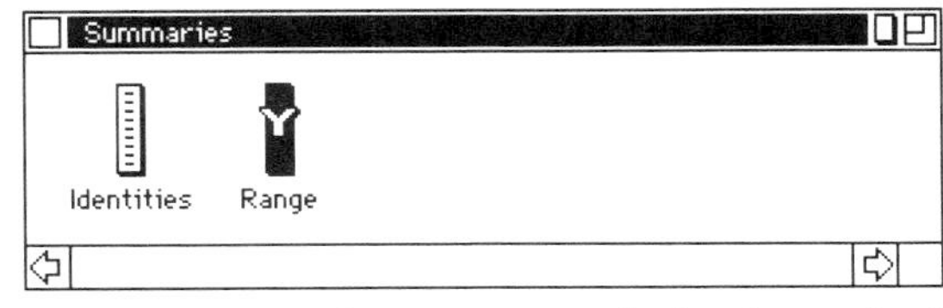

Figure 15-13. *Summary statistics for each simulated variable are collected into a new variable.*

The histogram of these values approximates the sampling distribution of the range for samples of eight cases drawn from a Uniform distribution. Figure 15-14 shows the histogram for one trial of these experiments.

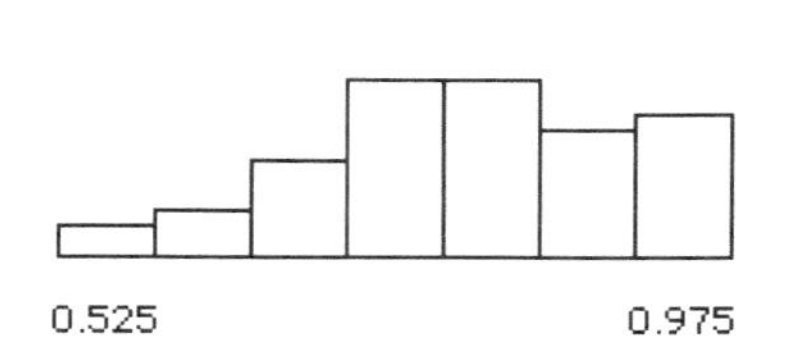

Figure 15-14. *A histogram of the ranges of one instance of random Uniform variables. If you repeat the experiment your histogram may look slightly different but should have the same general shape.*

15.12 *Central Limit Theorem*

Central Limit Theorem

The Central Limit Theorem states:

The theoretical sampling distribution of the mean of independent samples, each of size n, drawn from a population with mean μ and standard deviation σ, is approximately Normal with mean μ and standard deviation σ/√n. The approximation to Normality improves as n grows.

The theorem is fundamental to much of statistical inference, but the proof is beyond most introductory statistics courses. Data Desk can illustrate the Central Limit Theorem by approximating the sampling distribution of the sample mean by simulation.

For example, using the same methods as in the previous section, we can generate, say, 50 variables with 9 cases from a Uniform distribution. The Uniform distribution has a population mean, μ, of 0.5, and a population standard deviation, σ, of $1/\sqrt{12} = 0.289$.

Applying the {Calc ▶Summaries} **As Variables** command to the relation full of random variables generates a variable called *Means* containing the 50 means — one computed for each of the 50 Uniform variables. A histogram of Means shows how the means of these random samples are distributed and thus approximates the true sampling distribution of the sample mean.

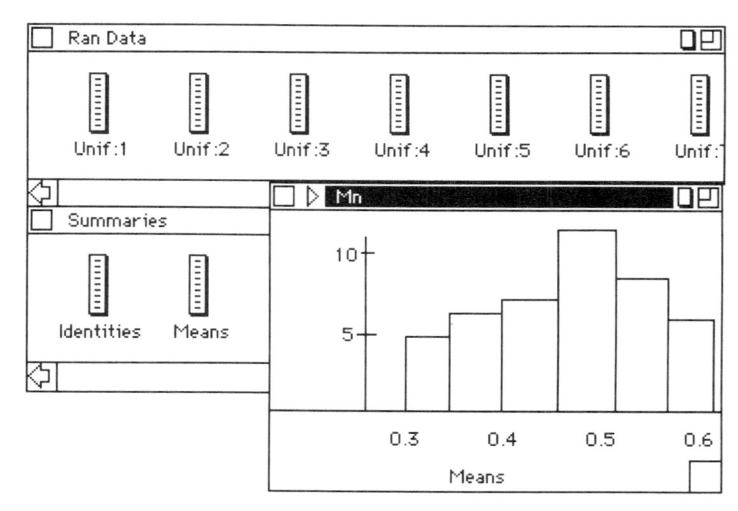

Figure 15-15. *We illustrate the Central Limit theorem by simulating the sampling distribution of the mean of samples from a Uniform population.*

As we expect from the Central Limit Theorem, the histogram of the means looks like a histogram of values from the Normal distribution even though the samples themselves were drawn from a Uniform population. A virtue of the Central Limit Theorem is that regardless of the underlying population from which samples are drawn, the sampling distribution of the sample mean is approximately Normal.

We can also check whether the mean and standard deviation of *this* distribution (that is, of the collection of 50 sample means in the variable Means) are as the Central Limit Theorem says they should be.

```
Summary statistics
for Means

NumNumeric = 50
Mean = 0.50576
Standard Deviation = 0.08103
```

The mean is near .5, the population mean for the Uniform distribution. The standard deviation we expect from the Central Limit Theorem is $\sigma/\sqrt{n}$, which equals .096 — a bit higher than the value .081 observed here, but still pretty close. In drawing only 50 samples, we can only approximate the true sampling distribution of the sample mean. Theoretically, if we were to draw an infinite number of samples, the mean and standard deviation of the corresponding Means variable would equal 0.5 and 0.96 exactly.

The simulation strategies in this chapter can help to make many aspects of statistical inference more intuitive. Many statistics perform well "on average" or "in the long run". For all of these methods, it can help your intuition to simulate "the long run" by generating independent random samples and applying the statistic of interest to those samples.

APPENDIX 15A
Details and Formulas

The pseudo-random numbers used by Data Desk originate from a random number generator built into the Macintosh. This generator is of a type known as a *multiplicative congruential* random number generator. Specifically, given the ith random number, x_i, in the range $0 < x_i < 1$, the next random number is generated as

$$x_{i+1} = (x_i * 7^5) \bmod (2^{31} - 1).$$

The initial random number (known as the "seed") is selected when Data Desk starts up by consulting an internal clock that counts 60^{th}s of a second since the machine was turned on, and is displayed at the bottom of the random numbers dialog. You can set the seed to be any large integer less than 2,147,483,647 by clicking in the seed window and typing the number you want. *If you start a simulation with the same seed and perform it in the same way, you will always get the same random numbers,* so you can repeat an experiment exactly if you wish. Usually, there is no need to adjust the seed.

Unlike other computations in Data Desk, random numbers are not generated to full 20-digit precision (because the largest number that can be generated by this generator is $2^{31} - 2$, which has only 10 digits).

This particular random number generator has been studied extensively and is known to perform well; sequences of random numbers show very little structure.

The random number generator itself generates random Uniform values. Random Normal values are computed by transforming pairs of random Uniforms using the Box-Muller transformation.

Random Bernoulli trials are generated by testing a random Uniform to see if it falls above or below a particular cut point (determined by the probability of a success). Random binomials are computed from repeated random Bernoulli numbers for large probabilities of success and with a wedge-tail approximation for small probabilities.

CHAPTER 16

Simple Inference

16.1	Confidence Intervals	*231*
16.2	Confidence Intervals for μ When σ is Known	*232*
16.3	Confidence Intervals for μ When σ is Unknown	*233*
16.4	Where Is the Randomness?	*234*
16.5	Multiple Intervals and the Bonferroni Adjustment	*234*
16.6	Testing Hypotheses	*235*
16.7	Hypothesis Test for μ When σ is Known	*235*
16.8	Hypothesis Test for μ When σ is Unknown	*236*
16.9	Multiple Hypothesis Tests and the Bonferroni Adjustment	*237*
16.10	Chi-Square Test of Individual Variances: Inference for Spread	*237*

APPENDICES

16A	Example: Simulating Confidence Interval Performance	*239*
16B	Example: Simulating Hypothesis Test Performance	*241*

STATISTICS PROVIDES WAYS to draw inferences about a population by examining a random sample from that population. While we don't expect to be able to make a precise statement about the entire population, we can use statistics to make a statement that is *likely* to be true and to understand how large an error we have probably made. In this way, we can control the error and draw conclusions from imperfect information.

Statistical inference usually considers the data at hand to be a representative sample from some population. We analyze the sample but make inferences about the population. We usually cannot examine the entire population, so we rarely learn whether the inferences we make are in fact correct. Simulation techniques, such as those of Chapter 15, allow us to control the characteristics of a special population to try to learn more about how inferential statistics work.

This chapter discusses the principles of statistical inference. If you are well-acquainted with these principles and wish to get to the business of constructing confidence intervals and testing hypotheses, you will probably need to read no farther than Section 16.2 to get a sense of how the **Test...** and **Estimate...** commands work. The Test and Estimate windows are interactive, and the pop-up menus guide you through all available choices.

16.1 *Confidence Intervals*

We saw in Chapter 15 that the sample means of many samples drawn from the same population fluctuate around the true mean of the population. When we wish to estimate the population mean, but have only one sample, we must account for the variability of the sample mean. Hence, we prefer to report a range of probable values for the population mean rather than the single sample average at hand. Confidence intervals provide a systematic way to construct a reasonable range of probable values.

Confidence intervals confront us with a tradeoff between precision and certainty. Ideally, we would like to compute a single, precise, value and be certain that we had found the true population value. For example, it would be nice to know with absolute certainty what the sales volume of a business will be tomorrow based on a sample of recent sales performances.

One alternative would be to acknowledge that we cannot predict unknown values with certainty and instead make probability statements about precise values. For example, we might say that the sales volume will *probably* be precisely $2745.12. Unfortunately, because it is so precise, the probability that this statement is true is so small as to make the statement worthless.

Another alternative would be to sacrifice precision to preserve certainty. For example, we could select a range of possible values but be certain that the true value is in the range. For example, we might say that while we cannot know the exact sales volume, it will certainly be between $0 and $5,000. Unfortunately, such certainty is only possible for ranges that are so large as to be uninteresting.

Ultimately, we are forced to forgo both certainty and precision by estimating both a range and a probability that the range encloses the true value. Thus we might say that while we do not know the true sales volume, we are 95% certain that it will fall between $2,500 and $3,000. By this we mean that in the long run, 95% of the intervals we might construct by similar processes would include the true sales volume (and 5% of them wouldn't).

We still have tension between precision and certainty. The more certain we want to be, the less precise we can be. The more precision we seek (as a narrow interval of values), the less certain we can be that the interval really encloses the true value.

16.2 *Confidence Intervals for μ When σ is Known*

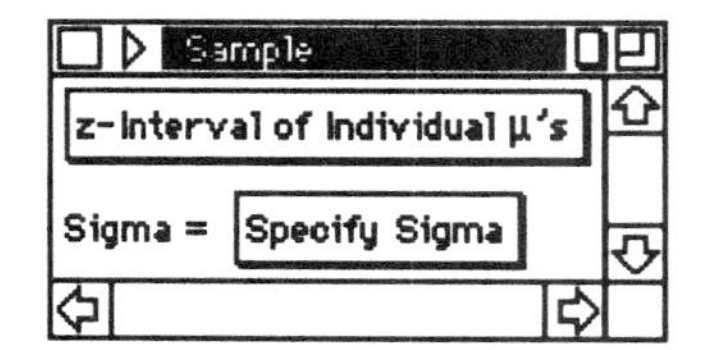

Figure 16-1. The Estimate Window.

To find a confidence interval for a population mean, μ, select a variable that contains a sample drawn from the population and choose **Estimate...** from the **Calc** menu. Data Desk opens an Estimate window. This window presents confidence interval results, expanding and recomputing according to the choices you make.

The first pop-up menu offers a choice of interval types. If the population standard deviation, σ, is known (not just estimated), a z-interval is appropriate.

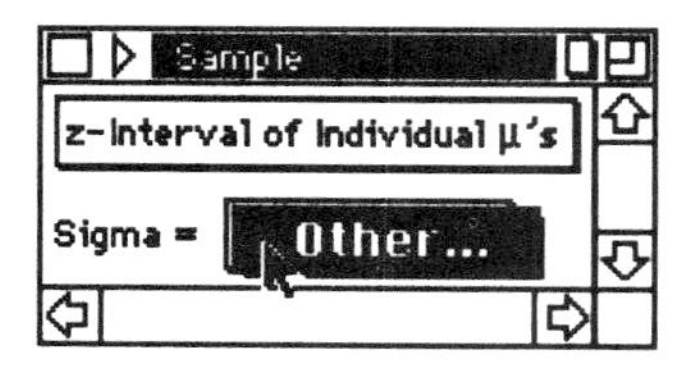

In the topmost pop-up menu which holds the different interval choices, select **z-Interval for individual μ's**. Click the **Specify Sigma** pop-up menu and type the population standard deviation value in the window provided. The Estimate window expands to look as in Figure 16-1.

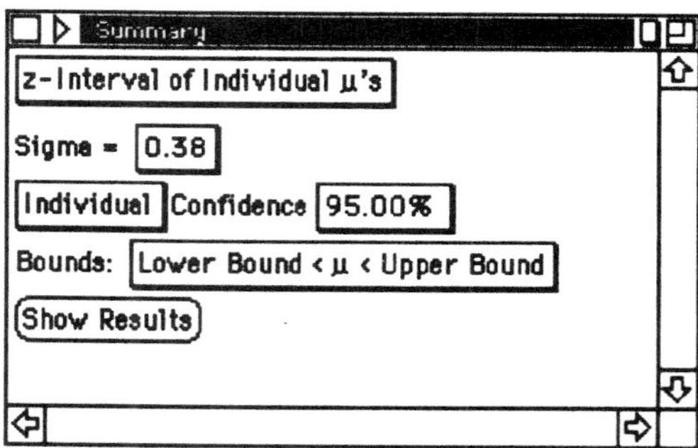

Figure 16-2. Specifying z-Interval for μ.

Data Desk computes z-intervals as

$$\bar{y} \pm z^{*} \sigma/\sqrt{n}$$

confidence level

where $\bar{y}$ is the sample mean, σ is the population standard deviation, n is the sample size, and z^* is the number of standard deviations one must move to either side of the mean in a Normal distribution to include the fraction of the distribution specified by the *confidence level*. The z^* values are also called the *percentage points* of a Normal distribution.

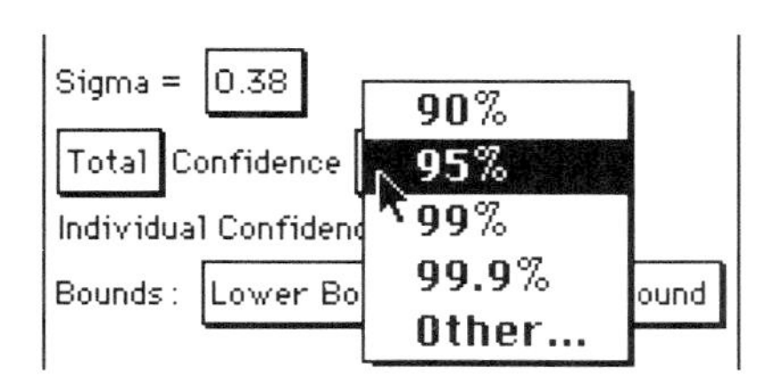

Figure 16-3. Click the ***Confidence*** *pop-up menu to specify the level of significance.*

The confidence level characterizes our choice in balancing between precision and uncertainty. If we imagine repeating independently many times the experiment that generated the data and computing, say, 95% confidence intervals with each sample, we would expect that in the long run, 95% of the computed intervals would in fact contain the true population mean and 5% of them would not.

Select the appropriate confidence level from the **Confidence** pop-up menu. While 95% is a very common confidence level, there is no correct value. If you want a level other than 95% choose **Other** from the pop-up menu and type a different confidence level in the window provided. The confidence level is specified as a percentage, so type 90 rather than 0.90 for 90%.

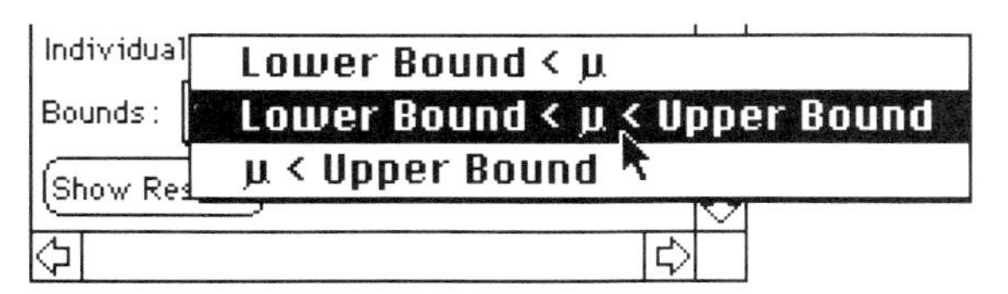

Figure 16-4. Click the ***Bounds*** *pop-up menu to select the type of confidence interval.*

One-Sided Confidence Intervals

Most confidence intervals are symmetric. We compute a statistic to estimate a population parameter and construct an interval that extends an equal distance above and below the value of the statistic. Sometimes however, we can be certain that the population parameter cannot be on one side or the other, or we are concerned only with a particular range of possible parameter values. In such cases, we may choose to construct a one-sided confidence interval. One-sided intervals can be either above or below the mean.

To finish creating the interval, select the appropriate bounds from the **Bounds** pop-up menu and click the **Show Results** button.

The pop-up menu to the left of the Confidence level holds the options **Total** and **Individual.** These options are important when computing two or more confidence intervals at once. See Section 16.5 for details.

16.3 *Confidence Intervals for μ When σ is Unknown*

t-distribution

Usually we do not know the standard deviation of the population and must estimate it from the sample. Confidence intervals based on the *t-distribution* use a sample-based standard deviation estimate. Like the Normal distribution, the t-distribution is symmetric about its mean. However, the peakedness of the t-distribution varies according to the size of the sample on which the standard deviation estimate is based, which determines the *degrees of freedom.* You need not specify degrees of freedom to construct a t-interval; Data Desk computes and reports degrees of freedom automatically.

degrees of freedom

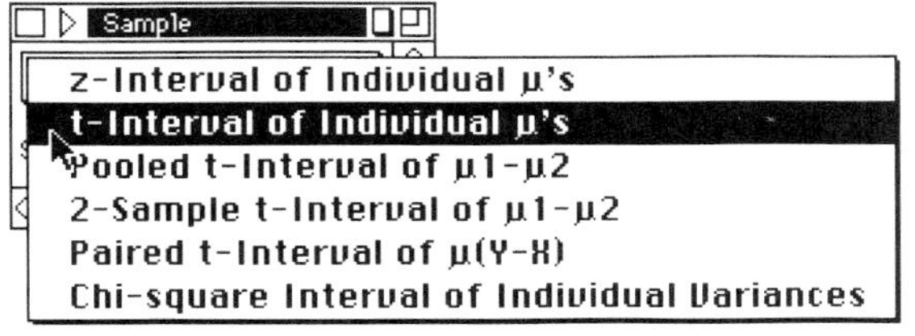

Figure 16-5. Select ***t-Interval of Individual μ's*** *from the pop-up menu.*

To construct a confidence interval for μ when σ is unknown, select **t-Interval for Individual μ's** in the Estimate window's top pop-up menu. The window will expand to show additional choices. Select the appropriate confidence level from the **Confidence** pop-up menu. Select the bounds parameters from the **Bounds** pop-up menu. The results appear at the bottom of the window when you click the **Show Results** button.

Data Desk computes t–intervals as

$$\bar{y} \pm t^*_{(n-1)}\, s/\sqrt{n}$$

where $\bar{y}$ is the sample mean, s is the sample standard deviation, n is the sample size, and $t^*_{(n-1)}$ is the appropriate percentage point of a t-distribution with $(n - 1)$ degrees of freedom. Unlike the Normal distribution, the t-distribution's shape, and thus its percentage points, depends upon degrees of freedom.

16.4 *Where is the Randomness?*

When constructing confidence intervals it is important to keep in mind that the confidence interval is the random quantity while the population parameter is fixed and unchanging. Interpretations of confidence intervals should reflect this distinction. When we say "With 90% confidence, $63.5 \le \mu \le 65.5$", we do *not* mean that "90% of the time μ will fall between 63.5 and 65.5", but rather that in the long run, 90% of the intervals we compute from independently drawn samples will include the true mean.

Of course, in practice we construct a single confidence interval from the sample we have at hand, and the interval can only be said to "probably" include the true population mean. The confidence level specifies our confidence that the interval in fact encloses the true population mean.

16.5 *Multiple Intervals and the Bonferroni Adjustment*

When we construct many confidence intervals, we expect that some of them will fail to cover the true population parameter value, but we don't know which ones. This is true whether the intervals estimate the same parameter or are completely unrelated, as long as the samples have been drawn independently. The *Bonferroni adjustment* lets us construct several intervals without increasing the likelihood of at least one erroneous interval.

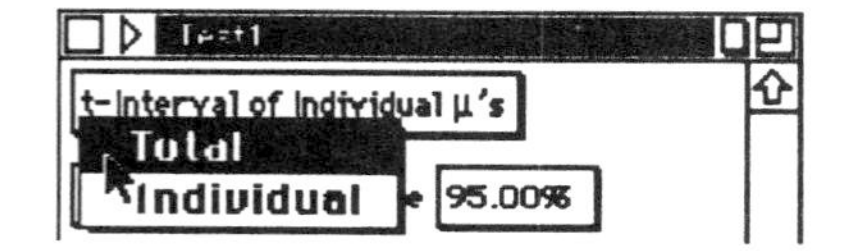

*Figure 16-6. Specify **Total** vs **Individual** confidence level for multiple µ's.*

If several variables have been selected, the Estimate window in Figure 16-6 offers the choice of a **Total** or **Individual** confidence levels. The **Individual** choice performs each test individually at the specified confidence level, and thus increases the chances that some of the intervals are in error. Choosing **Total** applies the Bonferroni adjustment to the individual confidence levels so that the probability of at least one erroneous interval equals the specified confidence level.

The Bonferroni adjustment considers the number of intervals requested and partitions the error probability equally among all the intervals. Thus, a request for 3 intervals at a total 95% confidence level generates each interval at the (100 – 5/3)% confidence level. This makes each interval somewhat wider and thereby reduces the probability that it is in error. Because each of the intervals now has a 5/3% chance of being in error, there is a total of no more than 5/3% + 5/3% + 5/3% = 5% chance of at least one error, as we would want from a 95% confidence level.

The Bonferroni procedure is conservative. That is, it makes slightly more allowance for multiple intervals than is necessary in many situations. However, it applies to a wide variety of situations and it is relatively easy to understand. The general problem of performing several inferential procedures together is called the problem of *multiple comparisons*.

multiple comparisons

16.6 *Testing Hypotheses*

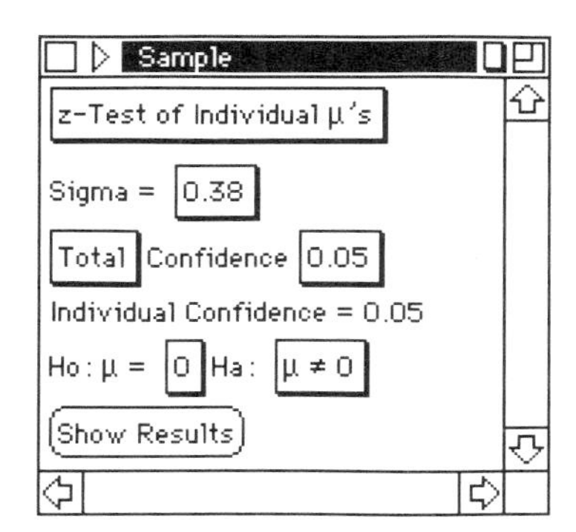

Figure 16-7. *Hypothesis test window.*

Hypothesis tests are closely related to confidence intervals. Hypothesis tests require that you specify four things:

- The test statistic
- The null hypothesis, H_0
- The alternative hypothesis, H_A
- The probability of rejecting a true null hypothesis, usually called the α-level.

The **Test...** command in the **Calc** menu opens a hypothesis test window. Pop-up menus in the hypothesis test window provide convenient ways to specify the components of hypothesis test.

16.7 *Hypothesis Tests for μ When σ is Known*

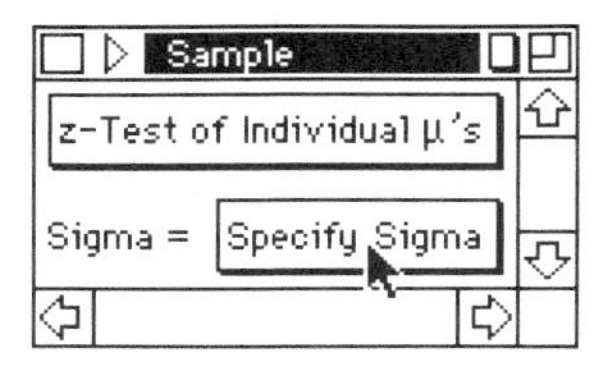

Figure 16-8. *Click the **Specify Sigma** button to type the value for population standard deviation.*

To test the null hypothesis that a population mean, μ, has some specified value, select a variable that contains a sample drawn from the population and choose the **Test...** command from the **Calc** menu. If the population standard deviation, σ, is known, then select **z-Test of Individual** μ**'s** from the topmost pop-up menu holding the different test selections.

Next, click on the **Specify Sigma** pop-up menu, select **Other...**, type the population standard deviation in the window provided and click the OK button. The window will look like Figure 16.9.

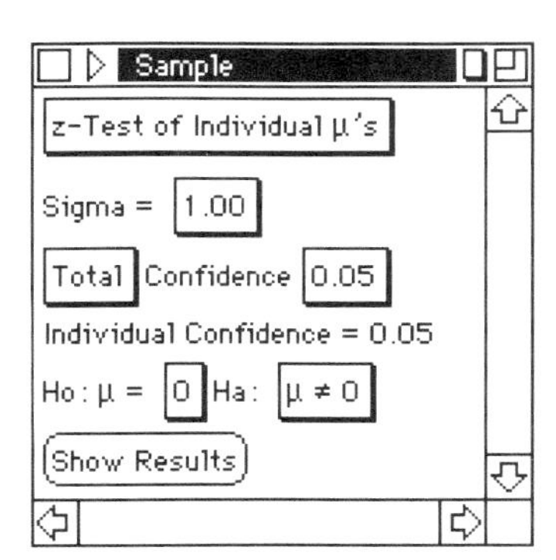

Figure 16-9. *A sample test window for z-*

Specify the appropriate confidence level in the **Confidence** pop-up menu where .05 is offered as the default. An α-level of .05 means that if we were to perform many independent repetitions of the experiment that generated the data and compute hypothesis tests *of a true null hypothesis* for each sample, we would expect that in the long run 5% of the tests would incorrectly reject the null hypothesis. Unlike the confidence level in the Estimate window, the α-level is *not* specified as a percentage. In general, for a given α-level, the corresponding confidence level is $100(1 - \alpha)\%$.

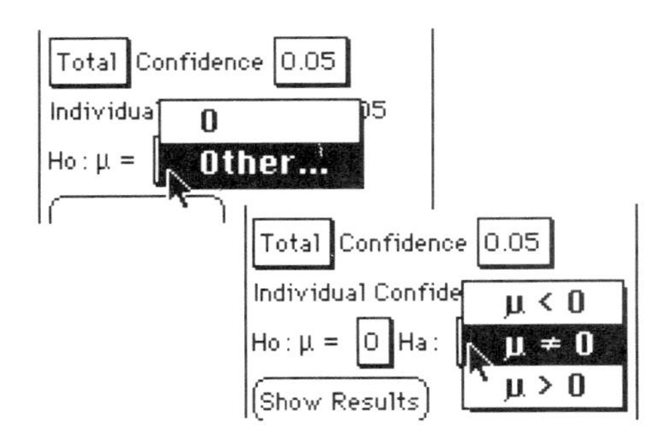

Figure 16-10. *Click the pop-up menu for* $\mathbf{H_0}$ *to specify the null hypothesis. Click the* $\mathbf{H_A}$ *pop-up menu to select the alternative hypothesis.*

Next, specify the null and alternative hypothesis. The null hypothesis is specified in the $\mathbf{H_0}$ pop-up menu. The default is $\mu = 0.0$, but you may specify any value.

The Test window offers three forms of alternative hypotheses. The alternative hypothesis is specified in the $\mathbf{H_A}$ pop-up menu. The default is a two-sided hypothesis test. Choosing a one-sided test improves your chance of rejecting the null hypothesis, but introduces another subjective decision that must be documented and defended.

For example, a one-sided test of whether the average pay for women in a company was lower than the corresponding pay for men might be open to the challenge that the analysis was blind to the possibility that the company discriminated against men.

Data Desk computes the z-test statistic as

$$z = \frac{(\bar{y} - \mu_0)}{\sigma / \sqrt{n}}$$

where $\bar{y}$ is the sample mean, μ_0 is the value of the population mean specified in the null hypothesis, σ is the specified population standard deviation, and n is the number of numeric cases in the variable. The test compares the observed z-test statistic to the percentage point of a Normal distribution that corresponds to the test's chosen α-level.

The pop-up menu to the left of the **Confidence** pop-up menu, holds the options **Total** and **Individual.** These options are used when testing hypotheses for two or more variables at once. See Section 16.9 for details.

16.8 *Hypothesis Test for μ When σ is Unknown*

Usually we do not know the standard deviation of the population and must estimate it from the sample. For this, we use a hypothesis test based on the t-distribution — a "t-test".

To construct a t-test, choose **Test...** command from the **Calc** menu. Data Desk opens a Test window. Select **t-Test of Individual** μ**'s** from the topmost pop-up menu and follow the steps as for the z-test described in Section 16.7.

From the **Confidence** pop-up menu, select the appropriate confidence level or choose **Other** and type the confidence value in the window provided. Click on the $\mathbf{H_0}$ pop-up menu to specify the null hypothesis. Specify the alternative hypothesis in the $\mathbf{H_A}$ pop-up menu. When you click the **Show Results** button the results will be displayed at the bottom of the window.

Data Desk constructs the t-test statistic as

$$t_{(n-1)} = \frac{\bar{y} - \mu_0}{s / \sqrt{n}}$$

where $\bar{y}$ is the sample mean, μ_0 is the value of the population mean specified in the null hypothesis, s is the sample standard deviation, and n is the number of numeric values in the variable. The test compares the observed t-test statistic to the percentage point of a t-distribution that corresponds to the test's chosen α-level. The t-distribution's shape, and thus its percentage points, depends upon the degrees of freedom, $n - 1$, in this case.

16.9 *Multiple Hypothesis Tests and the Bonferroni Adjustment*

The Bonferroni procedure available for multiple confidence intervals applies as well to hypothesis tests (see Section 16.5). It enables you to perform several tests without increasing the likelihood of rejecting at least one true null hypothesis. The *Total* confidence option performs a Bonferroni adjustment so that the probability of at least one erroneous test equals the α-level.

16.10 *Chi-Square Test of Individual Variances: Inference for Spread*

Data Desk also offers hypothesis tests and confidence intervals for the variance. For samples drawn from a Normal population, the statistic

$$(n - 1)\, s^2/\sigma^2$$

has a Chi-Square (χ^2) distribution with $(n - 1)$ degrees of freedom. Data Desk provides a confidence interval for σ^2 based on this fact. In particular, the interval is computed as

$$\frac{(n-1)s^2}{\chi^2_{\alpha/2,(n-1)}} < \sigma^2 < \frac{(n-1)s^2}{\chi^2_{(1-\alpha/2),(n-1)}}$$

where $\chi^2_{\alpha/2,\,(n-1)}$ is the $(100\alpha/2)$ percentage point of the χ^2 distribution with $(n - 1)$ degrees of freedom.

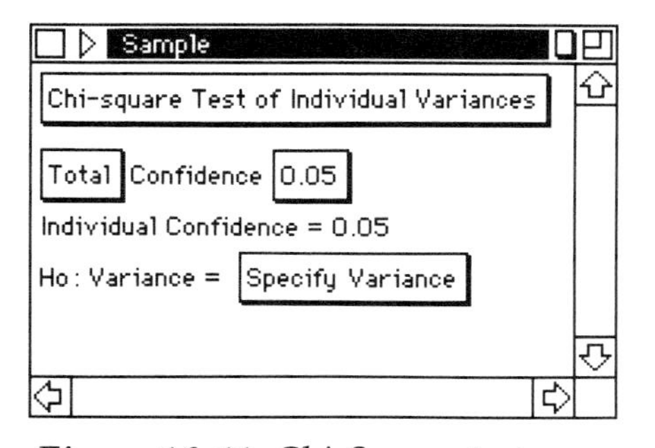

Figure 16-11. *Chi-Square test window.*

To compute the confidence interval select **Chi-square Interval of Individual Variances** from the topmost pop-up menu and specify the confidence level value and the confidence level type **(Total** vs **Individual)** of interval from the appropriate pop-up menus. Specify the bounds to be two-sided, one-sided upper or one-sided lower and click the **Show Results** button. The χ^2 distribution is not symmetric, so confidence intervals for σ^2 generally are longer on the side above the observed s^2 then on the side below s^2.

Data Desk also provides one and two-sided hypothesis tests for σ^2 based on the same distribution. Select **Chi-square Test of Individual Variances** from the topmost pop-up menu of the Test window and specify the confidence level value and the confidence level type from the appropriate pop-up menus. Click the **Specify Variance** button to enter the value for the null hypothesis; Data Desk automatically generates the term for the alternative hypothesis. You can still specify whether the test is to be two-sided, one-sided above the null value, or one-sided below the null value. Click on the **Show Results** button to perform the analysis.

All of the χ^2 -based tests are very sensitive to the distribution of the data. If the data are not Normally distributed or have outliers, then the statistic does not in fact follow a χ^2 distribution and should be used with great caution.

APPENDIX 16A

Example: Simulating Confidence Interval Performance

One way to get a better understanding of confidence intervals is to use Data Desk's ability to generate random values and simulate the collection of many samples from a known population. This appendix illustrates such an experiment.

Suppose that the true heights of the population of all women with soprano voices are Normally distributed with a mean of 64 inches and a standard deviation of 2 inches. Let an experiment consist of drawing a sample of n = 9 women from this population. First, invoke the **Generate Random Numbers...** command in the **Manip** menu. To draw 20 independent samples of size 9 from a Normal population with mean = 64 and standard deviation σ = 2, set the parameters in the Generate Random Numbers dialog.

Generate Random Data OK Cancel

Generate 20 variables with 9 cases/variable

Distribution

○ Uniform

◉ Normal μ= 64 σ= 2

○ Bernoulli trials: Prob(success)=

○ Binomial experiments

* Bernoulli trials / experiment

Prob(success)=

○ Poisson, λ =

Seed = 175805

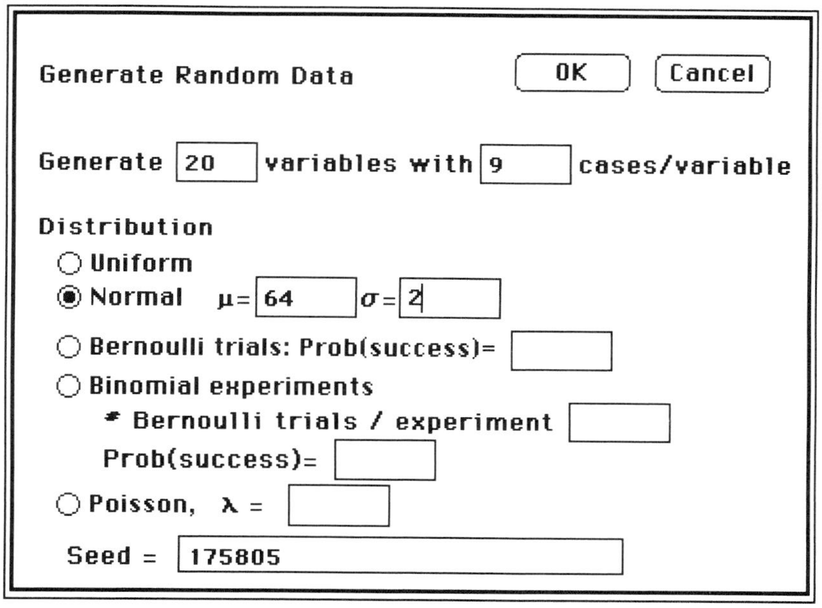

Figure 16-12. *To generate 20 random samples of n=9 from a Normal distribution with μ= 64 and σ = 2, set the parameters in the **Generate Random Numbers...** dialog.*

To construct a confidence interval for each of the 20 samples, select the icon of the *Ran Data* relation, choose the **Estimate...** command from the **Calc** menu, and specify *z-Interval of Individual μ's* and enter the sigma value specified in the Generate Random Number dialog.

For this experiment, we choose **Individual confidence level,** each with a 90% confidence level. We accept the default two-sided interval form. Clicking **Show Results** button performs the calculations and reveals an output record of confidence intervals. Because this is a simulation, you should get different numbers than these when you perform this experiment yourself. (The seed value shown in Figure 16-12 is for illustration; it is not the seed used to generate the example. In fact, you don't need to specify any seed; the Macintosh will do this for you automatically.)

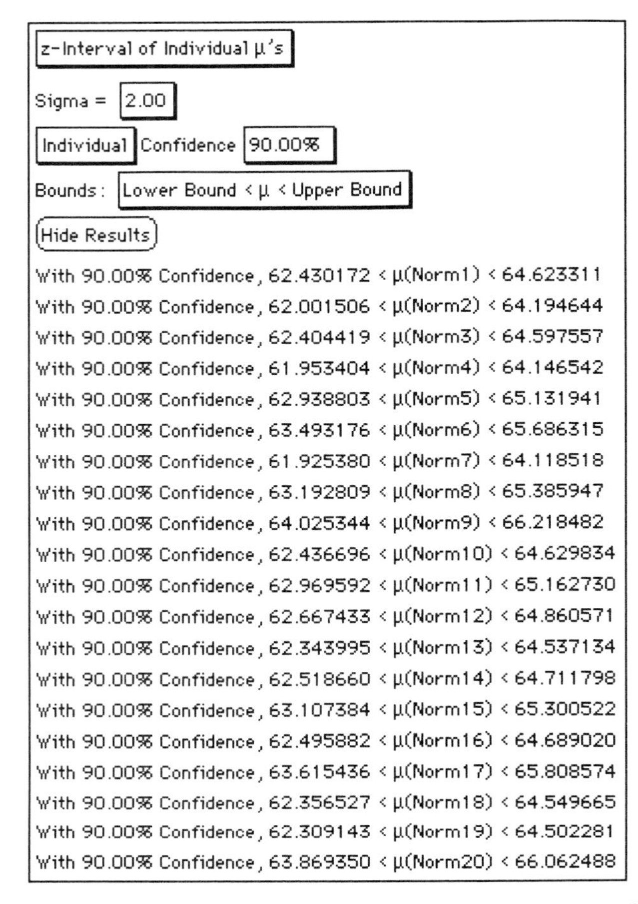

Figure 16-13. *The 20 simulated confidence intervals.*

When we examine these confidence intervals we find that 1 of them fails to include the true mean of 64 inches. Because we computed the intervals at a 90% confidence level, we might have expected 10% of them, or 2, to fail to cover the true mean. This discrepancy illustrates that statistical inference only specifies how statistics will behave in the long run.

APPENDIX 16B

Example: Simulating Hypothesis Test Performance

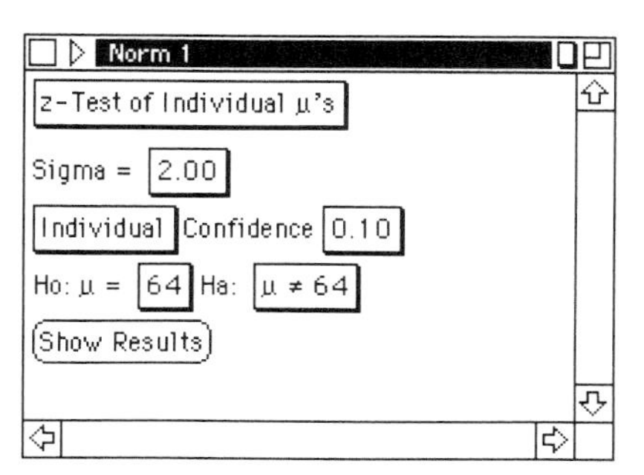

Figure 16-14. *Specifying z-test for 20 random normal samples.*

Suppose, as we did in the simulation of confidence intervals, that the true heights of the population of all women with soprano voices are Normally distributed with a mean of 64 inches and a standard deviation of 2 inches. Let an experiment consist of drawing a sample of $n = 9$ women from this population. For convenience, we use the same samples generated for the confidence interval experiment in Appendix 16A. Alternatively, we could have generated 20 new samples.

To perform a hypothesis test for each of the 20 samples, select the *Ran Data* relation icon, choose the **Test...** command from the **Calc** menu. Specify z-tests and set Sigma = 2.

For this experiment, we choose **Individual** confidence at α-level of .10.

We specify a null hypothesis that $\mu = 64$, and accept the default two-sided alternative hypothesis.

Clicking the **Show Results** button displays output record of hypothesis tests. Because this is a simulation, you should get different numbers if you perform this experiment yourself. There isn't room to show all 20 tests, but the first four we obtained appear in Figure 16-15.

```
Norm1:
Test Ho: μ(Norm1) = 64 vs Ha: μ(Norm1) ≠ 64
Sample Mean = 63.5   z-Statistic = -0.774
Fail to reject Ho at Alpha = 0.10
p<0.0001
Norm2:
Test Ho: μ(Norm2) = 64 vs Ha: μ(Norm2) ≠ 64
Sample Mean = 63.6   z-Statistic = -0.623
Fail to reject Ho at Alpha = 0.10
p<0.0001
Norm3:
Test Ho: μ(Norm3) = 64 vs Ha: μ(Norm3) ≠ 64
Sample Mean = 64.5   z-Statistic = 0.676
Fail to reject Ho at Alpha = 0.10
p≤0.4557
Norm4:
Test Ho: μ(Norm4) = 64 vs Ha: μ(Norm4) ≠ 64
Sample Mean = 64.7   z-Statistic = 0.995
Fail to reject Ho at Alpha = 0.10
p≤2.5741
```

Figure 16-15. *The first four simulated hypothesis tests.*

Because we computed the tests at an α-level of .10, we expect, on average, that 10% of the tests, or 2, will reject the null hypothesis even though it is true.

CHAPTER 17

Comparing Two Samples

17.1	Displaying Differences	*245*
17.2	Comparing Two Means When the Variances Are Assumed Equal	*245*
17.3	Example	*246*
17.4	Formulas	*247*
17.5	Confidence Intervals for Pooled Variance	*247*
17.6	Comparing Two Means When the Variances Are Not Assumed Equal	*247*
17.7	Paired Data	*248*
17.8	Example	*248*
17.9	Comparing Multiple Means When the Variances are Assumed Equal	*249*

MOST DATA ANALYSES deal with relationships. The comparison of two groups is one of the simplest relationships. The simplest comparison is a comparison of their centers. Such comparisons include questions like "Are women paid less than men for the same work?", "Is drug therapy A better than drug therapy B?", and "Will the Democratic or Republican candidate win more votes in the next election?".

Most of the methods in this chapter are strictly appropriate only when the underlying data are Normally distributed. But they can be applied to data for which this assumption is only approximately true. Unlike the multivariate methods in later chapters, these can compare unrelated groups of individuals, and thus can work with variables in different relations.

17.1 *Displaying Differences*

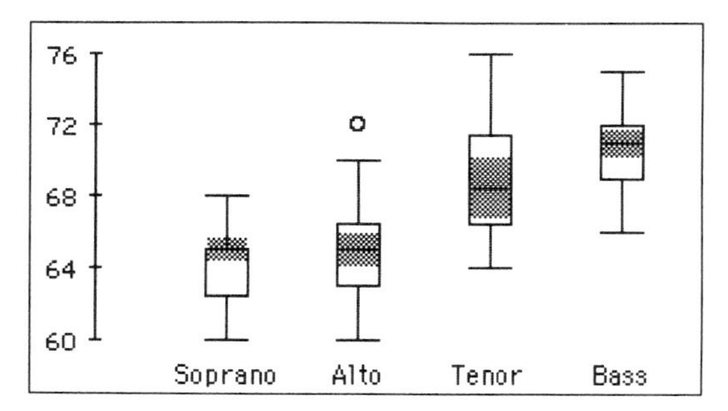

Figure 17-1. *Boxplots comparing the heights of singers by vocal part.*

Boxplots (see Section 8.7) are a useful way to display differences among variables because they focus on the features we most often compare. Boxplots show differences in the level or center in two ways – both as differences among medians (shown with a bar across the box) and as differences in the overall level of the boxes. They also show differences in spread through the size and extent of the boxes and of the whiskers. Finally, they explicitly exhibit outliers, which might affect numerical comparisons in unexpected ways.

The boxplots comparing the heights of singers in the Singers data provide a good example.

17.2 *Comparing Two Means When the Variances Are Assumed Equal*

Often the two groups to be compared are similar in most respects, except that their means might be different. When we are willing to assume that the population variances of the two underlying populations are equal, procedures that use *pooled variance* estimates are appropriate.

The assumption of equal variances, like the assumption of Normally distributed populations, is not tested in the data, but rather is asserted by the data analyst. You should be prepared to justify this assertion from your knowledge of the population separate from the values in the samples at hand.

The pooled t-statistic combines, or pools, the data from both samples to get a single estimate of variance. Because t-statistics usually derive their degrees of freedom from the variance estimate, a pooled t-statistic is based on more degrees of freedom than the corresponding two-sample t-statistic. (See Section 17.6). This makes the corresponding confidence intervals for $\mu_1 - \mu_2$ somewhat smaller and the corresponding tests for the difference between the means more powerful *provided the assumption of equal variances is correct.*

To perform a pooled t-test, select the two variables to compare, choose the **Test...** command from the **Calc** menu and select **pooled t-Test of μ_1 - μ_2.** from the topmost pop-up menu. Specify the *Confidence Level*, H_o, and H_a values in the corresponding pop-up menus and click on the **Show Results** button to see the statistics.

The pop-up menu that reads "Total" is useful only when computing several confidence intervals. Section 16.5 describes the Bonferroni adjustment in the context of simple hypothesis tests. The same adjustment can be made for tests that compare two means.

17.3 *An Example*

A glance at the boxplots in Figure 17-1 confirms that the basses in this sample are taller than the sopranos. But do the basses and tenors differ in height? To investigate this question, we select the two variables and choose the **Test...** command.

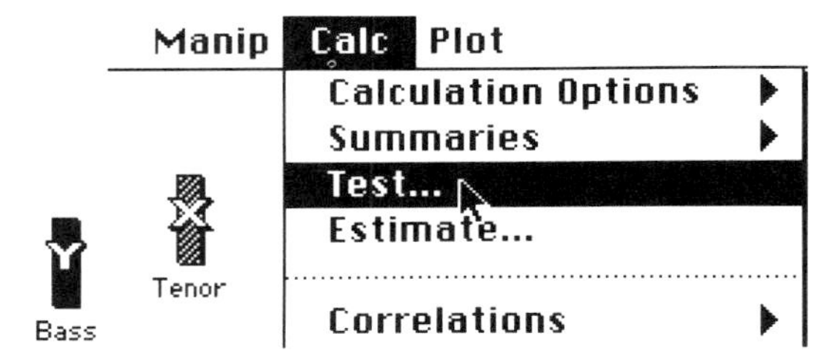

Figure 17-2. *Select Tenor and Bass and choose* ***Test...*** *from the* ***Calc*** *menu.*

(The Singers dataset is unusual. Each vocal part has different singers, so variables reporting the heights of Sopranos, Altos, Tenors, and Basses must each be in their own relation. The dataset is organized so that the different relations are kept in a folder named Relations, but the variables that hold the heights are in a separate folder named Heights. Even though the variables have been placed together in the same folder, they are still in separate relations. You can use them together only for those Data Desk operations that can work across relations.)

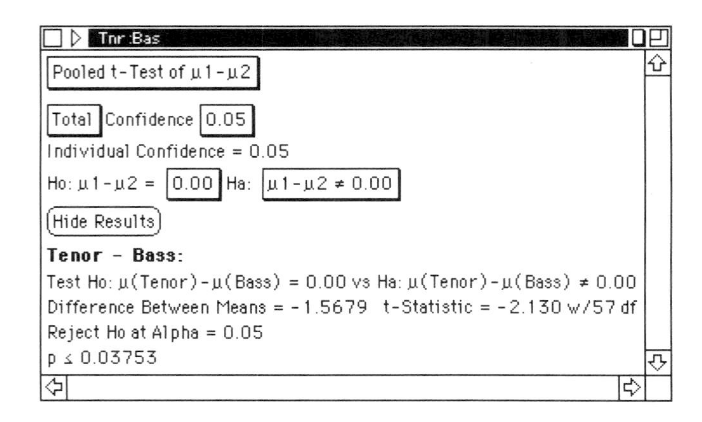

Figure 17-3. *The pooled t-test on heights of N.Y. Choral Society basses and tenors.*

We can assume that the tenors and basses in the N.Y. Choral Society are two independent samples from the population of men in the New York area, and thus that the variances of their heights ought to be equal. We request a pooled estimate of the underlying population variance of height.

The null hypothesis of no difference in height ($\mu_1 - \mu_2 = 0$) is appropriate here, as is the default two-sided alternative ($\mu_1 - \mu_2 \neq 0$). That is, we did not start out believing that either tenors or basses would be taller — a difference in either direction would be equally interesting.

The resulting test, in Figure 17-3, shows that at the .05 α-level, we can indeed discern a difference in the heights of N.Y. Choral Society basses and tenors.

17.4 Formulas

The pooled estimate of variance is computed as

$$s_p^2 = \frac{(n_1 - 1)s_1^2 + (n_2 - 1)s_2^2}{n_1 + n_2 - 2}$$

where n_1 and n_2 are the sample sizes of the two groups, and s_1^2 and s_2^2 are the sample variances of the two groups.

The pooled t-test statistic is

$$t = \frac{(\bar{y}_1 - \bar{y}_2)}{s_p\sqrt{1/n_1 + 1/n_2}}$$

where $\bar{y}_1$ and $\bar{y}_2$ are the sample means, $s_p = \sqrt{s_p^2}$ and n_1 and n_2 are the sample sizes. The t-statistic has $(n_1 + n_2 - 2)$ degrees of freedom.

17.5 Confidence Intervals for Pooled Variance

The **Estimate...** command offers confidence intervals for the difference in the population means, $\mu_1 - \mu_2$. Select **Pooled Interval of μ_1 - μ_2** from the topmost pop-up menu, specify the appropriate parameters in the other pop-up menus and click the **Show Results** button. Section 16.2 explains the parameters choices.

The confidence interval is computed as

$$(\bar{y}_1 - \bar{y}_2) \pm t_{df}^* s_p \sqrt{1/n_1 + 1/n_2}$$

where $\bar{y}_1$ and $\bar{y}_2$ are the sample means, $s_p = \sqrt{s_p^2}$ and n_1 and n_2 are the sample sizes, and t_{df}^* is the appropriate percentage point of the t-distribution with $(n_1 + n_2 - 2)$ degrees of freedom.

17.6 Comparing Two Means When the Variances Are Not Assumed Equal

two-sample t

When we are unwilling to assume that the underlying population variances are equal, but are still willing to believe that the populations are Normally distributed, the two-sample t procedures are often appropriate. The difference between the two-sample t procedures and the pooled t procedures is that the former uses two estimates of variance — one from each sample — and has fewer degrees of freedom associated with the resulting t-statistic.

In the **Test...** or the **Estimate...** windows, select the **2-Sample t-Test of μ_1 - μ_2**. The remaining steps are identical to those described in Sections 17.2 and 17.5.

The two-sample t-interval is computed as

$$(\bar{y}_1 - \bar{y}_2) \pm t_{df}^* \sqrt{s_1^2/n_1 + s_2^2/n_2}$$

where t^*_{df} is the appropriate percentage point from a t-distribution. The two-sample t-statistic is one of the few common statistics for which the calculation of degrees of freedom is complex. Data Desk approximates the degrees of freedom as

$$df = \frac{\left(s_1^2/n_1 + s_2^2/n_2\right)^2}{\frac{\left(s_1^2/n_1\right)^2}{n_1 - 1} + \frac{\left(s_2^2/n_2\right)^2}{n_2 - 1}}$$

The corresponding test statistic is

$$t_{df} = \frac{(\bar{y}_1 - \bar{y}_2)}{\sqrt{s_1^2/n_1 + s_2^2/n_2}}$$

17.7 Paired Data

paired-t

When each case in the first group is paired naturally with the corresponding case in the second group, we can take advantage of the additional structure in the data and compute *paired-t* statistics. Typical pairings are pairs of twins, measurements on the same patient before and after medication, or pairs of judgements made by the same individuals.

Comparisons of paired groups operate on the pairs. Paired t- statistics find the pairwise differences and then construct a t-statistic based upon these differences by treating them as if they were a univariate collection of data values.

When data are naturally paired, paired t-procedures are likely to produce smaller confidence intervals and more powerful tests. If the data are not naturally paired, paired t-procedures are invalid.

Paired variables are typically in the same relation.

17.8 Example

The *Labor Force* dataset contains the labor force participation rate (LFPR) of women in 19 cities in the United States in each of two years. These data might help us examine the growing presence of women in the labor force.

pooled t-Test

```
1968 - 1972:
Test Ho: µ(1968)-µ(1972) = 0.00 vs Ha: µ(1968)-µ(1972) ≠ 0.00
Difference Between Means = -0.033684  t-Statistic = -1.496 w/36 df
Fail to reject Ho at Alpha = 0.05
p ≤ 0.14341
```

Figure 17-4. *Testing for a change in labor force participation rate among women with a pooled t-test.*

We can compute a pooled t-test to see if the LFPR in 1972 was different from the LFPR in 1968. Pooling seems reasonable because the cities in the US did not change much between 1968 and 1972. The hypothesis test shown in Figure 17-4 does not discern any difference in the Labor Force Participation Rate between 1968 and 1972.

However, these data are naturally paired because the measurements were made in the same *cities* for each of the two years. If some cities naturally have a high LFPR and others

```
t-Test, paired samples

1968 - 1972:
Test Ho: μ(1968-1972) = 0.00 vs Ha: μ(1968-1972) ≠ 0.00
Mean of Paired Differences = -0.033684  t-Statistic = -2.458 w/18 df
Reject Ho at Alpha = 0.05
p ≤ 0.02435
```

Figure 17-5. The paired t-test, contrary to the pooled t-test in Figure 17-4 takes advantage of the natural pairing in the data, and rejects the null hypothesis of no change in LFPR among women.

a low LFPR (which would not be surprising), then the variability across all cities might have swamped any changes over the 4-year period we are studying. Moreover, because cities with a high LFPR in 1968 typically have a high LFPR in 1972, we should try to look beyond the city-to-city differences so we can concentrate on the year-to-year differences.

Thus to study changes in the LFPR, it is more appropriate to treat the data as paired, considering the change in each city between 1968 and 1972 and then looking at the collection of all these changes. The paired t-test shown in Figure 17-5 rejects the null hypothesis of no difference.

17.9 Comparing Multiple Means When the Variances Are Assumed Equal

F-Test

There might be a situation where we want to test if the means of more than two groups are equal. If we are willing to assume that the population variances of the underlying populations are equal, we can perform an *F-Test of Multiple μ's*. This test is the same test as a one-way Analysis of Variance (see Chapter 18). The decision to choose either command is based on how the data are entered in Data Desk.

When the values for each group are stored in individual variables, the F-Test of Multiple μ's is most appropriate. When they are stored as values in a single variable with groups named in a second variable, the ANOVA commands are more appropriate.

For example, we can compare the mean heights among the different singing parts in the Singers dataset. There are four parts — Sopranos, Altos, Tenors and Basses — each recorded in a separate variable. An F-Test tests the hypothesis "The mean heights of singers in all four vocal parts are the same".

To perform an F-Test, select the variables whose means you wish to compare and choose **Test...** from the **Calc** menu. Choose **F-Test of Multiple μ's** from the topmost pop-up menu, specify the confidence level and click the **Show Results** button. Data Desk computes an F-ratio and a Prob value.

The F-ratio is the ratio of an estimate of the variance of the means based on comparing the means of the different groups to an alternative estimate based on a pooled estimate of variance. When the null hypothesis is true, both values estimate σ^2, the population variance, so the F-ratio will be near 1.0. The pooled variance estimates σ^2 even when the treatment means differ, but the numerator will grow as the group means vary one from another. When the group means are different, the F-ratio grows.

The Prob value is the probability of observing an F-ratio as large as the one computed or larger, *if the null hypothesis were true.* The null hypothesis of equal treatment means can be rejected when the Prob value is smaller than the α-level you select for the test.

CHAPTER 18

One-way ANOVA

18.1	Example	*253*
18.2	Comparing Several Groups Graphically	*253*
18.3	One-way Analysis of Variance	*253*
18.4	The ANOVA Table	*255*

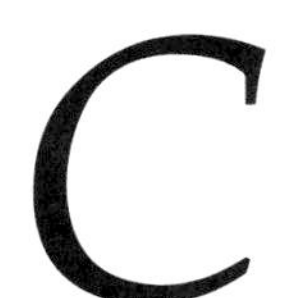

CHAPTER 17 DESCRIBES several methods for comparing the means of two groups. This chapter compares the means of several groups using a technique called analysis of variance, usually abbreviated as ANOVA or AOV.

18.1 *An Example*

Michelson's measurements of the speed of light in air were collected in five runs of 20 trials each. Although the trials were performed on the same equipment, the equipment was constantly being adjusted and tuned, so we might reasonably wonder whether there were differences among the five runs. Data to be analyzed with ANOVA is organized as one variable holding the responses (in this case, measurements of speed), and one or more variables naming categories or groups (in this case, trial numbers).

18.2 *Comparing Several Groups Graphically*

Boxplots, which were discussed first in Chapter 8, compare several groups. While boxplots show medians of the groups rather than means, they still make it easy to visualize the groups and compare their structure. In addition, boxplots reveal outliers, which might affect means and variances of the groups and thereby affect the analysis. The boxplots of the five runs in the Michelson dataset are shown in Figure 18-1.

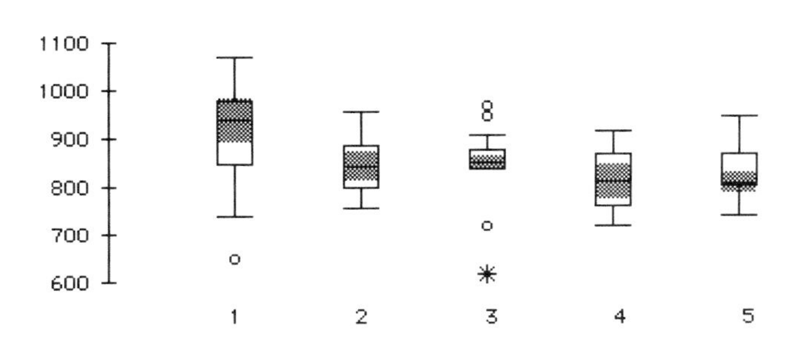

Figure 18-1. *Boxplot by Group of Michelson's measurements of the speed of light.*

It looks from the boxplots as if the runs were similar, but the first and third run show additional within-group variability. Of course, in any experiment we expect some variability in results from trial to trial, so each run should show some within-run variability. The main question here is whether the variability of the means *between* runs is large relative to the variability of measurements *within* runs. If the answer is yes, then we would conclude that there are differences due to the runs themselves.

18.3 *One-way Analysis of Variance*

One-way ANOVA attempts to answer the question, "Do all of the groups have the same mean?" Formally, to perform an ANOVA we are usually supposed to have random samples from Normally distributed populations, and we must be willing to assume that the groups being compared have the same underlying variance even if they have different means. In practice, ANOVA works well even when the Normality assumption is only roughly satisfied. The assumption of equal variance can also be violated, provided the groups are of similar size. When these assumptions are violated, it may be somewhat harder to discern differences among the means, but we are unlikely to falsely declare them different. The assumption of random sampling, however, is quite important.

Since ANOVA assumes that each of the groups being compared has the same underlying variance, it estimates this variability with a generalization of the pooled estimate of variance discussed in Chapter 17. It then computes the means of the groups and compares *their* variance to the pooled variance. The comparison statistic follows an F distribution — larger F-values suggest that the means differ more than we might have expected on the basis of the underlying sample-to-sample variability inherent among measurements within groups. Smaller F-values suggest that the means are not discernably different.

all meas...	Trial#
760	first
810	first
1000	first
1000	first
960	first
960	first
960	second
940	second
960	second
940	second

Figure 18-2. *The variable Trial# names the groups to be compared.*

To perform an ANOVA, Data Desk requires that the values being compared all reside in a single variable. A second variable supplies the category labels. If the measurements for each group are in separate variables, the **Append & Make Group Variable** command in the **Manip** menu will combine them into a single variable and create the required category variable, placing both in a new relation. The categories may be names or numbers.

For example, the Michelson measurements are in a single variable called *All Measurements*. The category variable is called *Trial#* and contains "first" for the 20 cases constituting the first run, "second" for the 20 cases of the second run, and so on.

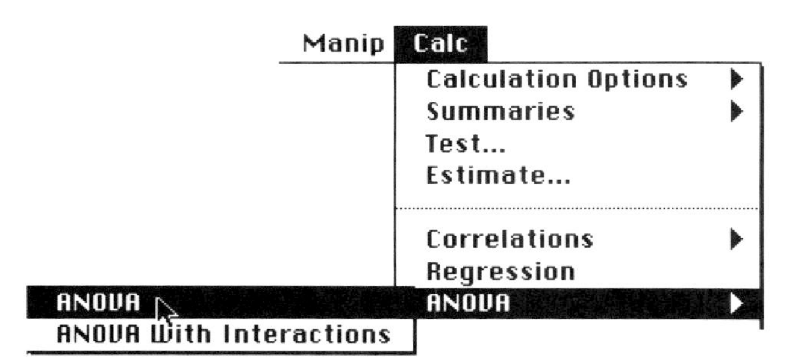

To compute a one-way ANOVA, select the response variable — the measurements whose group means are to be compared — as *y*, and select the category variable as *x*. Choose **ANOVA ▶ANOVA** from the **Calc** menu.

For the Michelson data, select *Speed* as y, then *Trial#* as x, and choose {Calc ▶ ANOVA} **ANOVA**. Analysis of variance output is almost always presented in a standard tabular form. Figure 18-3 shows the ANOVA table for the Michelson data.

Analysis of Variance For **speed**

Source	df	Sum of Squares	Mean Square	F-ratio	Prob
Tr#	4	94514.0	23628.5	4.2878	0.0031
Error	95	523510	5510.63		
Total	99	618024			

Figure 18-3. *The ANOVA table for comparing Michelson's trials.*

The F-statistic tests the null hypothesis $\mu_1 = \mu_2 = \ldots = \mu_p$ against the alternative that at least one mean is not equal to the others. When the null hypothesis is true, the F-statistic has an expected value of 1.0. Larger F values support rejecting the null hypothesis. The Prob value in the final column of the ANOVA table reports the probability of observing an F value at least as large as the one observed if the null hypothesis is true. A small Prob value says that although the observed F-value

The F-statistic tests the null hypothesis $\mu_1 = \mu_2 = \ldots = \mu_p$ against the alternative that at least one mean is not equal to the others. When the null hypothesis is true, the F-statistic has an expected value of 1.0. Larger F values support rejecting the null hypothesis. The Prob value in the final column of the ANOVA table reports the probability of observing an F value at least as large as the one observed if the null hypothesis is true. A small Prob value says that although the observed F-value could be a result of chance fluctuation when the null hypothesis is true, it is very unlikely. In fact, it says exactly how unlikely it is. The Prob value is equivalent to the smallest α-level at which we would reject the null hypothesis of equal means for the observed data.

In this example, the ANOVA confirms our suspicion that the runs did not all have the same mean. The *F*-statistic of 4.29 is sufficiently large to reject the null hypothesis of equal run means at any of the standard α-levels.

18.4 *The ANOVA Table*

degrees of freedom

sums of squares

treatment

Each column of an ANOVA table provides information about a different part of the analysis. The title of the table gives the name of the variable holding the measurements being analyzed — in this case, *Speed.* The Source column labels the major sources of variability under consideration — the differences in means of the groups defined by the category variable *Trial#* and the underlying measurement error. The row labeled Total holds values that refer to the data as a whole, but also are the totals for their respective columns.

The df column hold the *degrees of freedom* associated with each row of the

$$\sum_{i=1}^{\#\,groups} n_i(\bar{x}_i - \bar{\bar{x}})^2$$

table. The first df value is the number of categories or groups minus one. The second value sums the degrees of freedom associated with each group. The Total degrees of freedom is simply *NumNumeric – 1.*

$$\sum_i \sum_j (x_{ij} - \bar{x}_i)^2$$

The Sum of Squares column holds *sums of squares*. For the category variable, the sum of squares is the sum of squared differences of the group means (usually called *treatment* means in this context) from the mean of all the measurements:

$$\sum_i \sum_j (x_{ij} - \bar{\bar{x}})^2$$

Here, n_i is the number of observations in the ith group, x_i is the mean of the ith group, and x is the mean of all the measurements. For the error term, the sum of squares is a pooled sum of individual sums of squared deviations from means found within each group:

mean squares

The column labeled Mean Square contains *mean squares*. They are obtained by dividing the sum of squares values by the corresponding degrees of freedom. The mean square for Error is like the pooled estimate of σ^2 discussed in Chapter 17: it estimates the underlying variability of the data by combining data from different groups. Indeed, if there are only two groups, it is the same statistic.

The F-ratio is the ratio of the treatment mean square to the Error mean square. When the null hypothesis is true, both mean square values estimate σ^2, the population variance, so the F-ratio will tend to be near 1.0. The mean square for Error estimates σ^2 even when the treatment means differ, but the mean squares for treatments will grow as the treatment means vary. Thus, when the treatment means are different, the F-ratio will tend to be larger than 1.0.

The Prob value is the probability of observing an F-ratio as large as the one computed or larger, if the null hypothesis were true. The null hypothesis of equal treatment means can be rejected when the Prob value is smaller than the α-level for the test.

CHAPTER 19

Multi-way ANOVA

19.1	Notation	*259*
19.2	Interaction in ANOVA	*260*
19.3	One Observation per Cell	*261*
19.4	Working with ANOVA Tables	*262*
19.5	Notes on Computing ANOVA	*263*
19.6	ANOVA Options	*264*

treatment

ANALYSIS OF VARIANCE, commonly called ANOVA, is a general framework for analyzing experiments. In one-way ANOVA the groups whose means are compared are usually thought of as different categories of a single *factor* or treatment. Multi-way ANOVA introduces more factors, each specified by its own variable. The factors might affect the response variable both individually and jointly through some interaction.

level

Data Desk's ANOVA command works with any number of factors. Each factor is specified by a variable holding the names of the factor *levels* or categories. Factor categories are determined by the text rather than the numeric value of each case. Thus, for example, the factor levels "1" and "1.0" are different in Data Desk because their text is different even though their values are equal.

Two factors define a table with a row for each level of the first factor and a column for each level of the second. In Data Desk, the table they define may have any number of cases in each cell — that is, the analysis may be *unbalanced*. Cells may even be empty. The **Contingency Tables** command in the **Calc** menu can display a table that shows how many cases represent each factor-by-factor combination.

For example, the Eggs dataset reports the fat content of dried eggs. Two samples of dried eggs, labeled G and H, were drawn from a single well-mixed can of dried whole eggs and sent to six laboratories. At each laboratory, each sample was analyzed by two technicians. The measurements are recorded as %fat minus 41.40. We are interested in differences among the laboratories, which (because they were given equivalent samples) should show no differences in their reported measurements.

Analysis of Variance For **Fat Content**

Source	df	Sum of Squares	Mean Square	F-ratio	Prob
Lab	5	0.443025	0.088605	6.3106	0.0002
Tcn	1	0.004408	0.004408	0.31397	0.5783
Error	41	0.575667	0.014041		
Total	47	1.02310			

Figure 19-1. *The ANOVA table for the Eggs data.*

To compute the ANOVA, first select the response variable, *Fat Content,* as y, then the two factor variables, *Lab* and *Technician* as x (we will ignore *Sample* for now) and choose **ANOVA ▶ ANOVA** from the **Calc** menu. Figure 19-1 shows the ANOVA table.

balanced design

This example has the same number of cases for each combination of levels of the two factors, so it is a *balanced* design.

The F-ratio for Lab is quite large (and its Prob value is correspondingly small), so we can conclude that there is a difference in the mean fat content reported by the different laboratories indicating a difference in laboratories. The evidence of differences in the technicians is insufficient to reject the null hypothesis that they have measured the same values on average for fat content.

19.1 Notation

The algebraic notation that describes general ANOVA calculations is complex. Instead of presenting the most general form, this section presents the notation for two-factor ANOVA. The formulas generalize naturally for more factors.

We denote the response variable by y. The two factor variables are indicated by subscripts i and j, which count through the categories of the two factors, from 1 to I and from 1 to J, respectively. Each case has a response value, and a level name for each factor. If there are K_{ij} observations for the ij cell of the design, they are indexed with the replication subscript, k, which runs from 1 to K_{ij}. Thus, any response value can be denoted y_{ijk}, where i is the level of the first factor, j is the level of the second factor, and k is the replication count.

Factor levels are frequently nonnumeric. The values of i and j are integers assigned in an arbitrary order for notational convenience. Data Desk does not require factor levels to be coded as numbers. The values of the factor variables are read as text and interpreted as factor level names.

We replace a subscript by a dot (•) to indicate that an average has been taken over all values of that subscript. Thus $y_{11\bullet}$ is the average of all of the response values for the first level of the first factor and the first level of the second factor, while $y_{1\bullet\bullet}$ is the average of all of the response values for the first level of the first factor regardless of the level of the second factor. The grand mean, $y_{\bullet\bullet\bullet}$, is the average of all of the response values.

The elements of the ANOVA table are defined with this notation. The sum of squares for the first factor, or treatment, is defined as

$$JK\sum i\ (\ yi_{\bullet\bullet} - y_{\bullet\bullet\bullet}\ )^2$$

and has $(I - 1)$ degrees of freedom. The sum of squares for the second factor is defined similarly as

$$IK\sum_j (\ y_{\bullet j\bullet} - y_{\bullet\bullet\bullet}\ )^2$$

with $(J - 1)$ degrees of freedom. The Error sum of squares is defined as

$$\sum_i\sum_j\sum_k (\ y_{ijk} - y_{ij\bullet}\ )^2$$

with $IJ(K - 1)$ degrees of freedom.

The Total sum of squares is

$$\sum_i\sum_j\sum_k (\ y_{ijk} - y_{\bullet\bullet\bullet}\ )^2$$

with $IJK - 1$ degrees of freedom.

The mean squares are simply the sums of squares divided by their respective degrees of freedom. Each treatment's F-ratio is the ratio of the treatment mean square to the mean square for Error.

19.2 *Interactions in ANOVA*

interaction

In ANOVA, *interaction* refers to that part of the combined effect of two or more factors that is not accounted for by the effects of the individual factors. Interaction assesses whether the response variable, as measured for one of the factors, changes at different levels of the other factors. Data Desk's {ANOVA} **ANOVA With Interactions** command includes all possible two-factor interactions in the ANOVA model.

In the Eggs data, interaction might be present if the mean measured fat content in samples G and H depended on the laboratory. In a 2-way table, interaction can only be isolated when there are at least two observations for some combinations of factor levels.

In the notation terms of Section 19.1, the Interaction sum of squares is defined as

$$K\Sigma_i\Sigma_j (y_{ij\bullet} - y_{i\bullet\bullet} - y_{\bullet j\bullet} + y_{\bullet\bullet\bullet})^2$$

with $(I - 1 \times J - 1)$ degrees of freedom.

When there is no interaction between factors, omitting interaction terms increases the power of the hypothesis tests. However, if you omit interaction terms when there is in fact some interaction, the resulting analysis loses power and the consequent tests and confidence intervals may be invalid.

19.3 *One Observation per Cell*

It is not unusual to have only one observation for each cell of the design. In this case, it is not possible to compute the variability within each cell of the data table, and therefore to assess interaction. Instead, we predict a value for each combination of factor levels from the observed means for those factor levels. (Note that we now need only two subscripts because there are no replications.)

$$y_{ij} = y_{\bullet\bullet} + (y_{i\bullet} - y_{\bullet\bullet}) + (y_{\bullet j} - y_{\bullet\bullet})$$

The Error sum of squares is

$$SS_{error} = \Sigma_i\Sigma_j (y_{ij} - y_i)^2$$

For example, the Hearing dataset reports the percent of a list of 50 words understood correctly by each of 24 subjects when the words were spoken at low volume against a noise background. The word lists are ordinarily used to test hearing aids, and are designed to be equally difficult to perceive when heard with no noise. The experimenter was interested in determining whether the lists were still equally difficult to perceive when noise was present.

We could perform a one-way ANOVA comparing the four lists; however, there is variability in the hearing acuity of the people who were tested that is not relevant to the question being investigated. We might therefore prefer a two-way ANOVA with List and Subject as the two factors: List assesses the variability among the different lists, while Subject isolates the variability inherent among the people who were tested. In this design there is only one observation per cell. The resulting analysis appears in Figure 19-2.

Analysis of Variance For Hearing

Source	df	Sum of Squares	Mean Square	F-ratio	Prob
SbD	23	3231.63	140.505	3.8678	0.0000
LsD	3	920.458	306.819	8.4461	0.0001
Error	69	2506.54	36.3267		
Total	95	6658.62			

Figure 19-2. *The two-way ANOVA for the Hearing data.*

We can see that there is strong evidence that the lists are not equally difficult to perceive when noise is present.

19.4 Working with ANOVA Tables

Data Desk's ANOVA views are interactive, offering associated plots from HyperView popup menus and the ability to alter or correct the data and subsequently update the analysis.

The window's global HyperView offers overall diagnostic information and displays. The HyperView menu attached to the name of the response variable offers typical displays for a variable holding measured values. The HyperView menus attached to the factors offer typical displays for discrete or categorical variables and dotplots of the response variable against each factor. The window's HyperView offers overall diagnostic information and displays.

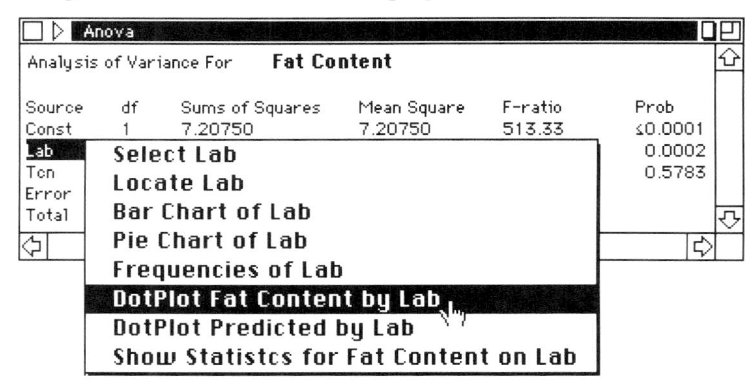

Figure 19-3. The HyperViews attached to each ANOVA factor offer displays that are appropriate for category variables and a dotplot of the response variable against the factor.

A dotplot of the response variable by the factor can be helpful in interpreting the ANOVA table. For example, in the Eggs data, a plot of Fat Content by Lab shows that much of the difference among the labs may be attributable to Lab I, which has a higher mean and is more variable than the others.

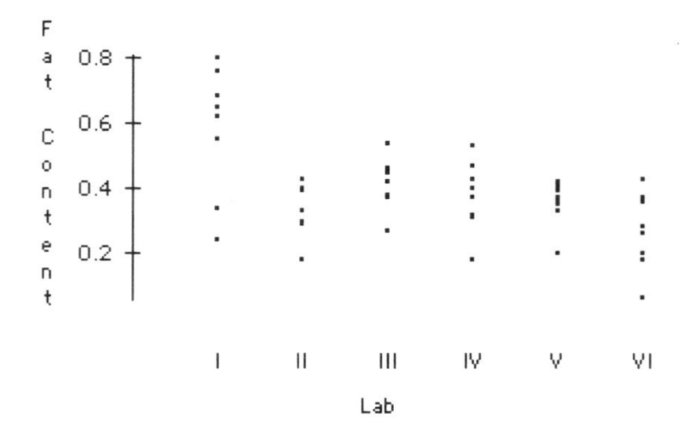

Figure 19-4. A dotplot of Fat Content by Lab. Here we can see that Lab I is different from the others, possibly accounting for much of the difference among the labs.

Data Desk's plot modification abilities can help us look beneath the surface of the ANOVA. For example, in the Eggs data, we can dotplot Fat Content by Lab. (The HyperView attached to the factor name Lab in the ANOVA table makes this easy.) Then select *Technician* and choose {Modify ▶ Symbols ▶ Add} **By Group**. Figure 19-5 shows that most of the difference among the labs is due to the results returned by Lab I, and that the two technicians in Lab I were entirely separate in their reported measurements. It is possible that a single technician in Lab I is responsible for most of the Laboratory effect in this analysis.

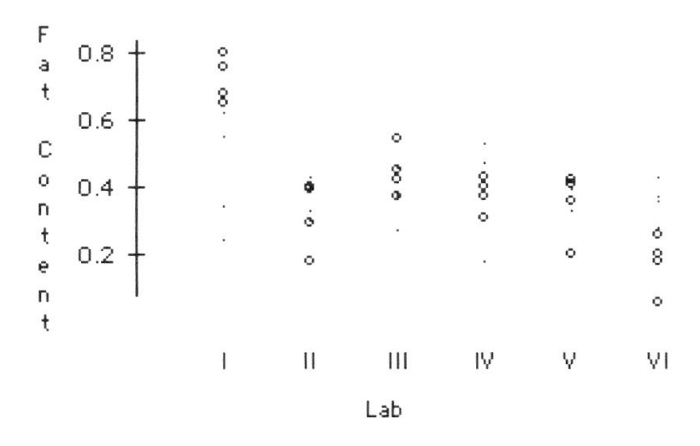

Figure 19-5. *Fat content in eggs by Lab with symbols added by group to show technician. Note that Lab I is different from the others and that the two technicians there are entirely separated in their measurements.*

If you modify any of the data underlying the ANOVA, the ANOVA table will offer to update in place or recompute in a new window. A ❢ symbol appears in the window's HyperView in the upper left of the title bar whenever the underlying variables have been changed.

19.5 Notes on Computing ANOVA

Analysis of Variance calculations can require more computer resources than many other Data Desk calculations. You may want to provide Data Desk with the maximum amount of memory possible to speed the calculations. Interaction terms in multi-way designs are especially demanding in these respects.

It is also possible that an ANOVA involving factors with many levels and using higher-order interactions will generate an intermediate structure that is too large for Data Desk to handle. Data Desk will alert you if this should happen. The only remedy for for this problem is to simplify the design, reduce the number of levels of some factors, omit higher-order interactions or increase the memory allocation to Data Desk.

You can abort a lengthy ANOVA calculation by pressing ⌘-period whenever the rotating cursor is visible.

19.6 ANOVA Options

Data Desk's ANOVA windows offer to compute statistics that can help you understand more about your data. The HyperViews found on both the design view and ANOVA table view of the analysis offer many helpful statistics and plots. This section discusses some of them.

The ANOVA options are closely related to the regression options discussed in Chapters 22 and 23. For most analyses of variance, the first two are the most useful.

PREDICTED VALUES

The ANOVA model predicts a value of the dependent (y) variable for each cell of the design. The prediction is the sum of an overall level plus contributions corresponding to the cell's level on each of the factors, plus contributions corresponding to each of the interaction terms. Data Desk computes the predicted values for each case and places them in a HotResult variable named *Predicted.*

RESIDUALS

The difference between the dependent variable value for each case and the value predicted for its cell by the ANOVA model is the residual for that observation. Data Desk creates a HotResult variable to hold the residuals, named *Residuals.*

DIAGNOSTIC STATISTICS

Other ANOVA options compute diagnostic statistics that can help you determine if you have an extraordinary case in your data, which might deserve special attention. These diagnostic statistics are derived from the corresponding diagnostics for regression analysis, which are discussed in Chapter 23.

LEVERAGE VALUES

Leverage is a measure of the influence of a case on the analysis due to an extraordinary combination of factor variables. Designed experiments rarely have cases with extraordinary leverage; in balanced experiments, all cases have the same leverage. Leverage is thus likely to be interesting primarily for complex unbalanced designs.

EXTERNALLY STUDENTIZED RESIDUALS
INTERNALLY STUDENTIZED RESIDUALS

The studentized residuals divide each residual by an estimate of its standard error. Chapter 23 describes the difference between internal and external studentizing and provides formulas. Histograms, boxplots, dotplots, and Normal probability plots of the studentized residuals can help you identify outlying or extraordinary points. Cases with large residuals tend to reduce the value of the F-statistics in an ANOVA; you may want to consider omitting them from the analysis and dealing with them separately, though you must be careful to note what you have done when reporting your results.

DFFITS
COOK'S DISTANCE

These diagnostic statistics combine information about leverage and residuals. Any case with large DFFits or Cook's distance deserves special attention. You may wish to remove it from the ANOVA and deal with it separately.

CHAPTER 20

Simple Regression

20.1	Coefficients in Regression	*269*
20.2	Least Squares Regression	*269*
20.3	Predicted Values and Residuals	*270*
20.4	Performing a Regression: An Example	*270*
20.5	Inference for Regression Coefficients	*271*
20.6	The ANOVA Table for Regression	*273*
20.7	R^2 and Adjusted R^2	*273*
20.8	Examining Residuals	*274*
20.9	Checking Assumptions	*275*

SCATTERPLOTS LET US EXAMINE the relationship between two variables, a *response*, y, and a *predictor*, x. A more formal description of the way in which y is related to x — especially when we have some reason to believe that the value of y might depend on the value of x — comes from estimating an equation relating y and x. The simplest useful equation to try is a straight line, and the method most often used to find a line to summarize the x-y relationship is called *regression*.

20.1 Coefficients in Regression

All of the points on any straight line fit an equation of the form

$$y = a + bx$$

for the appropriate choice of a and b. Once we have a and b, every pair of numbers (x, y) for which the relationship $y = a + bx$ is true will lie on the same straight line when plotted.

The numbers represented by a and b are called the *coefficients* of the regression equation. The coefficient a is called the *constant coefficient* because it is a base constant added to the value found as bx. It is sometimes called the *intercept coefficient* because it specifies the value of y when $x = 0$, where the line intercepts the y-axis. The coefficient b is called the *slope coefficient* because it specifies the steepness and sign of the relationship between y and x.

intercept coefficient

slope coefficient

The slope coefficient is best interpreted by noting the measurement units of y and x. The slope is measured in *y-units per x-unit*. Thus, for example, a regression of salary (in dollars) on experience (in years) has a slope estimated in dollars per year.

The intercept coefficient is measured in the same units as y, but may not be readily interpretable. Often the zero value on the x-axis is not meaningful, and thus cannot serve as a basis for interpreting a. For example, the intercept of a regression of house price on number of rooms would, under a naive interpretation, represent the value of a house with no rooms. To avoid such nonsense, we often interpret the intercept simply as a base constant for the equation.

20.2 Least Squares Regression

The plotted points (x, y) on most real-data scatterplots do not ordinarily lie on a perfect straight line, so the best we can do is to find a line that is close to the plotted points. The most common regression technique of closeness is *least squares regression*. We write the equation of the least squares regression as

least squares regression

$$\hat{y} = a + bx$$

where $\hat{y}$ denotes the predicted value of y for a specific value, x.

In least squares regression, the coefficients are determined by the condition that the sum of squared differences between each observed value, y_i, and its corresponding predicted value, $\hat{y}_i$ be minimized — the *least*

least squares criterion

squares criterion. This criterion, and its associated analysis, are so common that the technique is almost always referred to simply as *regression*, and that is how Data Desk refers to it as well.

The least squares criterion uniquely determines the values of a and b, and provides useful related statistics and plots. However, it does produce an analysis that can be greatly influenced by extreme data values, so it is a good idea to look at plots of the data as well as the summary table of the regression coefficients.

20.3 Predicted Values and Residuals

predicted values

Regression analysis estimates an equation that predicts a y value for any x-value. *Predicted values* are computed simply by substituting a value for x in the equation. The predicted values are usually denoted $\hat{y}$, to differentiate them from the observed y values. The regression options in Data Desk include saving a HotResult variable containing the predicted values for the regression. If a case has a missing value for x Data Desk reports a missing value for its predicted value. However, if a case is missing in y but not in x, Data Desk can still compute a predicted value for it, so you can compute predictions at x-values for which you have no y-value observation.

residuals

The differences between the predicted values and the observed values are known as *residuals*, and are denoted e_i. (The e is for the "error" the regression equation makes in describing each observed y-value.) There is one residual for each case:

$$e_i = y_i - \hat{y}_i$$

In a successful regression analysis the residuals are small relative to the original y's. Regression options in Data Desk include creating a variable holding the residuals for the regression. It is often useful to plot them in a scatterplot against the predicted values or against the x-variable. The regression summary table's global HyperView offers a scatterplot of residuals *vs* predicted values (y's). The HyperViews associated with the predictor's name offer the scatterplot of residuals (x's) *vs* the predictor.

20.4 Performing a Regression: An Example

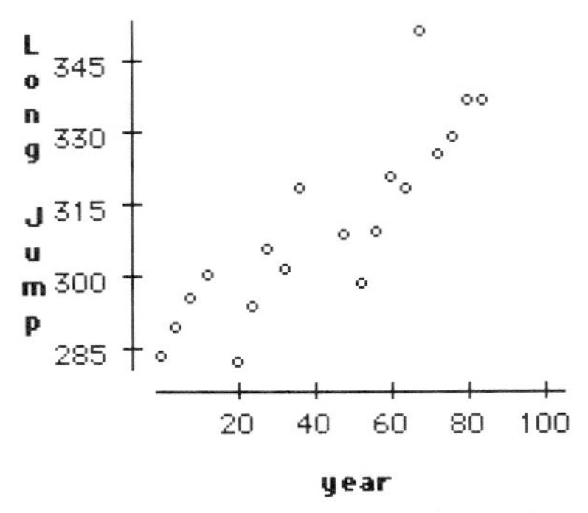

Figure 20-1. *Olympic long jump gold medal performance shows a strong trend over the years of the modern Olympic series.*

The Olympic Gold dataset contains the gold medal performance in the long jump for the modern Olympic games from 1900 to 1984. It is easy to see from the scatterplot in Figure 20-1 that long jump performances have been improving over time.

But exactly how fast have they been improving? How long a jump might we predict will be required to take the gold medal at the next Olympic games?

To perform the regression, select the response (*Long Jump*) as y, select the predictor (*Year*) as x, and choose **Regression** from the **Calc** menu.

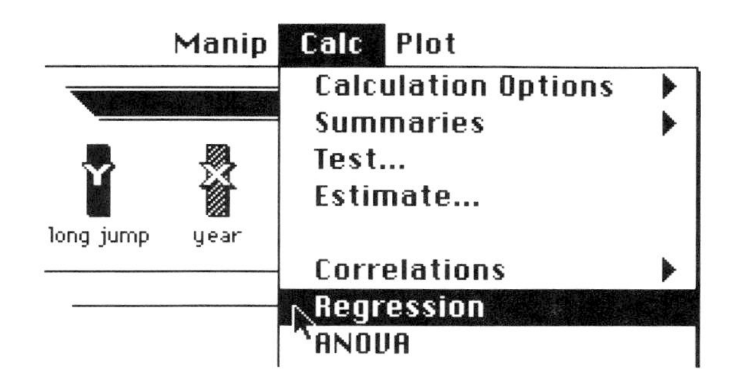

Figure 20-2. *Select the y-variable and the x-variable, then choose **Regression** from the **Calc** menu.*

Figure 20-3 shows the regression results table. Note that *Year* contains year since 1900.

Dependent variable is: long jump
20 total cases of which 1 are missing
R squared = 75.7% R squared (adjusted) = 74.3%
s = 9.757 with 19 - 2 = 17 degrees of freedom

Source	Sum of Squares	df	Mean Square	F-ratio
Regression	5054.89	1	5054.89	53.1
Residual	1618.33	17	95.1960	

Variable	Coefficient	s.e. of Coeff	t-ratio	prob
Constant	283.454	4.281	66.2	≤0.0001
year	0.613084	0.0841	7.29	≤0.0001

Figure 20-3. *The regression summary table for the regression of Long Jump on Year.*

The numbers in the column labeled "Coefficient" in the bottom part of the table are the values of a and b in the equation of the line. The constant coefficient, a, is about 283.5 and the slope coefficient, b, is about 0.61. Thus we write the regression equation as

$$\hat{y} = 283.5 + 0.61\,x$$

Or, in words and numbers,

predicted Long Jump distance = *283.5* + *0.61(years since 1900)*

Long jumps are measured in inches, so the equation says that in 1900 (when $x = 0$) the long jump distance was about 283.5 inches. (In fact, it was 282.875 inches, but the line is an estimated fit for all the data.) The slope coefficient says that long jump gold medal performance has increased by about 0.61 *inches per year.*

20.5 *Inference for Regression Coefficients*

The next column in the regression summary table, labeled "s.e. of Coeff" holds the standard deviation of the sampling distribution — the standard error — of each coefficient. Imagine drawing a multitude of random samples from the same underlying population, computing regressions for each sample, and collecting the coefficients. The standard error of the coefficient estimates the standard deviation of these collected coefficient estimates.

The statistical theory behind the estimate of the standard errors of the coefficients requires that the following things be at least approximately true:

- There is a true linear relationship between y and x in the underlying population. The equation of this line is usually written with Greek letters for the coefficients to indicate that they are the true population parameters:

  $$y = \alpha + \beta x + \varepsilon$$

  This assumption does not assert that there is a perfect linear relationship, but only that there is a true underlying relationship with errors (denoted ε) in the measurement of y. The assumption that the underlying relationship between y and x is linear should be at least approximately true.

- The true residuals, ε, are mutually independent. This assumption is sometimes violated by data measured sequentially over time, where a residual is likely to resemble the immediately previous or immediately subsequent residual.

- The true residuals have the same variance, σ^2_ε, for all values of x. Equivalently, the population variance of y is the same for all values of x. The square root of this variance is called the standard deviation of y about the regression line, and its estimate is denoted s on the second line of the regression summary table.

  This assumption is often the most difficult to satisfy. It is common, for example, for the variance of data values to increase as the value measured increases. If y grows with x, then the variance of the values may grow as well. Note also that this is a property of the true residuals, ε, rather than the observed residuals, e.

- The true residuals follow a Normal distribution with mean zero. This assumption is only required if we want to use statistical inference to interpret the t-ratios in terms of the t distribution.

If all of these assumptions are satisfied, then the ratio of the difference between a coefficient and a contemplated value (usually zero) to its standard error is a t-statistic. This fact can be used to construct confidence intervals and hypothesis tests for the true population value of each coefficient. These ratios for contemplated value zero are given in the right column of the regression summary table, labeled "t-ratio". The HyperViews attached to the t-ratios offer normal probability plots to look for extreme deviations from the normality assumption.

While the t-ratios can be used to test that the true coefficients have any specified value, the most common null hypothesis states that the true slope coefficient value is zero. This hypothesis is roughly equivalent to claiming that knowledge of x is not useful in linearly predicting y.

To find the value against which to compare the t-ratio, we need to select an α-level and to know the appropriate degrees of freedom to use. The regression summary table reports the degrees of freedom in the second line, at the top of the table. If we test the hypothesis $\beta = 0$ *vs* the alternative $\beta \neq 0$ at an α-level of 0.05 for the Long Jump data, we reject the null hypothesis if t is greater than 2.110. The observed t-value is 7.29, so we can reject the null hypothesis. The corresponding confidence interval,

$$0.61 \pm (2.11 \times 0.0841) = 0.433 \text{ to } 0.788$$

shows how precisely (or imprecisely) we have expressed the slope.

20.6 The ANOVA Table for Regression

The middle part of the regression output table is an Analysis of Variance table, much like the tables discussed in Chapters 18 and 19. The sums of squares here are defined as:

$$SS_{Regression} = \sum (\hat{y}_i - \bar{y}_i)$$

$$SS_{Residual} = \sum (y_i - \hat{y}_i)$$

$$SS_{Total} = \sum (y_i - \bar{y}_i)$$

As we might expect, $SS_{Total} = SS_{Regression} + SS_{Residual}$.

The mean square values are the sums of squares divided by their respective degrees of freedom. The F-ratio is the ratio of the mean square for regression to the mean square residual.

The mean square for regression is related to the steepness of the regression line. The mean square residual estimates σ_ε^2, the variance of the distribution of the residuals. (Its square root is the value of s given in the second line of the regression output table. The F-ratio is suitable for testing the hypothesis: "the true slope coefficient is zero". This is the same as the hypothesis tested by the t-ratio for the slope — in fact, the t-ratio for a regression with one predictor is the square root of the F-ratio. For regressions with more predictors (as described in Chapter 22), the F-ratio and t-ratio test different things.

20.7 R^2 and Adjusted R^2

The R^2 statistic, sometimes called the *coefficient of determination*, is an overall measure of the success of the regression in predicting y from x. The equation for R^2 can be written

R^2

$$R^2 = (SS_{Total} - SS_{Residual})/(SS_{Total})$$

R^2 measures the *fraction of the variability of y accounted for by its least squares linear regression on x*. Several of the words in this definition are important:

- R^2 is a *fraction* between zero and one. An R^2 value of zero indicates that y is not linearly predicted in any useful way by x. An R^2 value of one indicates that y is linearly predicted perfectly by x, so that each residual is zero.
- R^2 measures how successfully x *accounts for* or describes y.
- R^2 is most appropriate for *least squares linear regression*. It is usually not appropriate for other fitting criteria or for other functions relating y and x.

correlation coefficient

The square root of the R^2 statistic is the *correlation coefficient* for y and x. Chapter 21 discusses correlation and its relationship to regression.

adjusted R^2

The adjusted R^2 statistic is defined as

$$R^2(adjusted) = (MS_{Total} - MS_{Residual})/(MS_{Total})$$

where the MS, or *mean squares*, are

$$MS_{Residual} = SS_{Residual} / df_{Residual}$$

and

$$MS_{Total} = SS_{Total} / df_{Total}$$

It is most often used in multiple regression(see Chapter 22), where there are several predictors. In that case it helps to account for the number of predictors in the equation.

20.8 Examining Residuals

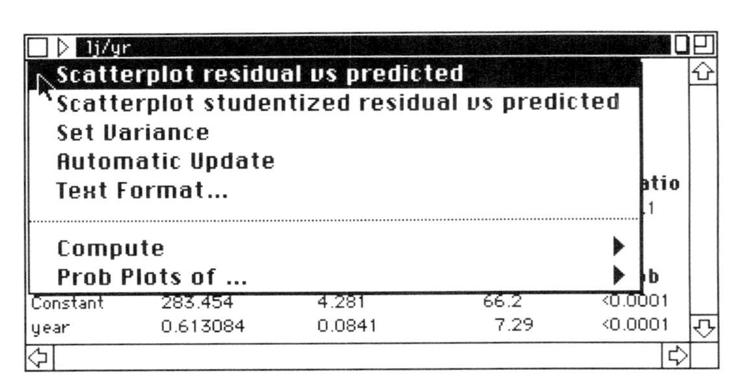

Figure 20-4. *The regression summary table's global HyperView offers to make a scatterplot of residuals vs predicted values. The arrow at the side of the popup menu indicates a submenu holding additional commands.*

A regression equation rarely accounts for all of the variability in the data, so you should examine the residuals — that part of y not accounted for by the regression. Removing the regression line from the x-y relationship often exposes less prominent patterns in the residuals. A scatterplot of the residuals against the predicted values is almost always useful, and the easiest way to make one is with the global HyperView menu on the regression summary table.

The scatterplot of residuals *vs* predicted values for the Long Jump data is shown in Figure 20-5.

Here, the largest residual (roughly equal to 25) stands out. It corresponds to Bob Beamon's record long jump performance at the 1968 Mexico City Olympics. You can confirm this by opening the *year* variable and clicking the extreme point with the **?** tool. This jump distance is considered one of the most remarkable sports achievements on record.

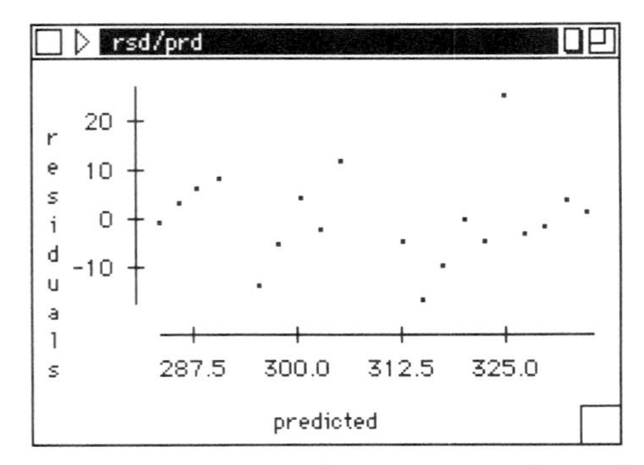

Figure 20-5. *The scatterplot of residuals vs predicted values for long jumps shows an extreme value.*

It is often useful to plot the residuals against the x-variable to see the part of the relationship between y and x not accounted for by the regression line. Once again, the regression summary table's HyperViews offer the plot directly. Alternatively, we could select the icon of Residuals and the icon of Year and choose **Scatterplot** from the **Plot** menu. In this example, it also helps to rescale the plot to emphasize the time sequence.

Here we can see a second pattern. The 1916, 1940, and 1944 Olympics were cancelled because of world wars. In the periods before World War I, between the two world wars, and after World War II, the residuals show an increasing trend. It seems that each world war set back long jump performances, perhaps by interrupting training and by killing many fine athletes. Between the wars, long jump performance advanced more rapidly.

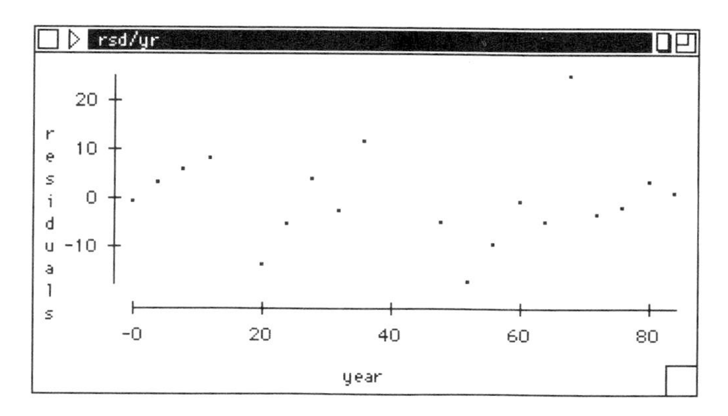

Figure 20-6. The scatterplot of residual vs year shows three distinct time periods separated by the two world wars.

It is not at all unusual for the residuals to reveal something new about the data. This example shows two of the kinds of things we might find; an outlying point and a pattern that was hidden in the overall trend. If we planned to use the regression equation for predicting future long jump performance, we would do better to take Beamon's extraordinary jump and the world wars into account.

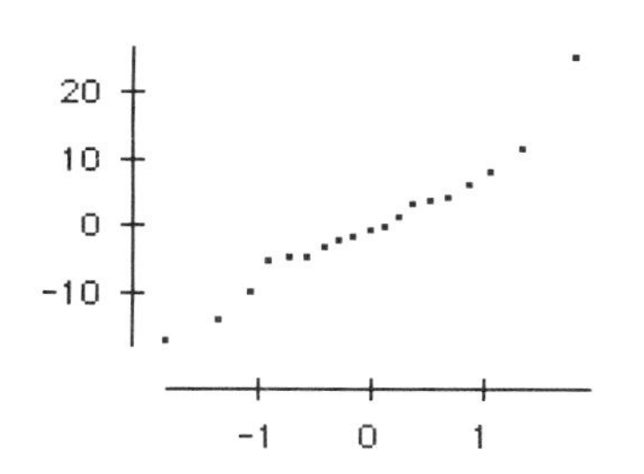

Figure 20-7. *A Normal probability plot of regression residuals for the Long Jump data shows that the largest residual is larger than would be expected if the residuals were truly Normal.*

20.9 Checking Assumptions

The scatterplot of residuals against predicted values is often an effective way to check whether the residuals seem to have constant variance as required by the regression assumptions. A Normal probability plot or histogram of the residuals can be used to check on the assumption that they are Normally distributed.

For example, the Normal probability plot of the residuals in Figure 20-7, shows a distribution that seems reasonably Normal except for Beamon's jump in 1968. See Chapter 8 for a discussion of Normal Probability Plots.

CHAPTER 21

Correlation

21.1	Pearson Product-Moment Correlation	*279*
21.2	Linear Association	*280*
21.3	Correlation and Regression	*280*
21.4	Correlation and Standard Scores	*281*
21.5	Correlation Tables	*281*
21.6	Spearman Rank Correlation (Rho)	*282*
21.7	Kendall's Tau	*283*
21.8	Covariance	*284*
21.9	Missing Values and Correlation	*284*
21.10	Extracting Values from Correlation Tables	*284*

CORRELATION COEFFICIENTS MEASURE the degree of association between two variables. Data Desk computes three kinds of correlation: Pearson product-moment correlation, Spearman rank correlation, and Kendall's tau. Each measures association differently and emphasizes different aspects of the relationship between variables. Data Desk also computes covariance, another measure of association between variables. All of these statistics are found in the **Correlations** submenu of the **Calc** menu.

The Pearson product-moment correlation is the statistic commonly called correlation. It is a multifaceted statistic that shows up in a variety of apparently unrelated places.

21.1 *Pearson Product-Moment Correlation*

The product-moment correlation of the variables x and y is

$$r = \frac{(x_i - \bar{x})(y_i - \bar{y})}{\sqrt{(x_i - \bar{x})^2 (y_i - \bar{y})^2}}$$

To compute the product-moment correlation between two variables, select both variables and choose **Pearson Product-Moment** from the **Correlations** submenu of the **Calc** menu.

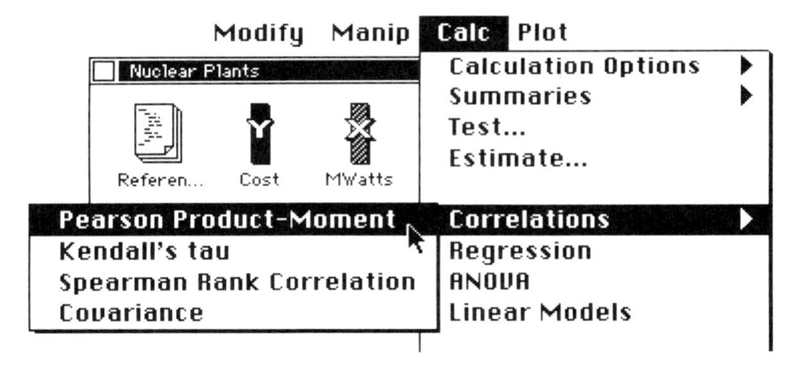

Figure 21-1. *To compute Pearson correlation, select two or more variables and choose* ***Pearson Product-Moment*** *from the* ***Correlations*** *submenu. The submenu may drop down to the right of the menu bar on larger screens.*

Data Desk creates a table of correlations. The correlation coefficient in the row for MWatts and the column for Cost is the correlation of those two variables.

Pearson Product-Moment Correlation

	Cost	MWatts
Cost	1.000	
MWatts	0.472	1.000

Figure 21-2. *The Pearson correlation of Cost and MWatts for the nuclear plants.*

Because the correlation of any variable with itself is always one, the diagonal entries of the table are always 1.000. Here the correlation of Cost and MWatts is 0.472.

21.2 *Linear Association*

Correlation measures *linear* association. Variables can be closely related by some non-linear function and still have a small or even a zero correlation. The correlation coefficient can have any value between –1 and +1. Negative values indicate a negative slope in the x-y scatterplot (that is, y decreases as x increases). Positive values indicate that x and y increase or decrease together. A value near zero means that any relationship between x and y is non-linear. A value near +1 or -1 means that there is a nearly perfect linear relationship between x and y.

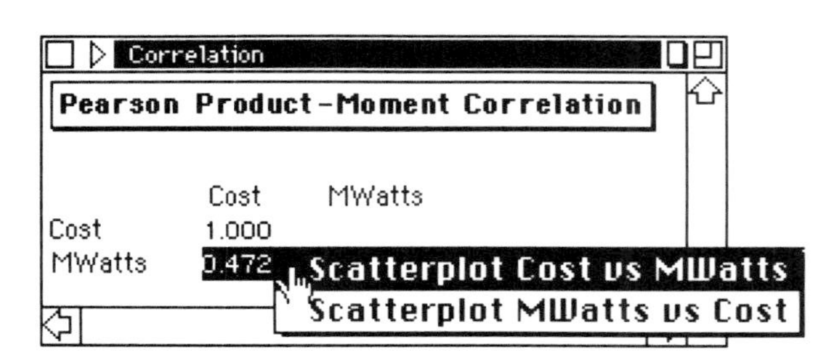

Figure 21-3. *The HyperView attached to each correlation coefficient offers the corresponding scatterplots. Correlation is symmetric, so either scatterplot is appropriate.*

Because product-moment correlation only measures linear association, it is possible for two variables to be closely related but have a small or even a zero correlation. For example, if $y = x^2$ for values of x equally spaced between –1 and 1, there is clearly a relationship between y and x, but their correlation is zero. Similarly, a single extraordinary datapoint can make any correlation either large or small (depending on its location). It is always a good idea to look at a scatterplot of y *vs* x before placing too much trust in a correlation value. The HyperView menus attached to correlation coefficients in Data Desk offer to draw the associated scatterplots.

21.3 *Correlation and Simple Regression*

Product-moment correlation and least squares regression are closely related. The R^2 statistic of the regression of y on x is the square of the correlation of y and x. This suggests another way to interpret the correlation coefficient: the squared correlation coefficient is the fraction of the variability of y accounted for by its least squares linear regression on x.

While regression treats y and x differently, correlation treats them symmetrically. That is, while the regression of x on y is different than the regression of y on x, the correlation of x with y is the same as the correlation of y with x. Thus, the square of the correlation between y and x is also the fraction of the variability of x that would be accounted for by its least squares linear regression on y.

The correlation of y and x is also the correlation of the regression's predicted values, $\hat{y}$, and the observed values, y. Thus, correlation measures the success of a regression. In this sense, behind every correlation coefficient is a regression analysis that might provide additional information about the relationship between y and x. Whenever a correlation is of particular importance, it is a good idea to make the corresponding scatterplot to check for linearity and to perform the corresponding regression. HyperViews help you follow this path easily.

21.4 Correlation and Standard Scores

Regression slopes are in *y*-units *per* *x*-unit, but correlation has no units. Sometimes this can be an advantage if the original units of the data are meaningless or if there is some reason to hide them (such as a promise not to report incomes or ages.)

For example, in a study relating income and age, a researcher might promise respondents that both numbers would be kept confidential. To establish alternative units to work with, she might standardize the values:

income/StDev(income)

age/StDev(age)

The resulting variables preserve information about the overall relationship of income and age, but report each individual in terms of standard deviation units above or below the mean. As long as the standard deviation is not reported, the original values are hidden.

The regression slope of *standardized income* on *standardized age* is the same as the correlation between the original income and age variables, so we can avoid standardizing and just use the correlation coefficient.

Returning to the nuclear plants data, we construct the derived variables

Cost/ StDev (Cost)

MWatts/ StDev (MWatts).

The regression of one on the other has a regression coefficient of 0.472, which is the correlation between Cost and MWatts.

Dependent variable is: StdCost
R squared = 22.3% R squared (adjusted) = 19.7%
s = 0.8963 with 32 - 2 = 30 degrees of freedom

Source	Sum of Squares	df	Mean Square	F-ratio
Regression	6.89931	1	6.89931	8.59
Residual	24.1007	30	0.803356	

Variable	Coefficient	s.e. of Coeff	t-ratio	prob
Constant	0.656832	0.7193	0.913	0.3685
StdMW	0.471761	0.1610	2.93	0.0064

Figure 21-4. *The regression on standardized scores has a slope coefficient equal to the correlation.*

21.5 Correlation Tables

To find all of the pairwise correlations among a collection of variables, select the variables and choose **Pearson Product-Moment**. Data Desk computes the correlations between each pair of variables and places them in a table. Because correlation is symmetric (the correlation of x with y is the same as the correlation of y with x), the numbers in this table are symmetric around the diagonal from upper left to lower right, so Data Desk so Data Desk only prints the lower half of the table. Because the correlation of any variable with itself is always one, the diagonal elements of the table are 1.000.

For example, a correlation table for several variables from the Cars dataset is shown in Figure 21-5:

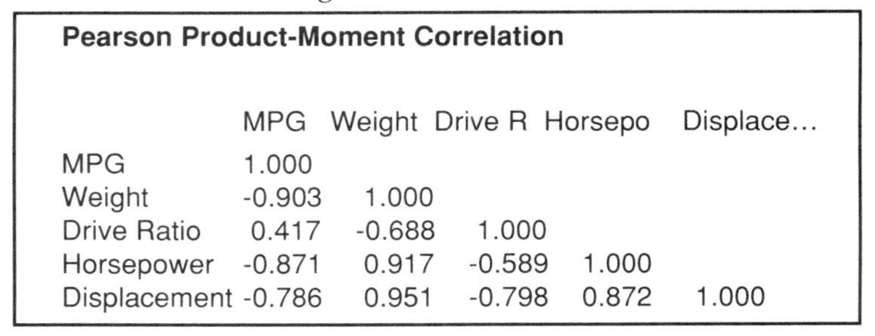

Pearson Product-Moment Correlation

	MPG	Weight	Drive R	Horsepo	Displace...
MPG	1.000				
Weight	-0.903	1.000			
Drive Ratio	0.417	-0.688	1.000		
Horsepower	-0.871	0.917	-0.589	1.000	
Displacement	-0.786	0.951	-0.798	0.872	1.000

Figure 21-5. *Correlation table for the Cars data.*

TIP

You can add variables to a correlation table simply by dragging their icons into the table. The correlation table recomputes automatically.

The correlation between any two variables is at the intersection of the row and column labeled with their names.

21.6 Spearman Rank Correlation (Rho)

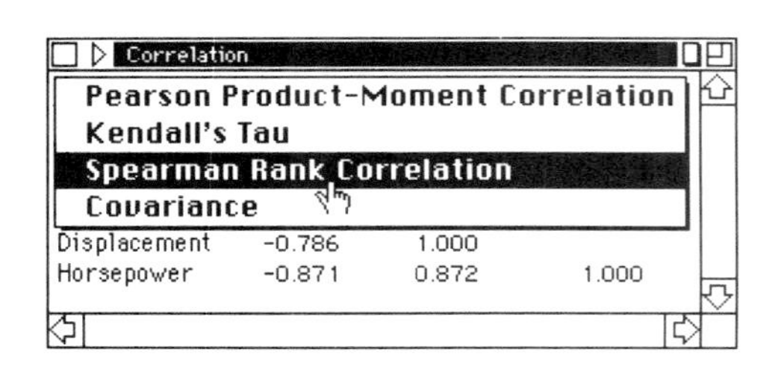

When x and y are not linearly related, but show a consistently increasing or decreasing trend, a nonparametric correlation such as *Spearman's rho* is appropriate. Spearman's rho is simply the correlation between the *ranks* of the two variables. To compute rho in Data Desk, choose **Spearman Rank Correlation** from the **Correlations** submenu in the **Calc** menu. Alternatively, click on the correlation table title to pop up a menu offering each of the correlation computations.

Spearman's rho

Because Spearman's rho is based upon ranks it has two special properties.

- It is not changed when either x or y is transformed by a *monotone* function. Monotone functions do not alter the order of values. The log, square root, and squaring functions are monotone for positive values. Because monotone functions preserve order, the ranks of *log(income)* will be the same as the ranks of *income*, so Spearman's rho will also be the same.
- It is less affected by outliers or extraordinary values than product-moment correlation. This, too, can be attributed to using ranks. If the largest value in one variable is extraordinarily large, then its correlation with another variable is likely to be strongly affected. However, the rank of the largest value depends only on the sample size and not on the magnitude of that value, so the value can be arbitrarily large without altering rho.

Rho can take on values between –1 and 1. If the value of rho is near 1 or –1, we can conclude only that there is a *monotone* relationship between x and y, rather than that there is a *linear* relationship between them. In a monotone relationship x and y are in the same order or are in exactly reversed orders relative to each other.

Conversely, if rho is near zero, we conclude that there is no evidence for a monotone relationship. For many types of data a monotone, but not linear, relationship exists between x and y. In such cases, rho is a more appropriate statistic than the product-moment correlation.

For example, we noted in Chapter 11 that the relationship between MPG and Displacement in the Cars dataset is not linear. Therefore, Spearman's rho is a more appropriate statistic.

Spearman Rank Correlation

	MPG	Displac...
MPG		1.000
Displacement	-0.838	1.000

Figure 21-6. *Spearman rho of MPG and Displacement.*

Transforming MPG to "Gallons per 100 miles" as -100/MPG makes the relationship more linear. Spearman's rho, however, is not affected by monotone transformations such as this.

Spearman Rank Correlation

	-1/MPG	Displace...
-1/MPG		1.000
Displacement	-0.838	1.000

Figure 21-7. *The Spearman correlation is unaffected by monotone transformations such as the negative reciprocal.*

21.7 Kendall's Tau

Kendall's Tau

Kendall's *tau* is similar to Spearman's rho in that it is based on the ranks of the data and is thus a nonparametric method. Kendall's tau measures the degree of monotonicity in the relationship between x and y by considering all *pairs* of datapoints in the scatterplot of y *vs* x. The slope between each pair of points is positive, negative, or zero. Kendall's tau is the difference between the number of positive slopes and the number of negative slopes, divided by the total number of pairwise slopes.

Tau ranges between -1 and 1. If it is near -1, then nearly all the pairwise slopes are negative, indicating a generally decreasing relationship between x and y. If tau is near 1, then nearly all the pairwise slopes are positive, indicating a generally increasing relationship. If tau is near 0, then the number of positive slopes is roughly equal to the number of negative slopes, indicating that x and y are not ordered similarly.

To compute Kendall's tau in Data Desk, select the designated variables and choose **Kendall's tau** from the **Correlations** submenu in the **Calc** menu.

21.8 *Covariance*

The covariance of x and y is computed as

$$\frac{\sum_{i=1}^{n}(x_i - \bar{x})(y_i - \bar{y})}{n - 1}$$

Covariance is a common measure of association that can be used in such calculations as Principal Components. Data Desk computes and displays covariances in the same manner as correlations.

21.9 *Missing Values and Correlation*

When Data Desk computes the correlation between two variables, x and y, it omits any cases that are missing on either x or y. That is, Data Desk omits cases in a *pairwise* manner. Data Desk thus regards a table of correlations as a table of two-variable statistics rather than as a part of a multivariate analysis for which cases missing on *any* of the variables included would be omitted.

This guarantees that any correlation in a table of correlations corresponds to the scatterplot of its component variables and does not change as variables are added to or removed from the correlation table. However, the correlations displayed in a correlation table of several variables may not be the same as the correlations Data Desk computes for some multivariate analyses. In a multivariate analysis Data Desk omits all cases with missing values on *any* variable.

21.10 *Extracting Values from Correlation Tables*

Some statistics calculations use correlations or covariances as parameters in a calculation. You can extract correlations from a Data Desk correlation table with the following steps.

- Compute the correlation table.
- With the table window frontmost, choose {Edit} **Copy Window**.
- Open a ScratchPad window and **Paste** the table into it.

The text in the ScratchPad window is now tab-delimited text containing the correlation table. You can select any correlation, copy it, and paste it in a derived variable expression, variable editing window, or program outside Data Desk.

If you delete the table title and labels, you can copy the body of the table, click on a folder window, and {Edit} **Paste Variables** to make each column of the table a variable. You can then select and re-order columns by selecting the variables and using {Edit} **Copy Variables**.

CHAPTER 22

Multiple Regression

22.1	Multiple Regression	*287*
22.2	Example	*287*
22.3	Interpreting the Regression Table	*289*
22.4	Predicted Values and Residuals	*289*
22.5	Missing Values in Regression	*290*
22.6	Regression Options	*290*
22.7	Working with Regression Summary Tables	*291*
22.8	Interactive Regression Model	*293*
22.9	Stepwise Regression	*295*

SIMPLE LINEAR REGRESSION, discussed in Chapter 20, describes the relationship between a response variable, y, and a predictor variable, x. Multiple regression extends simple regression to include more predictors in the prediction equation. Multiple regression describes data measured on several variables with a simple expression. The relative simplicity of the description, along with its usefulness as a basis for many other multivariate analysis methods, make multiple regression one of the most widely used statistical analyses.

22.1 *Multiple Regression*

Multiple regression describes the linear relationship between one dependent variable, y, and several predictor variables with a linear equation. In general, the regression equation for p predictors is written as:

$$\hat{y} = b_0 + b_1x_1 + b_2x_2 + \ldots + b_px_p$$

where the b_j are estimates of the true population parameters, denoted β_j. If $p = 1$, the equation is the same as for simple regression and the points that satisfy it fall on a straight line.

In Data Desk, multiple regression is specified in the same way as simple regression. Select the y-variable (response) first and then extend the selection to include the x-variables (predictors). To select the variables in any order, Option-click the y-variable and Shift-click the x-variables. You can add additional x-variables to the analysis by dragging their icons into the center of the regression summary table. You can replace an x-variable by dragging the new icon of the new variable over the label of the x-variable to be replaced. You can replace the y-variable by dragging the icon of the new variable over the label of the current y-variable.

22.2 *An Example*

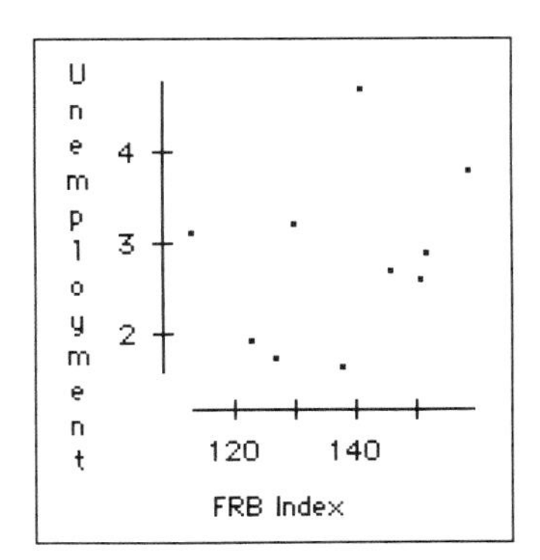

***Figure 22-1**. The scatterplot of Unemployment vs. the Federal Reserve Board Index.*

An example illustrates the similarities and differences between simple and multiple regression and points out some of the complexities of multiple regression. The Unemployment dataset contains the U.S. unemployment rate, Federal Reserve Board index of industrial production, and year of the decade for 1950-1959. A scatterplot shows the relationship of *Unemployment* to the *FRB Index.*

The simple regression of *Unemployment* on *FRB Index* shows two surprising things.

Dependent variable is: Unemployment
R squared = 9.8% R squared (adjusted) = -1.5%
s = 0.9719 with 10 - 2 = 8 degrees of freedom

Source	Sum of Squares	df	Mean Square	F-ratio
Regression	0.819310	1	0.819310	0.867
Residual	7.55669	8	0.944586	

Variable	Coefficient	s.e. of Coeff	t-ratio	prob
Constant	-0.035172	3.081	-0.011	0.9912
FRB Index	0.020690	0.0222	0.931	0.3789

Figure 22-2. *The simple regression of Unemployment on FRB Index.*

From the positive sign of the FRB Index coefficient, we might at first think that as industrial production rose during the 1950's, unemployment rose as well. (Rising industrial production would usually be expected to employ more people and reduce unemployment.) However, the t-ratio for this coefficient is quite small, which suggests that the coefficient's value is not reliably different from zero. Of course, even a value of zero could be surprising; the FRB Index and unemployment rate ought to be related. The R^2 for the regression shows that only 9.8% of the variability in unemployment is accounted for by its linear relationship with the FRB index, so the overall regression is rather unsuccessful.

Many aspects of the U.S. economy changed during the decade of the 1950's. The simple regression of unemployment on FRB index must ignore the impact of these changes because it is limited to a single predictor. Multiple regression introduces more predictors to account for some of the changes.

We can try to represent some of these changes with the variable *Year* that simply records the year of the decade. We can then include *Year* as a predictor in the regression. Click on the icon of *Year* and drag it into the predictor panel of the regression table. The resulting regression summary table will recompute to look like Figure 22-3.

Dependent variable is: Unemployment
R squared = 86.6% R squared (adjusted) = 82.7%
s = 0.4011 with 10 - 3 = 7 degrees of freedom

Source	Sum of Squares	df	Mean Square	F-ratio
Regression	7.24976	2	3.62488	22.5
Residual	1.12624	7	0.160891	

Variable	Coefficient	s.e. of Coeff	t-ratio	prob
Constant	13.4539	2.484	5.42	0.0010
FRB Index	-0.103339	0.0217	-4.77	0.0020
year	0.659417	0.1043	6.32	0.0004

Figure 22-3*. The multiple regression of Unemployment on FRB Index and Year.*

This regression is much more successful. The R^2 is now 86.6% and all three t-ratios are fairly large. After allowing for the changes during the decade represented by year, the coefficient associated with the FRB index is now negative, indicating decreasing unemployment with increasing industrial production.

22.3 *Interpreting the Regression Table*

Most of the multiple regression summary table is interpreted in the same way as the simple regression summary table discussed in Chapter 20, with the following exceptions:

- The null hypothesis tested with the F-ratio is now

  $$H_0: \beta_1 = \beta_2 = \ldots = \beta_p = 0$$

  which is a general hypothesis stating that the true coefficients are all zero. (Note that the constant coefficient, β_0, is not included in the hypothesis.) If the F-ratio is sufficiently large, we reject this null hypothesis.

- Multiple regression coefficients must be interpreted more carefully than simple regression coefficients. While each coefficient is still measured in y-units per unit of its associated x-variable, all the coefficients are now tied together and must be interpreted as a group. One way to describe this relationship in words is to say that a coefficient measures the change in y for a one-unit change in its associated x-variable *after removing the linear effects of all the other x-variables.* Section 23.2 discusses partial regression plots, which depict the basis for each regression coefficient.

- The t-ratios for the regression coefficients are interpreted as in simple regression, except that now they have, in general, $n - (p + 1)$ degrees of freedom (where n is the number of cases and p is the number of predictor variables). The third line of the regression table reports degrees of freedom.

- Adjusted R^2 is useful in multiple regressions because the adjustment accounts for the number of predictors in the regression equation. As new predictors are added to a regression equation, the value of R^2 can only increase, so even the addition of a predictor filled with random numbers can increase R^2. Adjusted R^2 accounts for this effect so that it does not generally improve if you add a useless predictor. Adjusted R^2 is more appropriate in a multiple regression, especially when alternative regression equations are compared.

22.4 *Predicted Values and Residuals*

Every regression computes *predicted* values and *residuals* from the Global HyperView of the regression summary table. The predicted values, usually denoted $\hat{y}$, estimate the observed values of y from a linear equation in terms of the predictors. There is a predicted value for each case in the regression variables.

The {Calc ▶ Calculation Options} **Set Regression Options...** command offers many choices, including saving predicted values. If selected, Data Desk saves predicted values as a HotResult variable named Predicted and places it in the regression folder.

Real data are almost never so structured as to fit a model as simple as the linear multiple regression equation. We expect the predicted values and the observed values to differ, although we hope that they differ relatively slightly. The differences between the observed and predicted values are the residuals. As for predicted values, there is one residual for each case. The Regression Options dialog and the Global HyperView offer to save residuals; they are placed in a HotResult variable named *Residuals*.

Least squares multiple regression minimizes the sum of the squared residuals, so we expect them to be small compared to the original data. For that reason, it is usually better to plot residuals directly rather than as part of a plot showing the original data. A scatterplot of residuals *vs* predicted values is often very useful for assessing the success of the regression. The window's regression HyperView offers this plot.

22.5 *Missing Values in Regression*

> HOW-TO
>
> To compute predicted values:
> • Add the new cases to the variables, leaving the y-values missing.
> • Compute a regression, requesting predicted values.
> • Data Desk finds predicted values for the new cases.

> TIP
>
> To see the numbers of a HotResult, open the HotResult and select Show Numbers from its HyperView.

Whenever any of the predictor variables in a regression has a nonnumeric or infinite value for any case, that case is omitted from the regression. This is sometimes called *casewise deletion*. To omit a case from the regression, make the value for that case in any of the variables included in the regression missing. Alternatively, use a Selector variable as described in Chapter 12. If you make the case missing in the dependent variable, a predicted value will still be computed for that case, based on the remaining points.

This feature provides a convenient way to compute predicted values for new datapoints. Enter the values of the predictors for the new points and mark the dependent variable missing. Then recompute the regression and examine the predicted values.

Data Desk tries to compute all requested statistics even for cases that are omitted. Omitting cases with a Selector variable results in estimates for those cases based on the regression computed without them.

22.6 *Regression Options*

The {Calc ▶ Calculation Options} **Set Regression Options...** command controls the regression, the computing of standard regression results, and the calculation of diagnostic statistics that can help you to understand your regression in greater depth. We postpone the discussion of regression diagnostics to Chapter 23, and discuss the other options here.

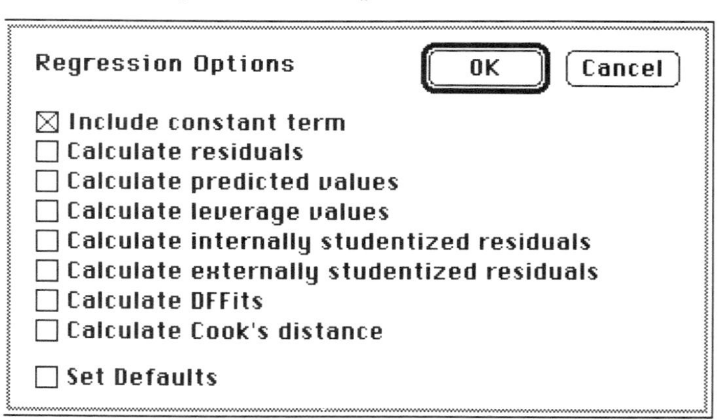

Figure 22-4*. The* ***Regression Options...*** *dialog.*

The regression options affect the computing of all subsequent regressions, but do not alter any that have already been computed. Data Desk will remember the options you set if you click on the box which says Set Defaults.

INCLUDE CONSTANT TERM

Turn this option off if you wish the regression to be computed without a constant or intercept term. This option must be used when you already have a predictor that is constant, but may also be used to compute a regression through the origin.

CALCULATE RESIDUALS

When this option is checked, all subsequent regressions create a HotResult variable named *Residuals,* and place it in the regression folder when created using the options dialog. If the residuals are created from the HyperView command, they are placed in the Results folder.

CALCULATE PREDICTED VALUES

When this option is checked, all subsequent regressions create a HotResult variable named *Predicted,* and place it in the regression folder.

22.7 *Working With Regression Summary Tables*

Data Desk summary tables are interactive in much the same way as Data Desk plots. They provide ways to alter and learn more about your analysis.

The simplest source of additional information is in the HyperView menus attached to the regression summary table. Many parts of the regression summary table have HyperView menus attached to them. You can tell when a HyperView is available because the cursor changes to when the mouse is over a HyperView. Press the mouse button to pop up the HyperView menu. Regression HyperViews offer plots to help you view important aspects of the regression analysis.

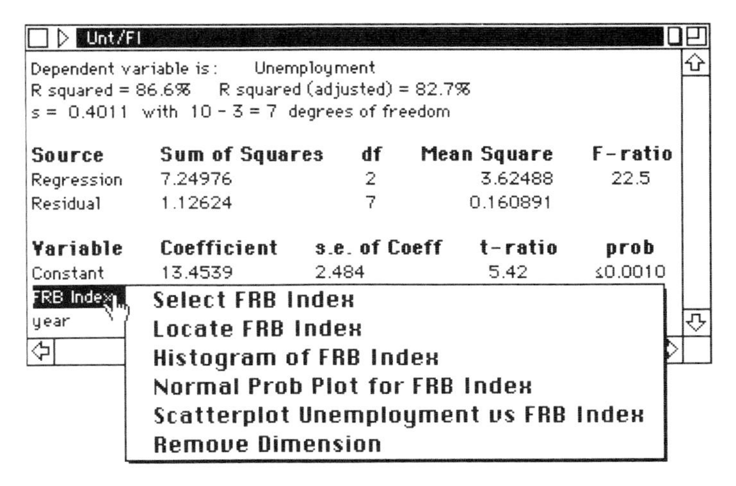

Figure 22-5. *The HyperView menu attached to the name of a predictor in a regression summary table.*

Some of these plots are basic background displays, such as histograms of the variables in the regression or scatterplots of y against one of the x-variables. It is always a good idea to examine such plots. Others offer more sophisticated views of the regression and are discussed in Chapter 23.

In addition to HyperViews, you can use the regression summary table to add new predictors to the regression model, remove predictors from the model, and update the regression calculation to reflect changes, corrections, or "what if?" alterations in the data.

The HyperView attached to the name of each predictor offers to remove that predictor from the model and recompute the regression. This provides a tool that can be useful in building regression models. Removing a predictor recomputes the regression, generating any additional statistics called for by the current setting of the regression options and placing them in the Results log.

To add new predictors to an existing regression analysis, select the icons of the predictors and drag them into the regression summary table. The border of the summary table highlights to acknowledge the new predictors. Drop the icons inside the table to initiate the new regression.

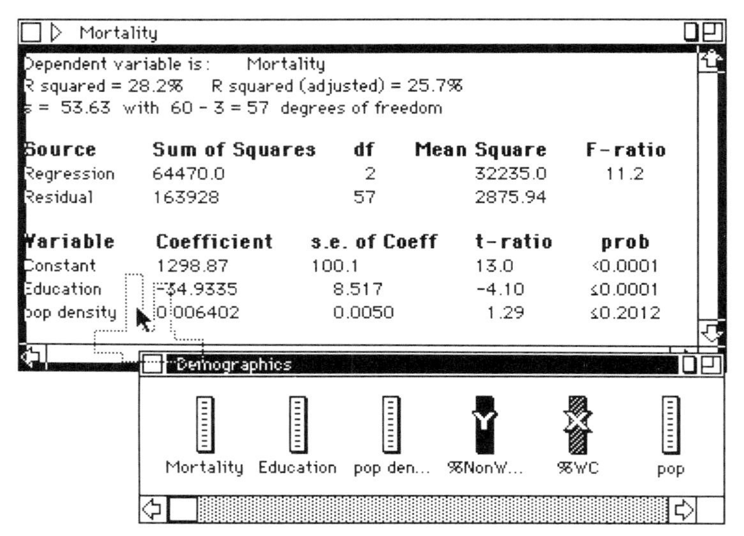

Figure 22-6. Drag additional predictors into the regression summary table to add them to the regression model. The window border becomes darker to acknowledge the drag.

The HyperView attached to a predictor's name includes the option to remove that predictor from the model. You can thus build and adjust regression models easily.

Data Desk also provides facilities for adjusting cases as well as variables. If you modify any of the data underlying the regression or the expression in any derived variable in the regression, the regression summary table immediately offers to update in place or recompute in a new window. A ! symbol replaces the window's HyperView button at the left of the window's title bar whenever the underlying variables have been changed. The menu attached to it offers to update the regression in place or redo it in a new window.

Often it is better to recompute the regression in a new window so that you can compare the "before" and "after" versions of the analysis. Regression analyses can change in subtle ways even when you make changes that seem trivial. Data Desk's updating capabilities also insure that you know immediately if the analysis you are using was computed from data values that have since been changed.

22.8 *Interactive Regression Model*

One of Data Desk's most powerful attributes is its ability to perform interactive analyses. Every Data Desk plot and table can have new variables dragged into it either to add those variables as predictors, or to replace predictors or the dependent variable. This process is especially valuable for exploring regression models. Not only can you change the regression model, but all plots and results computed from that regression will offer to update to reflect the new state of the model.

This section presents an example of interactive regression analysis. The example uses the dynamic function **Mix X and Y** and Data Desk's ability to replace predictor variables, dynamically update plots, and link all plots and windows together. We will create a predictor variable that mixes two potential predictor variables, and then slide dynamically from one to the other, viewing the effect on the regression. Data Desk provides direct dynamic control over the proportion of each predictor in the mixed variable with a slider. (For a refresher on sliders and dynamic parameters see Section 11.13.)

This example uses the Companies datafile. In particular, we examine a regression model predicting Market Value from either Sales or Assets. Let's start with a model that predicts Market Value from Sales.

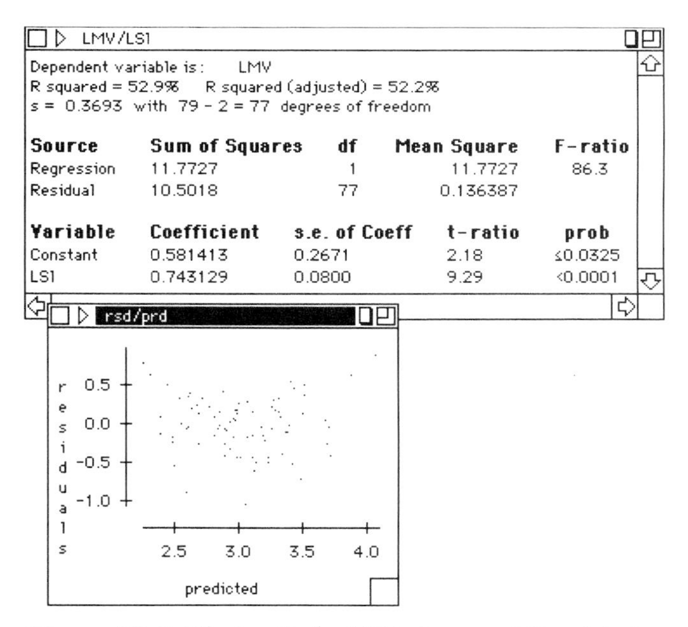
LMV/LS1

Dependent variable is: LMV
R squared = 52.9% R squared (adjusted) = 52.2%
s = 0.3693 with 79 - 2 = 77 degrees of freedom

Source	Sum of Squares	df	Mean Square	F-ratio
Regression	11.7727	1	11.7727	86.3
Residual	10.5018	77	0.136387	

Variable	Coefficient	s.e. of Coeff	t-ratio	prob
Constant	0.581413	0.2671	2.18	≤0.0325
LSl	0.743129	0.0800	9.29	≤0.0001

Figure 22-7. *The t-ratio for LSl is large and the plot of the residuals shows no particular pattern - it appears to be a pretty good model.*

Variables holding dollar amounts are usually better analyzed after taking logs. The variables *Market Value*, *Sales*, and *Assets* are all measured in dollars and should be transformed. Choose these variables and select {Manip ▶ Transform} **Log**. Data Desk creates derived variables named *LAss*, *LSl* and *LMV*. Select *LMV* as the *y*-variable, *LSl* as the *x*-variable, and choose **Regression** from the **Calc** menu. You should now have a regression summary table like the one in Figure 22-7. From the table's global HyperView, choose **Scatterplot residual vs predicted**. The results look pretty good — the t-ratio for *LSl* is large and the residuals show no particular pattern. Let's see what happens when we replace *LSl* with the new variable which is a mixture of Sales and Assets.

To create the mixed variable select LSl as y and LAss as x and choose {Transform ▶ Dynamic}**Mix X and Y** from the Manip menu. The Dynamic transformations create a slider (here, named p). The Mix X and Y command creates a derived variable named *pLSl+LAss* that contains the expression

(1-'p')(mean('LSl')+sdev('LSl')*('LAss'-mean('LAss'))/sdev('LAss'))+'p'*'LSl'*

While this expression looks complex, all it is really doing is combining the two variables *LSl* and *LAss* in a linear combination of the form

$$p * LSl + (1 - p) * LAss$$

The rest of the calculation standardizes the variables so that neither one can dominate the combination simply by having a large mean or standard deviation.

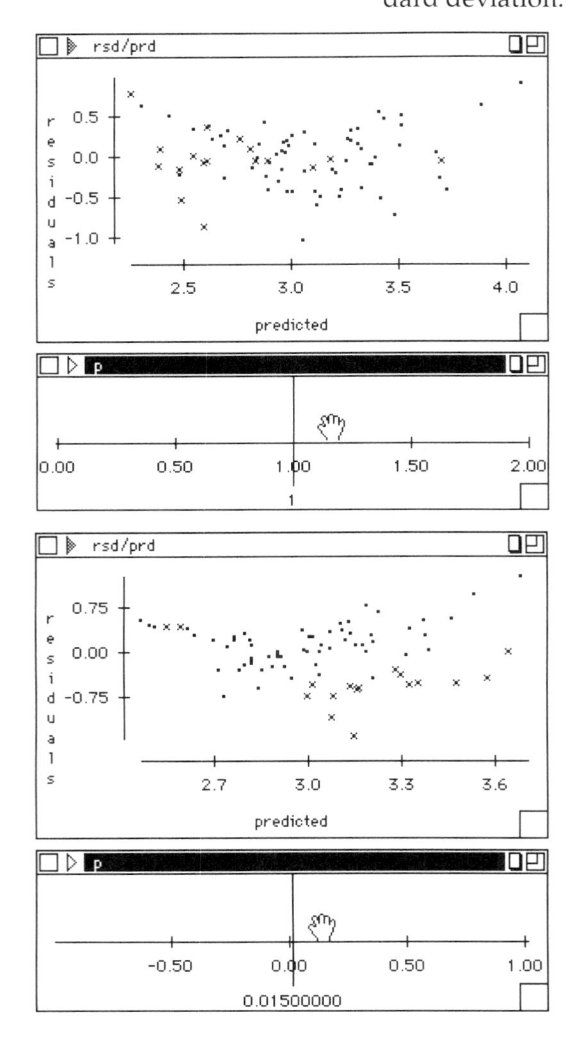

Figure 22-8. *As you move the slider's value closer to 0.0, the points plotted as x's appear to move against the grain of other companies. This pattern suggests that the Finance sector (plotted as x) deserves special attention.*

The slider is the weighting factor that determines the proportions, p and $(1 - p)$, of the variables in the mixture. When the slider value is 1.0, the mixture is completely *LSl*, when the value is 0.0, the mixture is all *LAss*. Any slider value between 1.0 and 0.0 is appropriate when using the Mix X and Y transformation. Any slider values outside that range are unlikely to be useful for the mixing derived variable.

Choose **Automatic Update** and **Manual Scale** from the scatterplot's global HyperView. Drag the derived variable's icon into the regression on top of the *LSl* label in the regression summary table to replace *LSl* with the derived variable. (The *LSl* label will highlight to acknowledge the proper positioning of the drag.) Because the initial slider value is 1.0, the initial value of the mixed variable is 100% LSl and 0% LAss, so there should be no immediate changes to the regression summary table or the scatterplot of residuals.

You are now ready to perform the interactive analysis. The slider and the plot of residuals are the most important windows, so make sure they are both visible. (See Figure 22-8, for example.) Grab the slider and drag the slider axis horizontally through the values from 1.0 to 0.0. As you move the slider's value closer to 0.0, the residuals become larger and start to form interesting patterns. Some of the points move in the opposite direction from the others.

Now try changing the symbol or color of companies in the Finance sector. (You could make a Barchart of *Sector* and select the Finance bar, or open the *Sector* variable and use {Edit ▶ Find} **Find** to select "Finance". Then choose {Modify ▶Symbols ▶ Show As} **x**.) Grab the slider and move it between the values 0.0 and 1.0. The x's appear to move against the grain of the other points. Perhaps the Finance sector is different from other sectors and should be analyzed separately.

If you have a large datafile and a small machine, you might find dynamic analyses slow. Remember that each time you change the value of the slider, Data Desk recomputes the regression, the residuals, the predicted values, and the scatterplot. One way to reduce the computing burden is to use the slider's "ratchet mode". First rescale the slider to place tick marks periodically between 0.0 and 1.0. Then slide the slider while holding down the option key. The slider value will change only when the hairline passes a tick mark, reducing the number of times Data Desk must recompute the regression.

You can also reduce computation by working with a representative subset of your data. Reducing the calculation to a few hundred cases often helps to improve the response (see Section 15.2).

22.9 *Stepwise Regression*

stepwise regression

Statistics packages designed for batch computing on mainframe computers often included a method known as *stepwise regression*. In stepwise regression the computer attempts to build a regression model by adding or deleting predictors according to a statistical rule of thumb.

Stepwise regression is useful when each regression calculation is expensive or time consuming because it returns more information about your data than you would ordinarily get from a single regression calculation. However, stepwise regression often leads to poor regression models because the computer cannot make intelligent choices among variables that are candidates for the regression model.

By contrast, Data Desk lets you add and delete predictors in such a way that *you* control the specification of the regression model. At each step, you can and should generate plots to check that an extraordinary case has not unduly influenced your regression, that the relationships are indeed linear rather than curved, and that other regression assumptions are not violated. The resulting regression models are likely to be more successful and more useful than those built without human intervention.

Data Desk provides tools that let you perform a stepwise regression by hand. Just follow these steps.

1) Select only the y-variable of your regression (as y) and choose **Regression** from the **Calc** menu. Data Desk makes a regression table with only the constant term.

2) From the global HyperView of the regression summary, choose **Compute ▶ Residuals**. Data Desk places a HotResult named *Residuals* in the Results folder.

3) Select *Residuals* first and then all the candidate predictor variables that you want to consider as possible predictors in the regression, and choose {Calc ▶ Correlations} **Pearson Product-Moment**. In the correlation table's HyperView choose **Automatic Update**.

   The first column of the correlation table shows correlations of the regression residuals with each of the candidate predictors. The predictor with the highest correlation is the one that a stepwise regression would add next.

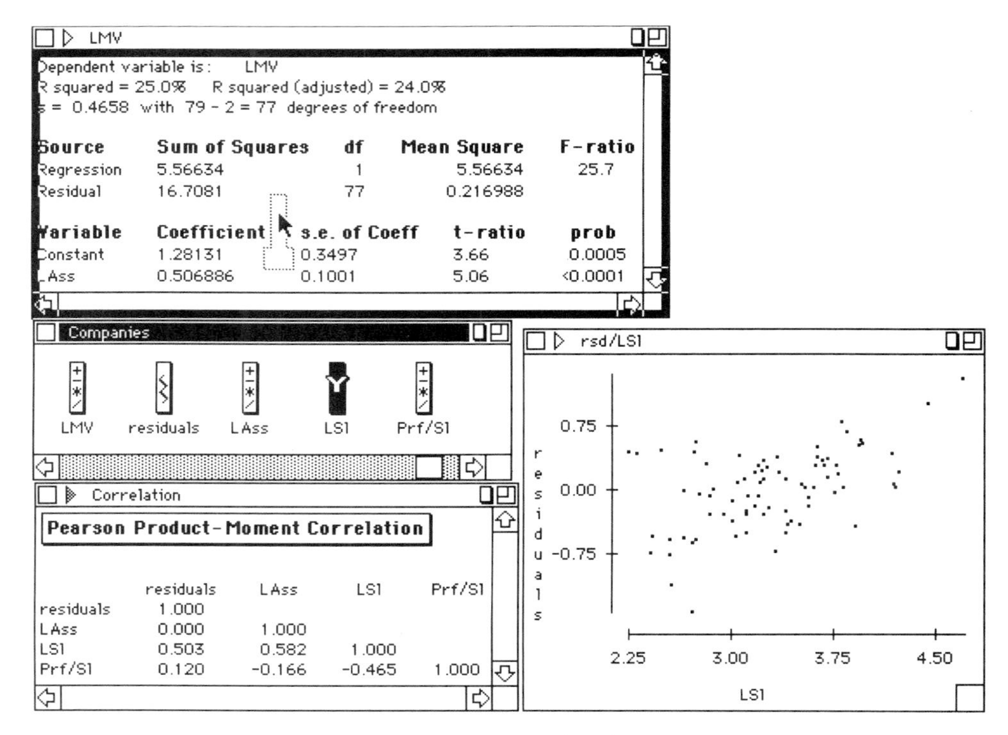

Figure 28-9. *Drag the icon of a variable that fits the assumptions of the regression model into the regression summary window. The regression will automatically recompute and the correlation of residuals vs the new predictor variable will change to 0.*

4) Select a predictor. If you wish to perform a standard stepwise regression, select the predictor with the highest correlation. The HyperView on the predictors' names in the correlation table offers to **Locate** the icon of the variable.

   A more responsible step is to scatterplot the residuals *vs* the contemplated predictor using the HyperView attached to the correlation value itself. Check the plot for outliers, nonlinearities, or heteroskedasticity (non-constant spread) and proceed only if the plot looks OK.

5) Drag the selected variable into the predictors section of the Regression table and drop it there. The regression will update, compute new residuals, and update the correlation table (because it is set to Automatic Update).

   Now return to step (4). All predictors already in the regression model must have correlation 0.0 with the residuals. Continue selecting predictors and dragging them into the regression until you wish to stop. One possible stopping rule is to stop when the added predictor has a t-ratio that is not significant.

   You can also remove predictors from the regression using the HyperView attached to the predictor name in the regression table. Stepwise regression often benefits from removing predictors whose t-ratios have fallen below significance. Each change in the model will be reflected in both the regression table and the correlation table.

- If you note an outlier in the scatterplots of residuals *vs* predictors, consider selecting the offending point and choosing {Modify ▸ Selection} **Record**. (See Chapter 9.) Then drag the resulting indicator variable into the regression table to isolate the outlying case. The t-ratio for the indicator tests the null hypothesis that the case is not in fact an outlier; a significant t-ratio argues that it is an outlier.

  Similarly, evidence of a nonlinear relationship might lead you to re-express the dependent variable. Select the icon of the dependent variable, choose a suitable re-expression from the Transform menus, and drag the new derived variable into the regression table over the name of the dependent variable to replace it. As before, the correlations will update automatically.

Note: If there are many missing values sprinkled throughout the candidate predictor variables then adding and deleting predictors may change which cases are included in the analysis. Changing the cases in the analysis and possibly the number of cases in the analysis can cause unexpected changes to other aspects of the regression — especially if one of the cases included or deleted is an outlier or an influential case.

CHAPTER 23

Regression Diagnostics

23.1	Checking Basic Assumptions	*300*
23.2	Partial Regression Plots	*301*
23.3	Leverage	*303*
23.4	Studentized Residuals	*306*
23.5	Distance Measures	*308*
23.6	Regression Options	*311*
23.7	A Note on Identifying Cases	*312*
23.8	Collinearity	*313*

Multiple regression analysis provides a deceptively simple description of multivariate data. The regression model is an effective basis both for understanding relationships among many variables and for predicting values of the dependent variable for new cases. But regression coefficients are more subtle than they may appear at first, and regression computations can be sensitive to anomalies in the data and to violations of the assumptions underlying regression analysis.

While the principles of regression analysis date back to the 19th century, statistics and plots for diagnosing the effects of individual cases on a regression are quite recent. Data Desk provides a wide range of plots and statistics designed to help you identify patterns in the data and individual data values that might unduly influence the regression analysis.

The diagnostic statistics discussed in this chapter can be requested with the {Calc ▶ Calculation Options} **Set Regression Options...** command or from HyperViews. Diagnostics requested by Options are placed in HotResults in the regression results folder. They are placed in the results folder when requested from HyperViews.

A NOTE ON NOTATION

Some of the diagnostic statistics defined in this chapter are best defined algebraically with matrix equations. But you need not know matrix algebra to understand how to interpret and use these statistics. We provide the equations to define the statistics.

Throughout our discussions of regression we denote the number of cases in the regression (after excluding any cases due to missing values or selector variables) by n and the number of predictor variables (not counting the constant term) by p. The predictors are sometimes gathered column-by-column into a matrix, $\mathbf{X}$. Values representing population parameters are denoted with Greek letters, while data-based estimates are denoted by placing a ^ over a term or with Roman letters. The notation (i) is read "not i" and denotes omitting the ith case from the calculation.

The examples in this chapter are based on an analysis of the SMSA's data and are set aside by printing in italics so that you can follow the story told by the analysis more easily. These data report properties of 60 Standard Metropolitan Statistical Areas (the standard Census Bureau designation of the region around a city) in the United States, collected from a variety of sources. The data include information on the social and economic conditions in these areas, on their climate, their geographic location, and some indices of air pollution potentials. The dependent variable of interest is age-adjusted mortality. We will illustrate diagnostic methods by working with the regression analysis shown in Figure 23-1.

Dependent variable is: Mortality
R squared = 60.4% R squared (adjusted) = 56.7%
s = 40.94 with 60 - 6 = 54 degrees of freedom

Source	Sum of Squares	df	Mean Square	F-ratio
Regression	137880	5	27576.0	16.5
Residual	90518.2	54	1676.26	

Variable	Coefficient	s.e. of Coeff	t-ratio	prob
Constant	1081.73	96.74	11.2	≤0.0001
Education	-21.9230	7.417	-2.96	0.0046
pop density	0.007931	0.0039	2.02	0.0482
%NonWhite	3.82884	0.6434	5.95	≤0.0001
Rain	0.601016	0.6042	0.995	0.3243
NOx	-0.002977	0.1359	-0.022	0.9826

Figure 23-1. *The regression of age-adjusted mortality on median education, population density, %nonwhite, mean annual rainfall, and pollution potential from nitrous oxides for 60 US Standard Metropolitan Statistical Areas.*

23.1 *Checking Basic Assumptions*

Most statistics texts recommend that you examine simple plots to ascertain that the variables in a regression do not have any wild values and that the relationships between the *y*-variable and each of the *x*-variables are essentially linear. Most data analysts will admit (if they are honest) that they often overlook this step because it can be difficult and time-consuming.

Data Desk makes it easy to examine these plots by providing HyperView menu commands to create them. Histograms of each of the variables in the analysis let us see at a glance any data value that is extraordinary. We can also check for bimodal distributions, which suggest that we separate the data into two groups and analyze each separately, and for variables with skewed distributions, which suggest that we transform the variable to improve its symmetry.

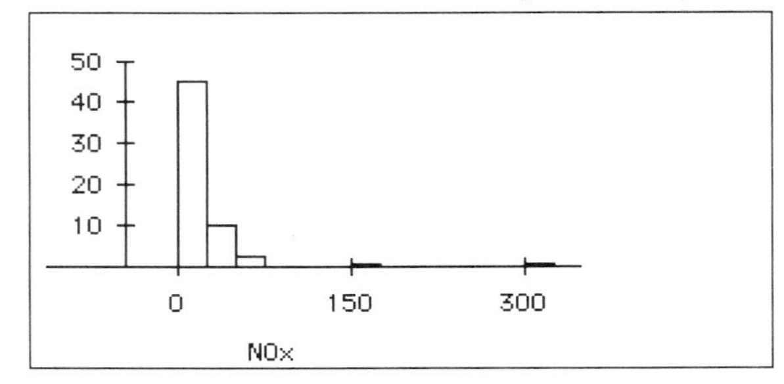

Figure 23-2. *A histogram of Nitrous Oxide pollution potential (NOx) reveals an extreme case that may deserve special attention.*

Scatterplots of the *y*-variable *vs* each *x*-variable let us check for linear relationships — a basic assumption of regression analysis. Nonlinear relationships can sometimes be improved by transforming one or both of the variables. Some nonlinear relationships suggest including a quadratic or cubic term in the model.

The extreme NOx values belong to Los Angeles and San Francisco. They might be telling us that air pollution in California is different from the rest of the U.S. However, the rest of the distribution is skewed, so we should first try re-expressing the variable. A histogram of log(NOx) is symmetric and unimodal, so there is no reason to treat the California cities separately. We perform a new regression using log(NOx) instead of NOx. Parts of the regression summary table are in Figure 23-3.

Dependent variable is: Mortality
R squared = 64.7% R squared (adjusted) = 61.4%
s = 38.65 with 60 - 6 = 54 degrees of freedom

Source	Sum of Squares	df	Mean Square	F-ratio
Regression	147753	5	29550.6	19.8
Residual	80645.6	54	1493.44	

Variable	Coefficient	s.e. of Coeff	t-ratio	prob
Constant	1031.73	93.36	11.1	≤0.0001
Education	-20.6236	6.992	-2.95	0.0047
pop density	0.003209	0.0040	0.799	0.4275
%NonWhite	3.16133	0.6481	4.88	≤0.0001
Rain	1.38444	0.5939	2.33	0.0235
LNOx	31.8577	12.39	2.57	0.0129

Figure 23-3. *Substituting log(NOx) for NOx improves the regression in several ways. Note the higher R^2, and several improved t-ratios.*

23.2 *Partial Regression Plots*

A good interpretation of the jth multiple regression coefficient, b_j, is that it reports the relationship between y and x_j *after the linear effects of the other x-variables have been removed from both.* One advantage of this description is that we can depict it with a *partial regression plot*. Partial regression plots are particularly useful because they provide a simple and intuitive way to see the influences of individual cases on the estimation of a multiple regression coefficient. Any intuition you might have about a simple slope in a scatterplot applies in the same way to the partial regression plot.

partial regression plot

Although it is now considered a modern computer-oriented method, partial regression plots and other plots related to them were widely used when regression was first developed — decades before computers were available to perform the calculations. They are given different names by different statisticians, including *Added Variable Plot, adjusted-variable plot,* and *individual coefficient plot.*

Each coefficient in a multiple regression can be depicted in its own partial regression plot. A partial regression plot graphs *y with the linear effects of the other x-variables removed* against *x with the linear effects of the other x-variables removed*. To remove the linear effects of the other x-variables from our chosen x, we perform another regression "on the side" and keep the residuals.

This extra regression estimates the linear effects of the other x-variables, and the residuals are what is left after removing these effects. The *partial correlation* of y and x is the correlation between the y and x adjusted in this way for the other x-variables. Some authors write of "partialing out" the linear effects of the other x-variables.

partial correlation

Coefficient	s.e. of Coeff	t-ratio	prob
1031.73	93.36	11.1	<0.0001
-20.6236	Partial regression plot of Education		
0.003209	0.0040	0.799	≤0.4273
3.16133	0.6481	4.88	<0.0001
1.38444	0.5939	2.33	≤0.0235
31.8577	12.39	2.57	≤0.0129

Thus, to construct a partial regression plot of y and a particular predictor, x_j, we compute the regression of x_j on the other x-variables and save the residuals, compute the regression of y on the same x-variables (that is, all of the predictors except x_j) and save those residuals, and then plot the y-residuals against the x-residuals. Data Desk provides a simple HyperView command to do all of this work. The HyperView menu attached to each regression coefficient in the regression summary table offers the partial regression plot for that coefficient. The **Partial regression plot** HyperView command automatically creates the y-residuals and x-residuals (naming them "<Varname> Resids") and places their icons in the Results folder.

The partial regression plot has several useful properties:

- The least squares slope of the partial regression plot of y and x_j is b_j, the least squares coefficient associated with x_j in the full multiple regression of y on all the x-variables.
- The residuals from the least squares line in the partial regression plot are the same as the *final* residuals for the full multiple regression. (We have removed from y the linear effects of all of the predictors except x_j to obtain the ordinate for this plot. If we now remove the linear effects of x_j we obtain the final residuals.)
- The influence of each individual data point as depicted in the partial regression plot for x_j, accurately reflects its influence on the regression coefficient b_j in the full regression model. Thus, for example, if one point is far from the others and evidently pulling the least squares line away from the slope indicated by the others, that case is exerting a similar influence on the coefficient of x_j in the multiple regression.

In short, you can transfer your intuition for how simple least squares regression relates to the standard y *vs* x scatterplot to how a multiple regression coefficient relates to its partial regression plot.

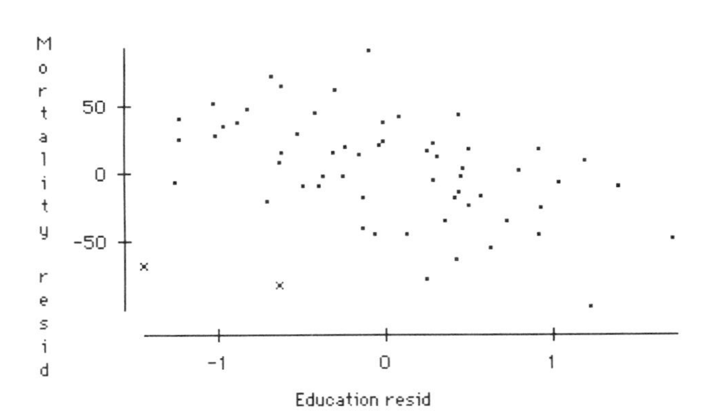

Figure 23-4. *The partial regression plot for Education shows a consistent negative trend and two points away from the others, indicated here by plotting them with an x.*

For example, the partial regression plot for Education in Figure 23-4 depicts the relationship between Mortality and Education in the full model given in Figure 23-3. The plot shows a generally consistent trend with two points away from the body of the data.

The two extraordinary points can easily be identified as York and Lancaster — SMSA's that are adjacent to each other geographically in the Pennsylvania Dutch country and that have a local culture that might lead to an unusually low median education level. Intuition from the partial regression plot suggests that these two points are forcing the coefficient of Education closer to zero than it might otherwise have been.

of individual cases on a regression analysis. Regression analyses can be strongly affected by even a single extreme value, so it is wise to be aware of any cases exerting such an influence. Influential cases may be in error, but it is more likely that they are extraordinary for some other reason. Whatever the reason, you can often learn more about your data by identifying the influential cases and giving them special attention.

leverage

The *leverage* of a data point measures how extreme it is on its x-variables, and consequently, how much influence it can exert on the regression. The leverage of the ith case, h_i, is the amount by which the ith predicted value, y_i, would change if the ith observed dependent variable value, y_i, were incremented by one and all other values in the regression data were unchanged.

Each h_i lies between 0.0 and 1.0. A leverage of zero indicates a data point with no influence on the regression — for example a value observed at $x = 0$ for a regression constrained to pass through the origin. A leverage of one indicates a data point guaranteed to have a zero residual.

In general, a case has higher leverage when it is farther from the center of the x-values. The appropriate physical model is a lever in which force applied farther from the fulcrum can move the lever more easily than force applied near the fulcrum. In least squares regression each point can be thought of as "pulling" on the line to try to make its residual as small as possible. Those farther from the center of the data can exert a stronger pull.

Leverage and residuals are related by the relationship

In a simple regression of y on a single x-variable, the leverage of the ith case is

In a multiple regression, the expression is more complex, and requires

$$h_i + \frac{e_i^2}{\sum e_i^2} < 1.0$$

matrix algebra.

$$\frac{(x_i - \bar{x})^2}{\sum (x_i - \bar{x})^2}$$

For the regression of the vector $\mathbf{y}$ on the matrix of predictor variables, $\mathbf{X}$, the leverage of the ith case is the ith diagonal element of the matrix

$$\mathbf{H} = \mathbf{X}\,(\mathbf{X}^{T}\,\mathbf{X})^{-1}\,\mathbf{X}^{T}$$

The matrix $\mathbf{H}$ is sometimes called the *hat matrix* because

hat matrix

so $\mathbf{H}$ "puts a hat on $\mathbf{y}$". Some authors call $\mathbf{H}$ the Prediction or the Projection matrix (because it projects any vector into the space of the $\mathbf{X}$-

$$\hat{y} = Hy$$

vectors), and denote it $\mathbf{P}$.

Cases with high leverage can be extraordinary either because they have an extreme value on one of the predictors or because the combination of val-

Cases with high leverage can be extraordinary either because they have an extreme value on one of the predictors or because the combination of values on some set of predictors is unusual. For example, the subject in a medical study who is 6'3" tall is not extraordinary. Nor is the subject who weighs 97 pounds. But a subject who exhibits both of these measurements is extraordinarily thin for his height.

Leverage values are fundamental building blocks of many regression statistics. For example, the variance of the ith predicted value is

$$Var(\hat{y}_i) = \sigma_\varepsilon^2 h_i$$

where σ_ε^2 is the variance of the regression residuals, and the variance of the ith residual is

$$Var(\hat{e}_i) = \sigma_\varepsilon^2 (1 - h_i)$$

high-leverage point

There are several rules of thumb for deciding that the leverage of a case is large enough for the case to deserve special attention. Some authors recommend that cases with leverage exceeding $2p/n$ or $3p/n$ be considered "high-leverage points". Others have proposed that leverages above 0.5 deserve attention. We have found that a histogram, dotplot, or boxplot of the leverage values is often better than inflexible rules. Any case whose leverage sticks out in such a plot deserves a second look.

You can make a histogram of the leverages by choosing **Compute ▶ Leverages** from the regression table's global HyperView and then choosing **Histogram** from the **Plot** menu. Figure 23-5 displays the leverages of the regression given in Figure 23-3 as a histogram. They reveal a high-leverage point.

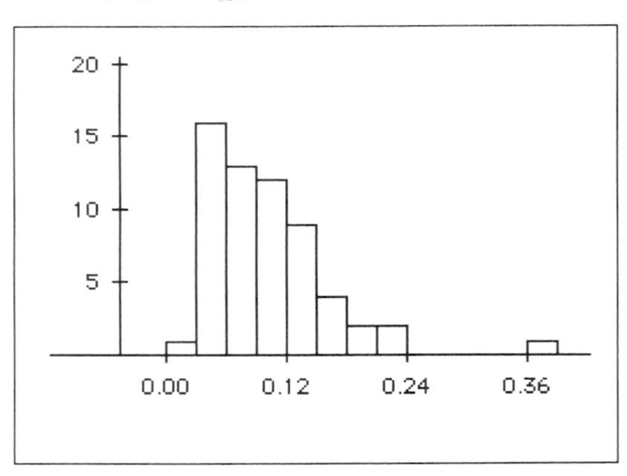

Figure 23-5. *A histogram of the leverages reveals a high-leverage point.*

The leverages are saved in the Results folder as a HotResult named *Leverages*. You can locate its icon conveniently with the HyperView under the name *Leverages* on the axis of the plot.

It is important to remember that leverage is a property only of the predictors. It takes no account of the y-values at all.

We have now seen in a few ways that York is a high-leverage point for this analysis. Along with the partial regression plot in Figure 23-4, we have substantial evidence that York is an extraordinary point that might be best analyzed separately from the other cases. We can isolate York from the analysis in a variety of ways. We choose here to generate an indicator variable that is 1 only for York and 0 elsewhere, and add it to the regression model.

To do this, select York alone in any plot, choose {Modify ▶ Selection} ***Record****, name the new variable "York", and drag its icon into the regression summary table. Figure 23-6 shows the results. The t-ratio associated with the indicator variable for York provides a test of whether York is an outlier for this analysis; the large t-ratio confirms that the datapoint is extraordinary.*

Dependent variable is: Mortality
R squared = 71.2% R squared (adjusted) = 67.9%
s = 35.25 with 60 - 7 = 53 degrees of freedom

Source	Sum of Squares	df	Mean Square	F-ratio
Regression	162551	6	27091.8	21.8
Residual	65847.3	53	1242.40	

Variable	Coefficient	s.e. of Coeff	t-ratio	prob
Constant	1040.19	85.19	12.2	≤0.0001
Education	-23.8313	6.445	-3.70	0.0005
pop density	0.010461	0.0042	2.48	0.0164
%NonWhite	2.82729	0.5990	4.72	≤0.0001
Rain	1.64921	0.5471	3.01	0.0039
LNOx	26.8664	11.39	2.36	0.0221
York	-155.443	45.04	-3.45	0.0011

Figure 23-6. *Isolating York by including a 0/1 variable improves the regression.*

A Note on removing cases

The question of when a case may be treated specially, such as we have done above with York, has been discussed by many authors but there does not seem to be any consensus of opinion. Certainly, you should not simply discard cases unless you have confirmed that they are wrong and beyond correcting. Nevertheless, often it is possible to obtain a very good fit to 90% or more of the data and deal specially with the remaining cases.

Dealing specially with selected cases means examining how they differ from the *pattern* described by the analysis, and reporting them as part of the analysis. In our experience, the occasional extraordinary case often can be understood best in terms of its deviation from a model or equation determined by the remaining cases, and this understanding can inform the entire analysis.

One response to an extraordinary case is to sample more data for similar situations. If the new data resemble the extraordinary case, then you have probably learned something new and important. If not, you may be able to correct or better understand the original extraordinary value. Unfortunately, we often do not have the luxury of being able to gather additional data.

23.4 *Studentized Residuals*

The residuals in a regression are the differences between the observed and predicted y-values:

$$e_i = y_i - \hat{y}_i$$

It is always a good idea to examine the residuals. The regression table's HyperView offers several appropriate plots including a probability plot of the residuals, scatterplots of the residuals and studentized residuals vs the predicted values (an effective way to check for some kinds of non-linear structure in the data). Context-sensitive HyperViews on each predictor's t-ratio value in the table offers scatterplots of the residuals and studentized residuals vs the predictor.

However, as we noted in the previous section, the residuals do not all have the same variance. Residuals for x-values near the mean of the x's are more variable than those for more extreme x-values. This follows from the observation above that the variance of the ith residual, e_i, depends, in part, on the leverage of the ith data point:

$$Var(\hat{e}_i) = \sigma_\varepsilon^2(1 - h_i)$$

This equation suggests that the raw residuals may be misleading, especially for assessing whether the residuals have constant variance.

A useful alternative is to standardize the residuals by dividing each by an estimate of its own standard deviation:

$$r_i = \frac{e_i}{\sqrt{Var(\hat{e}_i)}} = \frac{e_i}{\hat{\sigma}_\varepsilon\sqrt{1 - h_i}}$$

Residuals so standardized are called *studentized residuals* after the pseudonym of W.S. Gosset, the originator of Student's t.

internally studentized residual

To compute a studentized residual, we must estimate the standard error of the residuals, σ_ε, from the data. If we estimate the standard error by the residual standard deviation, s, we obtain the *internally studentized residual*. Internally studentized residuals have unit standard deviation, however, the ith residual participates in estimating the residual standard error so the numerator and denominator are not statistically independent.

externally studentized residual

The *externally studentized residual* estimates the residual standard deviation in the regression that omits the ith case, so it's numerator and denominator are statistically independent. We write the residual standard deviation omitting the ith case as $s(i)$. Note that this is not simply the standard deviation of all the residuals except the ith one, but rather is the residual standard deviation from a regression that omits the ith case. Data Desk uses a calculation method that does not require a new regression calculation to obtain $s(i)$.

Although the two kinds of studentized residuals are almost always very similar, some authors, including the authors of Data Desk, prefer the externally studentized residuals because they have several pleasant properties:

- The ith externally studentized residual is distributed as Student's t on $(n - p - 1)$ degrees of freedom — a distribution for which tables are readily available. By contrast, the internally studentized residuals follow a $Beta(1/2, (n - p - 1)/2)$ distribution. Of course, the studentized residuals in either form are not mutually independent.
- The ith externally studentized residual can be interpreted as a t-statistic for testing whether the ith case is an outlier in the regression.
- The estimate $s(i)$ is not inflated by gross errors in the ith residual.
- The externally studentized residuals are monotonic transformations of the residuals that may be arbitrarily large. They thus tend to exhibit outliers more dramatically.
- The ith externally studentized residual is the same as the regression coefficient of the indicator variable constructed to isolate the ith case, as we did with York in Figure 23-6.

The regression table's global HyperView offers probability plots of both the externally and internally studentized residuals and a scatterplot of the externally studentized residuals vs the predicted values. Context- sensitive HyperViews on each predictor's t-ratio value in the table offers a scatterplot of the externally studentized residuals vs the predictor.

NOTE:

The terms *standardized residual* and *studentized residual* are used in different ways by different statistics packages. Standardized residuals, in particular, are defined by some packages simply as e_i/s and in others as

$$\frac{e_i}{\sqrt{\sum e_i^2}}$$

These adjustments add little to the usefulness of residuals because the residuals still have different variances. It is always wise to determine exactly what is meant by these terms. The terminology used in Data Desk agrees with the most prominent recent books and articles on regression and regression diagnostics.

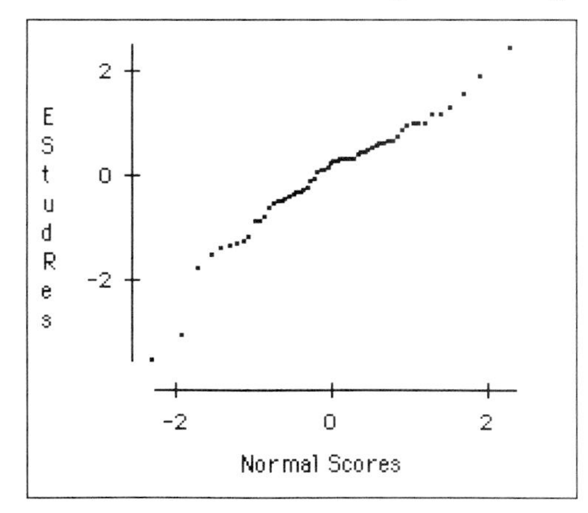

Figure 23-7. *A Normal probability plot of the externally studentized residuals from the regression given in Figure 23-6 is straight but for two cities, Lancaster and Miami.*

Working from the global HyperView menu for the regression in Figure 23-6, we can make a Normal probability plot of the externally studentized residuals. When $(n - p - 1)$ is large, the externally studentized residuals should be approximately Normal. The Plot shows a generally Normal distribution except for two points. These points turn out to be Lancaster and Miami.

This suggests that these two cities may deserve special attention. The partial regression plot for Education shown in Figure 23-4 provided a good explanation for why Lancaster should be influential and why it should be isolated from a regression intended to describe cities in general. We isolate it from the analysis here by creating a 0/1 indicator variable, naming it Lancaster, and adding it to the model.

```
Dependent variable is:     Mortality
R squared = 76.7%    R squared (adjusted) = 73.6%
s =  31.98  with  60 - 8 = 52  degrees of freedom
```

Source	Sum of Squares	df	Mean Square	F-ratio
Regression	175209	7	25029.8	24.5
Residual	53189.7	52	1022.88	

Variable	Coefficient	s.e. of Coeff	t-ratio	prob
Constant	1113.33	80.04	13.9	≤0.0001
Education	-29.7017	6.081	-4.88	≤0.0001
pop density	0.009130	0.0038	2.37	0.0214
%NonWhite	2.42814	0.5552	4.37	≤0.0001
Rain	1.68544	0.4965	3.39	0.0013
LNOx	28.8218	10.35	2.78	0.0075
York	-164.927	40.96	-4.03	0.0002
Lancaster	-120.331	34.21	-3.52	0.0009

Figure 23-8. *The regression model with a 0/1 indicator variable to isolate Lancaster. The large t-statistic for Lancaster confirms that it is an outlier relative to the remaining cases in this regression.*

23.5 *Distance Measures*

Leverage measures how extreme a case is in the predictors. Residuals and studentized residuals measure how extreme a case is in the dependent variable. Distance measures combine both of these concepts to measure the overall influence of a case on the regression.

DFFITS

The diagnostic statistic *DFFITS* is de fined as the change that would occur in the ith predicted value were the ith data point to be deleted, divided by the standard error of the ith predicted value:

$$DFFITS_i = \frac{\hat{y}_i - \hat{y}_i(i)}{s(i)\sqrt{h_i}} = \sqrt{\frac{h_i}{1-h_i}}\frac{e_i}{s(i)\sqrt{1-h_i}}$$

The calculation form of *DFFITS* shows it to be a product of a leverage-based term and the externally studentized residual. Although *DFFITS* does not in general follow a t-distribution, it is a t-like statistic, so cases with values of *DFFITS* greater than 1 or 2 may deserve special attention. Histograms, boxplots, and dotplots of the *DFFITS* values can highlight cases with extraordinary values regardless of the absolute magnitude of the *DFFITS* value.

Cook's distance

Cook's distance is similar to *DFFITS*, except that it uses the internally studentized residual and is squared relative to *DFFITS*. Specifically, Cook's distance is defined as

$$D_i = \frac{(\hat{y} - \hat{y}(i))^T(\hat{y} - \hat{y}(i))}{(p+1)\hat{\sigma}^2} = \frac{1}{p+1}\frac{h_i}{1-h_i}\frac{e_i^2}{s\sqrt{1-h_i}}$$

where the *(i)* subscript indicates values from a regression omitting the ith case. Cook originally proposed this statistic in an equivalent form that emphasizes its interpretation as a scaled change in the coefficients due to the omission of the ith case:

$$D_i = \frac{(\hat{\beta} - \hat{\beta}(i))^T \mathbf{X}^T \mathbf{X}((\hat{\beta} - \hat{\beta}(i))}{(p+1)\hat{\sigma}^2}$$

Cook suggests that D_i be compared to the percentiles of the F distribution with $(p + 1)$ and $(n - p - 1)$ degrees of freedom and interpreted relative to the confidence ellipsoids for β. Thus a D_i value that is approximately at the 95% point of its F-distribution can be interpreted to mean that removing the ith case would move the coefficient vector to the edge of its original 95% confidence ellipsoid.

The histogram of the DFFITS values for the regression in Figure 23-8 shows that Miami is influential in the regression.

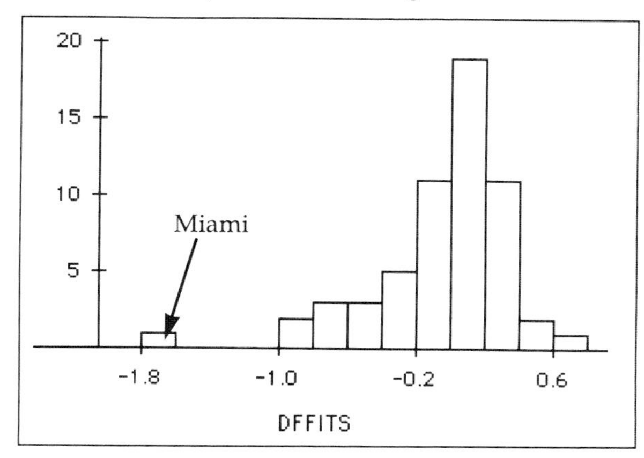

Figure 23-9. *A histogram of DFFITS for the regression in Figure 23-8 shows one influential point — Miami.*

*Having seen that Miami has a large studentized residual, we should not be surprised to find it influential according to this plot. We might want to identify the way in which Miami influences the regression so that we can better understand whether it is reasonable to treat Miami specially. One way to do this is to make Miami missing in one of the variables (for example, by typing a * in front of one of its data values) and compare the regression that includes Miami to the one that omits it.*

Variable	t-ratio Miami	t-ratio no Miami
Constant	13.9	13.9
Education	-4.88	-4.47
pop density	2.37	3.30
%NonWhite	4.37	4.89
Rain	3.39	4.32
LNOx	2.78	2.21
York	-4.03	-4.83
Lancaster	-3.52	-3.71

Figure 23-10. *Comparing the coefficient t-ratios with and without Miami shows that coefficients associated with population density and mean annual rainfall seem to have been affected most.*

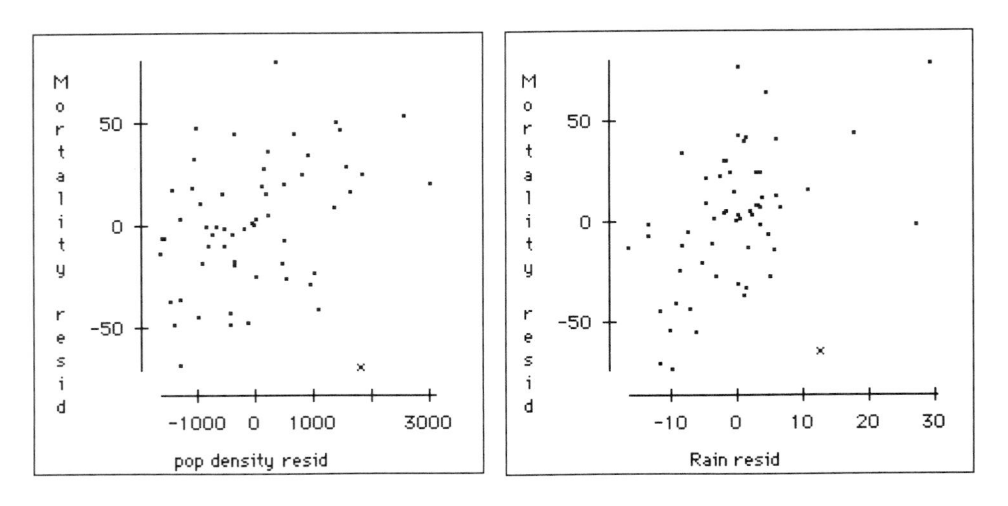

Figure 23-11. *Partial regression plots of Pop Density and Rain show that Miami (plotted with an x) influences both of these regression coefficients.*

It is easy to understand that Miami should have a climate different from other cities — Miami is classified as having a sub-tropical climate. It is harder to understand why Miami should be influential on the coefficient for population density.

To pursue the analysis further, we create a 0/1 indicator variable for Miami and add it to the regression model, obtaining the regression shown in Figure 23-12.

Dependent variable is: Mortality
R squared = 80.8% R squared (adjusted) = 77.8%
s = 29.32 with 60 - 9 = 51 degrees of freedom

Source	**Sum of Squares**	**df**	**Mean Square**	**F-ratio**
Regression	184550	8	23068.8	26.8
Residual	43848.3	51	859.770	

Variable	**Coefficient**	**s.e. of Coeff**	**t-ratio**	**prob**
Constant	1052.78	75.65	13.9	≤0.0001
Education	-25.5607	5.715	-4.47	≤0.0001
pop density	0.012006	0.0036	3.30	0.0018
%NonWhite	2.48953	0.5094	4.89	≤0.0001
Rain	2.01293	0.4659	4.32	≤0.0001
LNOx	21.5207	9.746	2.21	0.0318
York	-183.546	37.97	-4.83	≤0.0001
Lancaster	-116.452	31.38	-3.71	0.0005
Miami	-107.687	32.67	-3.30	0.0018

Figure 23-12. *The regression model with three cities isolated.*

A look at the diagnostics for this model suggests that Worcester, MA might also be extreme. Figure 23-13 shows the analysis with Worcester isolated

Dependent variable is: Mortality
R squared = 83.0% R squared (adjusted) = 79.9%
s = 27.87 with 60 - 10 = 50 degrees of freedom

Source	Sum of Squares	df	Mean Square	F-ratio
Regression	189567	9	21063.0	27.1
Residual	38831.3	50	776.625	

Variable	Coefficient	s.e. of Coeff	t-ratio	prob
Constant	1007.30	74.09	13.6	≤0.0001
Education	-23.0832	5.518	-4.18	0.0001
pop density	0.012236	0.0035	3.54	0.0009
%NonWhite	2.07282	0.5111	4.06	0.0002
Rain	2.59046	0.4977	5.20	≤0.0001
LNOx	23.6003	9.299	2.54	0.0143
York	-198.281	36.55	-5.42	≤0.0001
Lancaster	-120.636	29.87	-4.04	0.0002
Miami	-120.801	31.48	-3.84	0.0003
Worcester	-82.0973	32.30	-2.54	0.0142

Figure 23-13. *The regression with Worcester isolated as well.*

Diagnosis of the regression in Figure 23-13 reveals no additional extraordinary cases and general adherence to the regression assumptions. The studentized residuals appear close to Normal when plotted in histograms and Normal probability plots, and show no pattern when plotted against predicted values or individual predictors. The adjusted R^2 is large and the coefficient's t-ratios are all significantly different from zero.

23.6 *Regression Options*

The {Calc ▶ Calculation Options} **Set Regression Options...** command offers each of the diagnostic statistics discussed in this chapter. When you request a diagnostic statistic by checking a regression option, it is computed when the regression is computed and the results placed in a HotResult in the regression results folder.

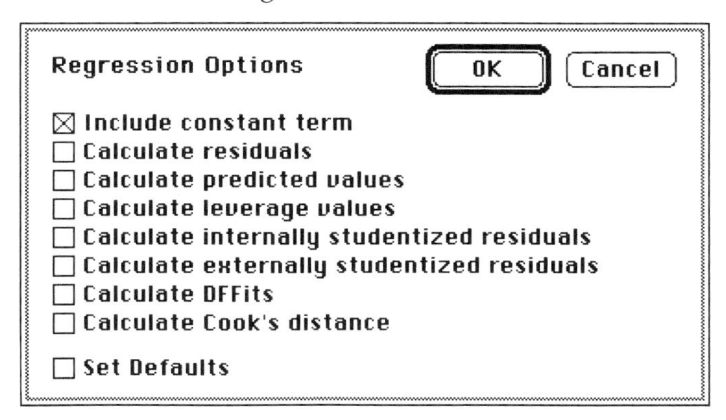

Figure 23-14. *The Regression Options dialog.*

The regression options affect all subsequent regressions, including those requested implicitly by updating, dragging variables into a regression or removing variables with a HyperView command. However, because the diagnostic statistics are generated as HotResults, they update automatically whenever the regression is updated. You can even update plots of diagnostic statistics without explicitly updating the regression summary table, but we do not recommend this because it can be confusing to see an out-of-date regression and up-to-date diagnostic plots.

The first three regression options have been discussed in Chapter 22. The remaining options calculate diagnostic statistics.

Calculate Leverage Values

When this option is checked, all subsequent regressions create a HotResult variable named *Leverages* that contains a leverage value for each case in the regression, and places it in the Regression folder.

Calculate Studentized Residuals

The options "Calculate internally studentized residuals" and "Calculate externally studentized residuals" each create a new HotResult variable in the Regression folder containing one appropriately studentized residual for each case. The created HotResult variables are named *Ext Stud Res* or *Int Stud Res* according to the kind of studentized residual requested.

Distance Measures

The options to calculate *DFFITS* and Cook's distance each create a new HotResult variable in the Regression folder that contains one value for each case. The variables are named *DFFITS* or *Cook's Distance.*

23.7 *A Note on Identifying Cases*

Regressions can be computed from a table of correlation coefficients. In the past, some statistics packages used this method because it is fast and requires relatively little computer memory. Modern regression methods, however, require that the individual cases be available for examination and correction during a regression analysis.

The diagnostic statistics discussed in this chapter are *casewise* diagnostics. That is, they generate an entire variable full of values, one for each case. The HotResults holding diagnostics are in the same Relation as the data because they hold a value for each case. The easiest way to use them is to simply make probability plots, histograms, or dotplots of them and identify any cases with extraordinarily large diagnostic values. A large diagnostic value typically means that the case in question is influential in the regression. The case might be entirely correct, but you will want to know that it is influential anyway.

Because all Data Desk plots link together, you can identify extraordinary points with the **?** tool. Alternatively, you can select them and see where they lie in other plots. Plotting several diagnostic statistics together as the variables making up a scatterplot can help you to identify extraordinary or influential cases.

23.8 Collinearity

A multiple regression in which two or more predictors are highly correlated is said to be *collinear*. Data Desk checks for severe collinearity and aborts the regression computation before the rounding errors that collinearity causes can ruin the value of the regression output.

Sometimes a case can be influential because it is the only point that stands between your analysis and collinearity. For example, in a regression of y on two predictors, if the predictors are collinear but for one point, the regression analysis is balancing the regression plane like a table top on a knife edge and one other point. That point will have leverage very near to 1.0; if it is moved up or down, the table top will follow. If you omit the point from the regression, the underlying collinearity shows through and the table top is no longer stable.

One way to assess collinearity is to perform a partial regression of one predictor on the others. The R^2 for this regression measures the collinearity of the multiple regression model as it affects the coefficient of that predictor. R^2 values near 1.0 indicate strong collinearity.

Appendix

Special Keys

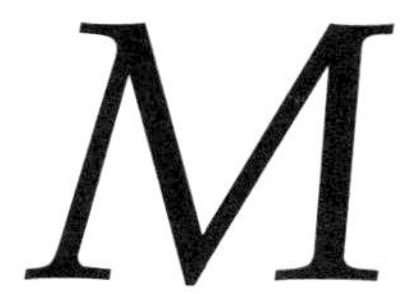ANY DATA DESK COMMANDS are available from the keyboard with ⌘-key equivalents. This section first lists the key assignments and then discusses other special keys. The command-key combinations marked with an asterisk (*) are standard assignments for Macintosh programs.

A Select **A**ll*

Works on text, icons, and plotted points.

B New **B**lank Variable

Creates a new blank variable, opens it and places it on the Data Desk desktop.

C **C**opy*

Copies text, variables, and cases.
The Copy command changes to reflect what is being copied, but you cannot see the change when you use a command key.

D **D**uplicate*

Duplicates any icon. Duplicating related or linked icons duplicates the relationship as well.

E New D**E**rived Variable

Creates a new derived variable
(⌘-D was already assigned to the Duplicate command)

F **F**ind

Searches for text in variables.

G **G**o To Next Selected Case

Helps you scan through the selected cases.

H Toggle **H**idden Points

Shows the currently hidden points and hides the currently visible points on the frontmost plot.

I **I**nfo...*

Shows the information records for all selected icons.

J **J**ot Note

Creates a new Jot Note and places it on the Data Desk desktop.

L **L**ocate...

Prompts for the name of an icon and seeks the icon of the specified name.

M Plot Tools

Places the Plot Tools palette on the Data Desktop.

N **N**ew Folder

Creates a new folder and places it on the Data Desk desktop.

O **O**pen*

Opens all selected icons.

P **P**rint*

Prints the window frontmost on your screen.

Q Quit*

Quits Data Desk.

R **R**eplace All Selected Cases

Substitutes specified text in all selected cases of front most variable editing window.

S **S**ave*

Saves any changes to a permanent copy of the file.

T **T**op Selected Case

Useful with ⌘-G. Start at the top and scan through the selected cases.

V Paste*

Think of an editor's insertion mark.

W Close*

Closes the window frontmost on your screen. A shortcut to clicking in the window's close box.

X Cut*

Think of crossing it out.

Y Show/Hide Axes

Hides the axes in a scatterplot to provide a clearer view of the data.

Z Undo*

Undoes the effects of the previous typing or editing command in most circumstances. The menu changes to reflect what action can be undone.

= Evaluate Scratchpad

Evaluates expression in frontmost scratchpad window. If a derived variable is selected, evaluates it and creates an ordinary variable holding the evaluated values.

- Hide Lines

Hides any added lines from the frontmost plot.

[or{ Select all in Groups

Selects all points with the same plot symbols as the selected points. The key is mnemonic of the bracket and brace used in Pascal programming or in mathematics to start a list of items in a set.

Other Active Keys

Shift

Datafile Window

Extends the selection of icons, adding new icons to the selection. All added icons are *x*-selected. The cursor changes to an X-cursor.

Shift-selection of already selected icons changes them to *x*-selection.

Editing Window

Editing within a case, extends the selection of characters from the current insertion point to the mouse point.

Editing several cases, extends the selection of cases from the case with the current insertion point to the mouse point.

Plots

Extends the selection of points and uses "Toggle mode" selection (de-selecting selected points and selecting de-selected points). Works with the ☜, lasso, or selection rectangle. Shift-brush sets Toggle selection so points flash on and off to help locate them easily in other displays.

Turns the ? tool into the ☜ tool working in toggle mode. This makes it easy to select a point once you have identified it.

Option

Datafile Window

Extends the selection of icons, selecting new icons with y-selection. The cursor changes to a Y-cursor.

Selecting already selected icons changes their selection to y-selection.

Holding Option while opening an icon "forgets" the old location of the icon's window and places the window neatly on the screen next to other open windows. For variable editing windows, this means aligning the windows into a table.

Plots

Alters the current plot tool to a default tool defined according to the type of plot. Option also enables the modification of the plot tool by the Shift and ⌘ keys, so that those keys in combination with Option select still other plot tools.

In Histograms, holding the Option key while dragging the window's size box rescales the histogram to the number of bars accommodated by the new window size.

Any Window

Option-Click the close box of any window to close all Data Desk windows.

Option-Shift

Datafile Window

De-selects any selected icon.

Command (⌘)

Variable Editing Windows

Enables discontinuous selection. Places a ✥ cursor on the mouse to indicate discontinuous selection.

Plots

In brushable plots, sets the size of the brush rectangle when the brush tool is selected.

For the **?** tool, displays the case ID number rather than any identifying text.

Tab

Datafile Window

When editing an icon name, Tab advances to the next icon to the right (if any) and selects its name for editing. This makes it easier to rename several icons at once.

Variable Editing Window

Tab ordinarily advances editing to the next variable in the editing sequence. Its function can be altered by the Preferences dialog.

Dialogs

In dialogs in which several boxes may be filled in, Tab advances to the next box.

Enter

Variable Editing Window

Enter ordinarily advances to the next case in the current editing window. Its function can be altered by the Preferences dialog.

Dialogs and Alerts

Most dialogs and alerts have a default button indicated by a black outline. Usually this is the OK or the Cancel button. Enter is the same as pressing this button. When no button is outlined, Enter is the same as pressing the OK button.

Return

Variable Editing Windows

In variable editing windows, Return advances to the next case in the same variable.

Dialogs and Alerts

Most dialogs and alerts have a default button indicated by a black outline. Usually this is the OK or the Cancel button. Return is the same as pressing this button. When no button is outlined, Return is the same as pressing the OK button.

Index

A

α-level, 235-237, 241, 246, 249, 255-256, 272
aborting, 46
About Data Desk, 14
active window, 16, 59, 66, 68, 205
Add Lines, 127, 134-135
  (see **Lines** submenu)
Add to Editing Sequence, 57
Align Editing Windows, 56
alphabetic sorting, 180
alternative hypothesis, 235
  for chi-square test, 210
  for one-sample mean, 235-237
  for one-way ANOVA, 254
  for regression coefficients, 271
  for two-sample means, 246
  in tables, 211
Alternative Delimiters, 76
amounts, 32-33
analysis of variance, 30, 157, 181, 249, 253-254, 263, 273
  ANOVA, 253-255
  ANOVA options, 264-265
  assumptions, 253
  balanced vs unbalanced, 259
  computing resources, 263
  Contingency Tables, 259
  displays for, 262-263
  HyperViews, 262-264
  levels, 259
  multi-way (see multi-way ANOVA)
  one-way (see one-way ANOVA)
  treatments, 256, 259
ANOVA (see analysis of variance)
ANOVA, 253, 259
ANOVA with Interactions, 260
Append & Make Group Variable, 183-184
  in ANOVA, 254
appending variables, 183
Apple menu
  About Data Desk, 14
archiving data, 43
area principle, 95, 107
arithmetic functions
  index, 169
arithmetic operators
  in derived variables, 167
  precedence order, 167
arrow keys, 57
arrow marker
  in HyperViews, 197
ASCII file, 76
assumptions
  for ANOVA, 253
  for pooled-t, 245
  for regression, 272, 275, 299-300, 311
Automatic Update, 39, 48, 156
average (see mean)
axes, 118
  Axis Name, 79
  Axis Scale, 79
Axes submenu,130
  Hide/Show Axes, 130
  Hide/Show Axis Names, 130

B

Backspace key
  in deleting cases, 54
  in Layout window, 80
Backup copy, 43, 46
balanced design
  in ANOVA, 259
balances, 32-33, 161
bar charts, 30, 33, 99, 106, 117, 120, 130, 206
Bar Charts, 99, 106
batch computing, 5, 295
before and after views, 47-48, 199
Bernoulli distribution, 221
  drawing simple random samples, 221
  in simulation, 228
  relationship to Binomial, 221
Beta distribution
  internally studentized residuals, 306
bimodal distribution, 100
Binomial distribution, 221
  in simulation, 228
  relationship to Bernoulli, 221
biweight, 84, 90
Black on White, 98
  for printing, 98
blunder, 30
Boolean expressions, 148, 151-152
Bonferroni adjustment

in confidence intervals, 234
in hypothesis tests, 237, 246
bounds, 16, 105, 132, 161, 219, 233, 237
Box-Cox Transformation, 159-161, 172
Box-Muller transformation, 228
Boxplot Side by Side, 104
Boxplot Y by X, 104
boxplots, 88, 104-105
by many groups, 105
components, 104
confidence intervals, 105
definitions, 113
displaying differences, 245-246, 253
in ANOVA, 253
inference, 105
Set Boxplot Options..., 105
braces
comments in expressions, 167
in derived variables, 146, 149
Brackets, 153, 163, 167, 173
Brush tool, 122
resizing, 122
brushing and slicing, 122-123, 141, 155
Brush tool, 122
Knife tool, 122
brushing
principles, 141-142
button presser
in HyperViews, 198

C

Calc menu,14
Cars data, 31, 83, 106-107, 112, 282
case, 27
copying, 57-58
cutting, 57-58
observation, 28
pasting, 57-58
period, 28
record, 28
respondent, 28
selecting, 20-22
shifting, 60
subject, 28
case identities, 200
case insertion point, 55
casewise functions, 167
index, 169
categorical variables, 106-108
Append & Make Group Variable, 183-184
cell (of a contingency table), 206
center (see measure of center)
Central Limit Theorem
definition, 226
illustrating, 226-227
Chi-Square distribution
inference for variance, 237
Chi-Square Interval of Individual Variances, 237
chi square statistic, 206, 213
chi square test for independence, 210, 213
alternative hypothesis, 210
degrees of freedom, 210-211, 213
expected values, 210, 213
null hypothesis, 210-211, 213
standardized residuals, 206, 210, 213
Chi-Square Test of Individual Variances, 237
Clear Colors, 129
Clear Editing Sequence, 57
Clear Lines, 134
Clear Selector, 186
Clear Symbols, 128
click, 8
Command, 9
Double, 9
Option, 8
Shift, 8
Clipboard, 40, 73-76
Copy Window, 40, 212
Show Clipboard, 40
Close All..., 68, 97, 187
Displays, 97
Variables, 68
Icon Windows, 68
Siblings, 187
close box, 16
Close Window, 54
coefficient of determination (R-square), 273
coefficients
regression, 269, 271-273, 289, 301-303, 306-307
cold objects, 47
collinearity in regression, 313
collapsing functions, 163, 168
function list, 175
Colors, 9, 47, 103-104, 108, 118-120, 122, 127-129, 133, 135, 189
palette, 9, 102, 118-119, 128-129
Show As, 128
Record, 129
Set Group, 129
Add by Group, 129
Add by Rank, 129
Add Linear, 129
Add by Indices, 129
Clear, 129
color screen, 135
column percent
in tables, 208
Command key
in brushing, 122
in selecting cases, 55
to move points in plots, 123
Command-hyphen

to show lines, 134
command-key equivalents, 9, 46
Command-Option
to identify and select, 120, 123
Command-period
to abort, 46
comments,
in derived variables, 146
Compact, 43
conditional distributions
in brushing and slicing, 142
confidence ellipsoid
Cook's distance, 308-309
confidence intervals
confidence level, 232
Estimate…, 232-233
for boxplots, 105
for one-sample mean & variance known, 232-233
for one-sample mean & variance unknown, 233
for regression coefficients, 271
for variance, 237
one-sided, 236-237
percentage points, 232-234
simulating behavior, 239-240
t-interval, 233
two sample means common variance, 245
two sample means unequal variances, 247
z-interval, 232-233
confidence level, 232
individual, 234
total, 234
Configure Editing, 69
confirmatory analysis, 217
context sensitive Hyperview, 197
contingency tables, 207
column percent, 208
contents, 207
Contingency Tables, 207
count, 208
expected values, 210
HyperViews, 211-212
in ANOVA, 260
margin, 208
row percent, 209
table percent, 209
test for independence, 209-210
control functions, 172
Cook's distance, 308, 312
in ANOVA, 265
internally studentized residuals, 308-309
Copy, 40, 46, 47, 58, 73
Copy Cases, 40, 58, 61, 76
Copy Front Window, 212
Copy Variables, 40, 74, 76, 284
Copy Window, 22, 40-41, 78, 80, 96, 190, 212, 284
copying cases
Copy Cases, 40, 58, 61, 76
copying data, 78
copying results, 78, 190
special symbols, 78
tables, 212
copying variables, 74
Clipboard, 73-76, 78
Copy Variables, 40, 74-76, 284
data table, 74
delimiters, 75-76
Scrapbook, 78
variable names, 74
counted fractions, 32, 172
counts, 32
in tables, 211
correlation, Chapter 21
Correlations submenu, 279
extracting from tables, 284
HyperViews, 280
Kendall's tau, 283
Pearson Product-Moment, 279-281
regression, 273, 295
Spearman Rank Correlation, 282-283
table, 280-281
correlation coefficient, 279-281
and predicted values, 280
and R-square, 280
linearity, 280
partial in regression, 301
Pearson Product-Moment, 282
standard scores, 281
table, 281-282
Covariance, 284
cropping tool (see Refocus tool)
cross-beam, 54
CumSum(.), 171
cursor control keys, 57
Cut, 40, 46, 58
Cut Cases, 58, 68

D

data, Chapter 3
data analysis, 3, 5
data analysis displays, 95
data context, 197
Data Desk Limits, 49
Data Desk menus, 14
Data Desk submenus, 14
Data Desk desktop, 13, 14, 47-48
data entry and editing, Chapter 5
data importing, Chapter 6
Data menu, 14, 17, 63
data table, 34, 73, 179
using the Return key, 73
using the Tab key, 73
datafile, 15, 43

Compact, 43
limits, 49
New Datafile, 15
Open Datafile, 15
Revert To Saved..., 43
vs **Revert,** 63
Save Datafile, 43-44, 63-64
Save Datafile As..., 43-44, 63
vs **Store,** 62
Defaults, 22, 79, 83, 220, 224, 291
degrees of freedom
ANOVA, 255
chi square, 210-211
Cook's distance, 309
externally studentized residuals, 307
inference for one-sample mean, 233
inference for two sample means, 247-248
inference for variance, 237
internally studentized residuals, 306-307
multi-way ANOVA, 260-261
one-way ANOVA, 255
paired t-test, 249
regression, 273, 289
tables, 210, 213
delimiters, 75-76
Set Delimiters button, 76
dependency, 149-150, 186
among duplicate icons, 189
and selectors, 191
circular, 150
derived variables, 149-150
discarding icons, 199
locating dependent icons, 42, 150, 199
underlying variable, 199-200
when variables have identical names, 154
derived variable, 39, Chapter 11
arithmetic functions, 169
arithmetic operators, 167
as hot objects, 47
avoiding errors, 164
braces for comments, 146
capitalization, 149, 166-167, 169
casewise functions, 167, 169
collapsing functions, 163, 168, 175
comments, 145-146, 149, 167
control functions, 172
dependencies, 149
dynamic parameters, 155
efficiency, 159, 162
evaluation to a variable, 147-148
expressions, 146, 148-149
formatting, 149
HotResults, 168
icon, 37, 145
ladder of powers, 158, 160, 169
locating dependent icons, 149
logical functions, 169
manipulating text, 152, 166
manipulation functions, 173
miscellaneous functions, 171
names of variables, 166
New Derived Variable, 145-146
numbers, 166
open, 150
parentheses, 149, 154, 167
re-expression, 145, 157-161, 172
recoding, 172
relational functions, 166, 168, 173
rounding functions, 170
Scratch Pads, 150
Slider, 155-160, 172
spaces in, 149
text constants, 148-149, 152
Transform submenus, 145-146
trigonometric functions, 171
underlying variable, 156
transformations, Chapter 11
variable names, 147, 149, 151, 154-155, 164-166
desktop metaphor, 8
desktop
resizing, 13
DFFITS, 312
externally studentized residuals, 312
in ANOVA, 265
diagnostics
distance measures, 308, 312
for ANOVA, 264
for regression, Chapter 23
importance in regression, 311
leverage in regression, 303-306, 311
studentized residuals, 306-308, 311-312
discarding folders, 18-19
discarding icons
with dependencies, 199
discontinuous selection, 55-56
discrete data, 106
distance measures, 308, 312
Cook's distance, 265, 308, 312
DFFITS, 265, 308-309, 312
distribution, 219
Bernoulli, 221, 228
Binomial, 221
Chi Square, 210, 237
Normal, 87, 112, 222-224
parameter, 219
percentage points of, 232, 236
Poisson, 219, 221-222
sampling, 217-218, 224-227
shape, 99-100
standard Normal, 222
Student's t, 233, 237, 247
Uniform, 222
Dotplot Side by Side, 103
Dotplot Y by X, 103
dotplots, 103
by many groups, 103

  - in ANOVA, 262-263
  - **Dotplot Side by Side,** 103
  - using the Grabber tool, 123
- double quotation marks
  - in derived variables, 149, 152
- dragging icons, 17, 46
- **Duplicate Icons,** 14, 189
- dynamic displays, 95
- dynamic parameters, 155, 293

E

- Edit menu, 14, 61, 63
- editing data, 59
  - **Align Editing Windows,** 56
  - example, 64-68
  - Backspace key, 56
  - case insertion point, 55, 57, 65, 67-68
  - Configure Editing, 69
  - **Copy,** 58
  - **Copy Cases,** 58, 61, 76
  - **Copy Window**, 78, 80, 96, 190, 212, 284
  - **Copy Variables,** 74, 284
    - setting delimiters, 76
  - correcting cases, 59
  - cross-beam, 55-56, 60
  - cursor control keys, 57
  - **Cut,** 58
  - **Cut Cases,** 58, 68
  - deleting cases, 60
  - discontinuous selection, 55-56
  - editing sequence, 56-60, 65, 69
  - Enter key, 57, 69, 150, 160, 164, 200
  - extended selection, 55
  - finding cases, 61-62
  - finding typos, 61
  - **Go To...,** 59, 61-62, 120
  - I-beam, 8, 55
  - inserting cases, 67
  - linking, 48, 60, 95, 122
  - **Paste,** 58
  - **Paste Cases,** 58, 68
  - recoding, 62
  - replacing cases, 61
  - Return key, 54, 57, 60, 69
  - **Revert,** 62-63
  - several variables, 56
  - shifting cases, 60
  - **Store,** 59, 62-63
  - Tab key, 56-57, 59-60, 65, 69
  - text insertion point, 54-56
  - **Undo,** 55, 58, 68
- editing sequence, 56-60, 65, 69
  - example, 64-68
- Eggs data, 259-260, 262-263
- **Empty Trash,** 18, 41
- engineering notation, 23
- Enter key, 57, 69, 150, 160, 164, 200
  - in editing data, 69
- entering data, 53, 65
  - example, 64-68
  - one variable, 53
  - several variables, 56
- equal variance in regression, 271, 275
- **Estimate...**
  - one-sample mean, 232
  - two-sample means & common variance, 247
  - two-sample means & unequal variances, 247
  - variance, 237
- Evaluate, 31, 41, 167
  - **Evaluate Derived Variable,** 148, 150
- exclamation mark, 38, 199
  - in ANOVA, 263
  - in regression, 292
- expected values, 117, 206, 210, 213
  - in tables, 206, 210, 213
- Exploratory Data Analysis, 5, 113, 157, 197
- exploratory analysis, 217
  - removing cases, 305
- **Export...,** 76
- exporting data, 76
  - delimiters, 75-76
  - text files, 76-77
  - variable names, 73-76
- expressions
  - precedence order in derived variables, 166
- externally studentized residuals, 306-307, 312
  - advantages over internally studentized residuals, 306-308
  - coefficient of indicator variable, 306-307
  - DFFITS, 308
  - in ANOVA, 262
  - in Regression, 306-307
  - regression options, 306-307
  - t-distribution of, 307
- extracting values from tables, 284

- F-test
  - and Cook's distance, 312
  - one-way ANOVA, 253-254, 256
  - regression, 273, 289
  - two-way ANOVA, 260
- **F-Test of Multiple μ's,** 249
- factor, 206-207
- Fiber data, 208-211
- field, 28, 111
- File menu, 14, 23, 43-44, 49, 63-64
- Find submenu, 61

Find..., 59, 61
Find Same, 61
Finder Desktop, 4, 8, 13, 15, 19, 49, 77
finding cases, 61-62
Fisher's Exact, 211
  two-by-two contingency table, 211
fixed notation, 23
folder, 19-20, 34,
  discarding, 20
  icon, 19
  New Folder, 19
  Results folder, 20, 201
  selecting, 20
Font
  changing, 79
Frequencies Options dialog, 206
Frequency Breakdown, 205
frequency table, 205-206

G

Gauss C.F., 222
Gaussian distribution (see Normal distribution)
Generate Patterned Data..., 110, 134, 156, 181-182, 223-224
  drawing simple random samples, 222
Generate Random Numbers..., 179, 220-221, 224, 239
generating random samples (see simulation)
GetCase(y, x), 173
Global Hyperviews, 198
Go To..., 59, 61-62, 120
  Next Selected Case, 59, 61-62
  Previous Selected Case, 62
  Top Selected Case, 59, 62
  Bottom Selected Case, 62
  Case #, 59, 62
Grabber tool, 124
  in plots, 124
  in histograms, 100, 102
grades, 32-33
Graduation data, 133-134
gray icon
  as a derived variable, 150
Greater(y, x), 169
Group submenu
  analyses by group, 186-187
  Group Button, 187
  Set Group, 128-129, 187

H

hat matrix, 303
Hearing data, 261
Hide/Show Axes, 130
Hide/Show Lines, 134
Hide/Show Axes Names, 130
hinges in boxplots, 113
histograms, 6, 18, 87, 99-102, 106, 109, 117, 120, 131, 142, 157, 161, 183, 219, 222, 264, 292, 300, 308, 311-312
  centering, 100-101
  scaling, 101-102, 131
  using the Grabber tool, 102
Histograms, 18, 99
  rescaling with the Option key, 101
Hoaglin, David, 113
hot objects, 47
HotResults, 21-22, 40, 168, 299, 312
Huff, D., 96
HyperViews, 16, 37-38, 95-96, 197-201
  and selectors, 186
  as warm objects, 48
  button presser, 198
  context-sensitive, 198, 306
  cursor, 38, 198
  global, 198
  in ANOVA, 262-263
  in correlation, 280
  in regression, 270, 272, 274, 291-292, 299, 306-307
  in **Summary Reports,** 87-88
  in tables, 211-212
  marker, 29, 39
  updating windows, 38, 199
hypothesis tests, 235-238
  α-level, 235-237
  alternative hypothesis, 235
    for one-sample mean, 235
    for two-sample means, 246
  assumptions for two sample means, 245
  Cook's distance, 308
  degrees of freedom for regression, 289
  DFFITS, 308-309
  F-test, 249
  for μ when σ is known, 232, 235
  for μ when σ is unknown, 233, 236
  for outliers in regression, 306-307
  for regression, 289
  for regression coefficients, 271-273
  for two-sample means and common variance, 246-247
  for two-sample means and unequal variances, 247-248
  for variance, 237-238
  null hypothesis, 235, 289
  null hypothesis for one-sample mean, 235
  null hypothesis for two-sample means, 246

paired samples, 248
percentage points, 232, 234, 237
pooled procedures, 245
several means and common variance, 253-255
simulating the behavior of, 241
t-test, 236-237, 245-249
test statistic, 235
Test..., 235-236, 241, 246-249
z-test, 235-236, 241

I

I-beam, 8, 55
icon, 3, 8, 17
datafile, 15
de-selecting, 8
derived variable, 39, 145
discarding, 8
duplicating, 127, 189
folder, 19
gray, 17
HotResult, 22
locating dependencies, 149-150
opening, 17
output record, 22
plot, 22
renaming, 18
Results folder, 20
ScratchPad, 17
text file, 15
Trash, 8, 19
variable, 17
x-variable, 18
y-variable, 18
icon alias, 16, 20, 80, 96, 98, 126, 132-133, 150, 190, 201
icon window, 15-16
identifier text
in plots, 123
identifying points, 123
Identifier tool, 123
IF/THEN/ELSE, 152-153, 158, 159-160, 164-165, 172-173
Import, 41, 73, 75-77, 184
importing data, Chapter 6
delimiters, 75-76
from several files, 77
text files, 73, 76-77
independence, 209-211, 213
indicator variable, 121, 136, 162, 211
in regression, 297, 305, 307-308, 310
individual confidence level, 234, 239
inference
boxplots, 105
confidence intervals, 231-232
for many-sample means, Chapter 18
for regression, 271-272
one-sample mean, Chapter 16
two-sample means, Chapter 17
variance, 237-238
infinity, 30, 166, 175,
in sorting, 180
Option-5, 85-86
influential points (see leverage)
Info..., 42, 44, 49
Information record, 27, 42
inserting cases, 67
interaction,
in ANOVA, 260-261
interactive displays, 95
intercept coefficient, 269
removing the intercept, 291
internally studentized residuals, 264, 306-307, 312
Beta distribution of, 307
Cook's distance, 308
in ANOVA, 264
interquartile range, 85-86
interval data, 32
interval size, 132
inv(y), 169
Iris data, 128
isolating points, 125
Refocus tool, 126
Scale to Selected Points, 125, 130
isolating subgroups
using indicator variables, 162

JK

JotNotes, 41

Kendall's tau, 279, 283
kinds of data, 32-33
amounts, 32
balances, 32
counted fractions, 32, 172
counts, 32
grades, 32
interval, 32
names, 32
nominal, 32
ordinal, 32
ranks, 32
ratio, 32
Knife tool, 122
kth %ile difference, 86

L

Labor Force data, Chapter 17
ladder of powers, 158-161, 169
    example, 159-161
lagged variable, 60
    Lag(y k), 173
    shifting cases, 60
Lasso tool, 120
Law of Large Numbers, 223-224
Layout, 80
    Layout windows, 80, 132, 190
least squares criterion,
    in regression, 270
leaving Data Desk, 23
Lesser(y x), 169
level, 206
    in ANOVA, 259
leverage
    and predicted values in regression, 303-304
    and residuals in regression, 303-304
    Cook's distance, 308
    definition, 303
    DFFITS, 308
    example, 304-305
    formula, 303-304
    hat matrix, 303
    high-leverage points, 304
    HyperViews, 304
    in ANOVA, 264
    in regression, 303-305
Line tool, 133-134
linearity
    in correlation, 280
    in regression, 269, 272, 287, 300
lineplot, 110-111, 128, 135, 223-224
    by adding lines to scatterplot, 135
Lineplots, 110-111
lines in plots, 133-135
Lines submenu, 133-135
    Add Lines, 134-135
    by Series, 134
    by Group, 134-135
    by From/To, 133
    Clear Lines, 134
    Hide/Show Lines, 134
    Record Lines, 134
linking, 48, 60, 95, 199-200 122
    discarding icons, 200
listwise deletion, 192-193
Locate, 19
Locate button, 154, 200
Locate Arguments of, 43, 150
Locate Selector, 186
Locate...Users Of, 42, 150
logarithm, 32-33, 39-40, 145, 157-161, 169, 181, 191, 200, 223
logical expression, 162
logical functions
    list, 169-170
Long Jump data, Chapters 8 and 20
LookUp(y, k), 163, 173-174
LookUpFirst, 153, 163, 174
LookUpLast, 153, 163, 174
Lower Bound, 132

Make Data Table, 74, 88-89, 189
managing windows, 59
Manip menu, Chapter 12
manipulation functions
    function list, 173
    Ranking, 181
    Sorting, 179-180
Manual Update, 39, 156
margins
    in tables, 208
matrix, 34
mean, 83-85
mean squares
    multi-way ANOVA, 260
    one-way ANOVA, 256
    regression, 274
measures of association (see correlation)
measure of center, 83
    average, 84
    biweight, 84, 90
    mean, 84
    median, 84
    midrange, 84
    robust, 84, 90
measure of central tendency, 83
measure of dispersion, 84
measure of location, 83
measure of spread, 84
    interquartile range, 85
    range, 85
    standard deviation, 85
    variance, 85
measure of variation, 84
median, 21-22, 84-86, 90, 104-105, 113-114, 153, 171, 300, 302
memory requirements, 44
memory
    limits, 49
menu, 14
    submenu, 14
Michelson data, 254-255
mid k, 86
midrange, 84
miscellaneous functions, 171

missing values, 29-30, 61, 85-86, 175, 191-193, 297
in correlation, 284
in ranking, 181
in regression, 290
in sorting, 180
Option-8 to type •, 29
pairwise and listwise deletions, 192
selectors, 193
Mix X and Y, 172, 293-294
mode
in distributions, 100
Modify menu, 15, 118, 120, 127-128, 130, 136
moments, 87
coefficient of excess, 87, 91
coefficient of kurtosis, 87, 91
coefficient of skewness, 87, 91
Mosteller F., 32
moving points in plots, 123-124
Grabber tool, 124
multimodal distribution, 100
multi-way ANOVA, 259, 261, 263, 265
ANOVA, 259
ANOVA table, 259-260, 262-264, 273
degrees of freedom, 260-261, 272-273
F-test, 260
interaction, 259-261, 263-264
mean squares, 260
notation and formulas, 260-261
one observation per cell, 261
sum of squares, 260-261
multiple comparisons, 235
Bonferroni adjustment, 234, 237, 246
Multiple Lineplots, 110-111, 135

N

names, 33
renaming icons, 17
NaN (Not a Number), 29
New Blank Variable, 53-54, 64
New Datafile, 15
New Derived Variable, 145-146
New Folder, 19-20, 158
New JotNote, 41
New Layout, 80
New ScratchPad, 40
New Slider, 155-156
New Variable, 18
nominal data, 32
nonadditivity, 157
nonconstant variance, 157
nonlinearity, 157
Normal distribution, 91, 112, 114, 171, 219, 222-224, 227, 232-234, 239, 272, 307
Central Limit Theorem, 226-227
in ANOVA, 253
in regression, 272
inference for means, 232, 245
kurtosis, 87
parameters, 219
percentage points, 232, 234, 237
shape, 219
simulation, 219
standard Normal, 222
underlying inference for variance, 237
Normal probability plot, 38, 112, 114, 171, 275, 307
definition, 114
for regression, 275
Normal Prob Plots, 112
normal score, 112, 114
normality, 112, 226, 253, 272
assessing, 112
in regression, 275
Central Limit Theorem, 226-227
Not a Number (see NaN), 29
notched boxplots (see boxplot inference)
Nscores, 112, 114, 171
null hypothesis
in tables, 210- 211, 213
for one-sample mean, 235-237
for one-way ANOVA, 254-256
for regression, 272-273, 289, 297
for two-sample means, 246
number, 30-31, 166
 # NonNumeric Cases, 86-87
 # Numeric Cases, 86-87
numeral, 30
in sorting, 180

O

object-oriented analyses, 39
observation, 28
Olympic Gold data, 110-111, Chapter 20
one-way ANOVA, Chapter 18
ANOVA, 253-255, 259-265
alternative hypothesis, 254-255
ANOVA table, 254-255
degrees of freedom, 255-256
F-test, 253-256
mean squares, 256
notation and formulas, 255
null hypothesis, 254-256
sum of squares, 254-256
Open Datafile, 15
Open, 17, 64, 77
Option-8 to type •, 166
Option key
in aligning variables, 56
in histograms, 101

in selecting icons, 21
Options
in frequency tables, 205-206
in ranking, 181
in regression, 290-291
in sorting, 180
in two-way tables, 207-209
order statistic, 114, 171
kth %ile difference, 86
maximum, 85-86
mid k, 86
minimum, 85-86
percentile, 85-86
quartile, 86
rank, 86
ordinal data, 32
outliers, 30, 238, 245, 253, 282, 296, 307
in boxplots, 104
in regression (also see leverage), 303
output window, 16

P

paired t-test, 248
example, 248-249
null hypothesis, 249
pairwise deletion, 192-193
Page Setup..., 63, 79-80
Palette, 9
Plot Colors, 9, 118, 127
Plot Symbols, 47, 118, 120, 122, 127-128
Plot Tools, 9, 21, 102, 109, 117-118, 120, 136
Parallel Append, 77-78, 183-184
parameter, 219
parentheses
in comments, 167
in derived variables, 149, 167
partial correlation, 301
partial regression plots, 289, 301-302, 310
assessing collinearity, 313
interpreting the slope, 301
residuals, 301
Paste Cases, 40, 43, 58, 68
Paste Variables, 40, 43, 74-76, 151, 284
pasting variables, 74-76
Clipboard, 74
delimiters, 75-76
variable names, 75
patterned data, 110, 134, 156, 181-182, 223-224
Pearson product-moment correlation (see correlation coefficient)
Pearson Product-Moment, 279, 281-282, 295
percentage points, 232, 234, 237
percentile, 85-86, 113
period, 28
pi, 156, 166
PICT form, 79, 212
in copying and printing results, 190
pie chart, 31, 107-108, 120, 122, 136-137, 197
Pie Charts, 107
plot actions, 119
plot aspects, 118
plot axes, 130
Plot Colors palette, 9, 118
Plot dimensions, 132
Plot menu, 14, 96-98, 118
Plot Options submenu, 118
Plot Symbols palette, 118, 128
plot tools, Chapter 9
command-key equivalents, 137
Plot Tools palette, 9, 102, 118
plot window, 16, 119, 125-126, 130
plots, Chapter 8
actions, 119
aspects, 118
axes, 130
axis labels, 37, 198
brushing and slicing, 119, 122-123, 141
comparing categories, 106-108
conventions, 96-98
depicting distributions, 100-105
depicting relationships, 109-112
icons, 22
identifying points, 123
isolating points, 124-125
lines, 127, 133-135
linking, 121-122
moving points, 123-124
recentering, 124
recording group variables, 129
rescaling, 100-101, 109, 111
resizing, 4, 16, 101, 105
scale, 130-133
selecting, 120-121
Show As submenu, 128
showing case numbers, 123
slicing (see brushing and slicing)
specify plot scale, 131
substituting, 126-127
symbols, 127-129
tools, Chapter 9
visibility submenu, 124, 133
plotting
Black-on-White, 79, 98
White-on-Black, 79, 98
Pointer tool, 120, 123
Poisson distribution, 219, 221-222
pooled procedures, 245
ANOVA, 253-255
estimate of variance, 247
population, 29
Precision, 23, 78, 132, 212, 228, 231-233
predicted values, 168, 193
and correlation, 280

and leverage in regression, 303
in ANOVA, 264
in regression, 270, 274-275, 289-290, 295, 306-307, 311
Prediction matrix, (see Hat matrix)
Preferences, 69
preferences, 14, 22-23, 57, 60, 69
presentation displays, 96, 107, 127-128, 130
Print Front Window, 63, 79-80, 190
for tables, 212
Print Variables, 63, 79
printing, 14, 63, 79, 98, 190, 212, 299
Page Setup..., 63, 79-80
PICT form, 79, 212
Print Front Window, 63, 79-80, 190
Print Variables, 63, 79
Save Desktop Picture, 80, 96
tables, 212
variables, 63, 79
printing results, 190
pseudo-random-numbers, 228

Q

Q-Q plot, 114
Quantile-Quantile plot (see Q-Q plot), 114
quartile, 86
question mark (see Identifier tool)
Quit, 23, 43-44, 63, 76

R

R-square, 273-274, 289
adjusted, 273-274, 289
and collinearity, 313
and correlation, 274, 280
random sample, 188, 217-218, 221, 223, 231
random number generator, 228
randomness, 217, 219, 234
in confidence intervals, 234
range, 85
Rank, 181
rank(•), 171
rank correlation, 279, 282-283
Kendall's tau, 279, 283
Spearman rank correlation, 279, 282
ranking, 30, 85-86, 129, 181
missing values, 181
ranking options, 181
Ranking Options dialog, 181
ratio data, 32
reciprocal, 157-158
recoding, 62, 172
IF/THEN/ELSE, 152-153, 158-160
Replace..., 62
record, 27-29
Record Line, 134
Rectangle tool, 120
rectangular data set, 28, 73
Redo in New Window, 38, 199
re-expression, 145, 157-161, 172, 297
example, 159
ladder of powers, 158, 160, 169
list, 169-174
Refocus tool, 126
regression, Chapters 20, 22, 23
adding or removing predictors, 292
adjusted R-square, 273-274, 289
adjusting cases, 292
alternative hypothesis, 272
ANOVA table, 273
assumptions, 271-272, 275, 295-296, 299-300, 311
coefficients, 269-272, 279-280, 289, 299,
collinearity, 313
degrees of freedom, 273, 289
DFFITS, 265, 308-309, 312
diagnosing residuals, 274-275
diagnostics, Chapter 23
distance measures, 308, 312
estimating residual variance, 273
example, 270-271, 287-288
externally studentized residuals, 306-307, 312
F-test, 273, 289
hat matrix, 303
high-leverage points, 304
hypothesis tests for coefficients, 271-273
importance of diagnostics, 312
indicator variable, 297, 305, 307-308, 310
internally studentized residuals, 306-307, 312
least squares criterion, 270
leverage, 303-304, 306, 308, 312-313
mean squares, 274
missing values, 290, 297, 299
null hypothesis, 272-273, 289, 297
options, 289-291, 299, 311-312
partial regression plots, 289, 301, 310
predicted values, 270, 274-275, 289-291, 295, 306-307
R -square, 273-274, 289
Regression, 270-271
residuals, 270, 272-275, 289-291, 293-297, 301-304, 306-308, 311-312
Results folder, 291, 295, 299, 302, 304, 311
selector variable and, 290
Set Regression Options..., 289-290, 299, 311
standardized residuals, 307
stepwise regression, 295-296
studentized residuals, 306-308, 311-312
sum of squares, 273-274
summary table, 270-272

updating, 292
relation, 20-21, 27,28, 34, 135
importing, 77
Multiple Lineplots, 135
relational functions, 166, 168, 173
removing cases, 305
Remove From Editing Sequence, 57
Repeat Variables, 111, 182
Replace..., 62
replacing cases, 61-62
residuals, 112, 159, 168, 193, 206, 210-211, 213, 264-265, 270, 272-275, 289-291, 293-297, 301-304, 306-308, 311-312
and leverage in regression, 303
diagnosing in regression, 274-275
estimating variance in regression, 273
in ANOVA, 264
in regression, 269, 290
standardized, 307
studentized, 306-308, 311-312
Resize tool, 126
resizing plots, 100-102, 126
respondent, 28
Results folder, 20, 163, 201, 218-219, 295, 299, 302, 311
in regression, 299, 311
Results log, 20, 201, 292
Set Results Log, 20
Return key, 54, 57, 60, 69
in a data table, 73
in editing data, 56-57, 69
Revert, 43-44, 58-59, 62-63
vs **Revert to Saved...,** 63
Revert To Saved..., 43-44, 63
rho (see Spearman Rank Correlation)
rogue, 30
rounding functions, 170-171
RoundEven, 171
RoundUp, 171
RoundDown, 171
row percent
in tables, 209

S

Sample..., 188, 218
sample, 29
simple random, 217
sampling distribution, 114, 171, 224-227, 271
sample mean, 226-227
Save, 43, 44
Save as, 44
Save Datafile, 43-44, 63-64
Save Datafile As..., 43-44, 63
Save Desktop Picture, 80, 96
saving data, 43-44, 62-63
Compact, 43
Revert, 43-44, 58-59, 62-63
Revert To Saved..., 43-44, 63
Save Datafile, 43-44, 63-64
Save Datafile As..., 43-44, 63
Store, 43, 62-63
saving screen images, 80
scale, 79, 101-102, 110-111, 118, 125, 130-132, 145, 155-157, 187, 294
Scale submenu, 130-133
Home, 130
Scale to Selected Points, 125, 130
Specify Plot Scale, 102, 131-133
scatterplot, 109-110
Scatterplots, 109
scientific notation, 23, 132, 166
Scratch File, 22-23
ScratchPad, 22-23, 40-41, 61, 75, 77, 80, 150-151, 201, 284
extracting values from correlation tables, 284
icon, 40
scrolling, 16, 59
seed for random number generator, 220, 228
Select All, 21
Select Summary Statistics, 22, 83, 88-90, 225
selecting, 21
selecting cases, 21, 55, 136
discontinuous selection, 55-56
indicator variable, 121
extending selection, 55-56
Selection submenu, 120
selecting icons, 13, 17, 21
selecting folders, 20
selecting points, 120, 141
Edit menu (Select All), 120-121
Lasso tool, 120
Pointer tool, 120
Rectangle tool, 120
selecting variables, 18, 37, 198
x-selection, 18, 145
y-selection, 18, 145
selection criteria, 185
Select Symbols, 128
Selection submenu, 120
Add Selection, 121
Clear Selection, 121
Record, 120
Toggle Selection, 121
Selector button, 121, 185-187, 191, 193, 198, 212, 218, 221
selector variable, 73, 121, 186, 191, 193, 198, 221, 290
criteria, 185
setting and clearing, 185-186
selectors, 169, 185-186, 191
in HyperViews, 198
Selector submenu
Set Selector, 121, 221
Clear Selector, 186
Locate Selector, 186
Send Window Behind, 17
sequence (see editing sequence)

sequence box, 56-57, 59
Sequence submenu
    Clear Editing Sequence, 57
    Remove from Editing Sequence, 57
Set Boxplot Options..., 105
Set Frequencies Options..., 205
Set Group, 128-129
Set Plot Scale..., 102, 131-132, 155, 157, 187
Set Ranking Options..., 181
Set Regression Options..., 289-290, 299, 311
Set Results Log, 20
Set Selector, 121, 221
Set Sorting Options..., 180
Set Table Options..., 207
Shift Cases Down, 60
Shift Cases Up, 60
Shift key
    in selecting cases, 55
    in selecting icons, 21
    in selecting multiple windows, 45
Shift-Option-Command
    to resize plots, 126
shifting cases
    Shift Cases Down, 60
    Shift Cases Up, 60
Show all Points, 125, 133
Show As submenu, 128
Show Black on White, 98
Show Clipboard, 40
Show White on Black, 98
simple random sample, 188, 217, 221
    drawing, 218-219, 220
simulation, Chapter 15
    and the Central Limit Theorem, 226-227
    behavior of confidence intervals, 239-240
    behavior of hypothesis tests, 241
    Bernoulli distribution, 221
    Binomial distribution, 221
    Box-Muller transformation, 228
    details and formulas, 228
    Generate Random Numbers..., 179, 220-221, 224, 239
    generating random samples, 220
    Poisson distribution, 221-222
    pseudo-random numbers, 217, 228
    random number generator, 228
    sampling distribution, 224-227
    seed, 220, 228, 240
    standard Normal distribution, 222, 224
    Uniform distribution, 222, 225-227
Singers data, 104-105, 245
single quotation marks
    in derived variables, 149, 166
size box, 16
skewed distribution, 100, 112
skewness, 87, 91, 114
slicing (see brushing and slicing)
slider, 42, 132, 155-160, 172, 199, 293-295
slope coefficient, 269
SMSA's data, Chapter 23
sort key, 179-180
Sort on Y Carry X's, 110, 179
sorting, 179-180
    alphabetically, 180
    infinities, 180
    missing values, 180
    numerals, 180
    sort key, 179-180
    sorting options, 180
    stable, 179
    Unsort Indices, 179
Sorting Options dialog, 180
Spearman Rank Correlation, 279, 282-283
Special menu, 16, 43,
    Locate Users of, 42, 150
    Locate Arguments of, 43, 150
Specify Plot Scale, 102
splitting variables, 183
spread (see measure of spread)
spreadsheet
    importing and exporting, Chapter 6
square root, 145, 157-158, 161, 169,
stable sorting, 179
Stack Windows, 97
standard deviation, 84-85
standard Normal distribution, 222-223
standard scores and correlation, 281
standardized residuals
    in regression, 307
    in tables, 210, 213
statistic, 219
statistics context and HyperViews, 197
stepwise regression, 295-296
Store, 62-63
    vs **Save Datafile,** 63
straight line equation, 269
Student's t-distribution (see t-distribution)
studentized residuals, 264, 306-308, 311-312
    Beta distribution of, 307
    Cook's distance, 308
    DFFITS, 308
    HyperViews, 307
    regression options, 311-312
    t-distribution, 307
subgroups, 95, 100, 106-107, 162, 186-187
subject, 28
submenu, 9, 14-15
Subscripting, 153-154
subset (see selector button)
substitute, 119
sum, 87
sum of squares,
    multi-way ANOVA, 260-261
    one-way ANOVA, 254-256
    regression, 273-274
Summaries as Variables, 83, 88-89,
    simulating a sampling distribution, 225-227

Summaries as Variables By Group, 89
Summary functions, 175
Summary Reports, 78, 83, 87, 192
summary statistics, Chapter 7
    example, 84
    Select Summary Statistics, 22, 83, 88-90, 225
    Summary Reports, 78, 83, 87, 192
symbols, 128
symmetric distribution, 87,100
    skewness, 87, 91, 114
System 7, 13, 17, 19, 44, 74, 80
Systematic Sample, 188, 218

T

t-distribution
    DFFITS, 308
    for one-sample mean, 233-234, 236-237
    for paired samples, 248
    for two-sample means, 247
    externally studentized residuals, 307
t-interval
    for one-sample mean, 233-234
    for two-sample means, 245, 247
t-test
    paired samples, 248-249
    one sample mean, 236-237
    regression coefficients, 272
    two-sample means and common variance, 245-246
    two-sample means and unequal variances, 247-249
Tab key, 17-18, 56-57, 59-60, 65, 69, 220
    in a data table, 73
    in editing data, 56-57
Table Options dialog, 207
table percent
    in contingency tables, 209
tables
    multi-way, 214
    two-way (see two-way tables)
tails, 87, 100
test statistic, 235, 248
Test...
    one-sample mean, 236-237
    paired samples, 248-249
    two-sample means and common variance, 245-246
    two-sample means and unequal variances, 247-249
test for independence in tables, 210-211, 213
text files, 73, 76-77, 184
    delimiters, 75-76
    importing, 75-77
text form, 78, 190, 212
    in copying and printing results, 190
Text Format..., 63, 73, 79, 212
text insertion point, 54-56
TextOf, 148, 152, 162, 165, 172-174, 185
Tile Windows, 16, 97
title bar, 16
Toggle Hidden Points, 133
total confidence level, 234
Total # Cases, 87
Transform submenus, 39, 145-146, 158, 169
transformation (see derived variable)
    to improve analyses, 33, 157
Transpose, 159, 184-185
Trash, 8, 13
    Empty Trash, 18
trend
    equally-spaced intervals, 109
    in brushing and slicing, 142
    in plots, 133
    unequally-spaced intervals, 110
trigonometric functions, 171
Tukey John W., 5, 33
Tukey's Lambda, 172
two-way tables, 206
    column percent, 208
    contingency tables, Chapter 14
    HyperViews, 211-212
    margin, 208
    row percent, 209
    Table Options dialog, 207
    table percent, 209
    test for independence, 209-210

U

underlying variable, 156, 199-200
Unemployment data, Chapter 22
Uniform distribution, 222, 225-227
    drawing simple random samples, 224-225
    in simulation, 228
Undo, 40, 46, 55, 58, 68, 80, 146
unimodal distribution, 100
unit, 29
univariate statistic, 83
Unsort Indices, 110, 179
Untitled datafile, 68
Update This Window, 199
updating, 38-39, 47, 58, 148, 164, 186, 199-201, 293, 312
    exclamation mark, 38, 199
Upper Bound, 132
Use colors (in Pie Charts), 108

V

variable, 17-19, 27
- appending and splitting, 183
- category, 106-108
- cold objects, 47
- column, 28
- copying, 74
- derived (see derived variable)
- discarding, 18-19
- editing, Chapter 5
- editing window, 54,
- field, 28
- finding cases, 61-62
- finding typos, 61
- icon, 17
- lagged, 60, 173
- **New Blank Variable,** 18, 53-54, 64
- recoding, 62, 172
- replacing cases, 61

variance, 85
- pooled estimate, 246-247

Visibility submenu, 124, 133
- **Show All Points,** 125, 133
- **Show Only Selected Points,** 124, 133
- **Toggle Hidden Points,** 125, 133

W

warm objects, 48
whiskers, 104, 113, 245
White-on-Black, 79, 98
window, 8, 15-17
- active, 16, 59
- aligning editing windows, 56
- arranging on Desktop, 16-17
- avoiding duplicates, 201
- avoiding overload, 201
- closing, 16
- exclamation mark, 38, 199
- limits, 49
- size box, 16
- Windows submenu, 14, 16
- zoom box, 16,

Windows submenu, 14, 16
word processor,
- importing and exporting, 74, 78

XYZ

x-cursor, 18
x-selection, 18, 37

y-cursor, 18
y-selection, 18, 37

z-interval, 232, 239
- for one-sample mean, 232

z-test, 235-236, 241
- for one-sample mean, 235

zoom box, 13, 16, 97, 99

•, 166 (see missing values)
∞ , 166 (see infinity)
π , 166 (see pi)

Exercises

SIMPLE SUMMARIES

NAME____________________ DATE ____________________

CLASS____________________ INSTRUCTOR ____________________

1. a) Retrieve the Singers dataset. Create boxplots comparing the heights of all 4 voice parts. Use **Summary Reports** in the **Compute** menu to obtain the statistics, and record them in the table:

	Mean	**Median**	**Standard deviation**	**Interquartile range**	**Upper Quartile**	**Lower Quartile**
Soprano	______	______	______	______	______	______
Alto	______	______	______	______	______	______
Tenor	______	______	______	______	______	______
Bass	______	______	______	______	______	______

b) Sketch the boxplot of the Altos and indicate on it the values of the median, upper and lower quartiles, maximum, and minimum.

2. Some of the order statistic summaries are in fact measures of center or spread. Find each of these for the Altos and enter it in the appropriate column:

	Center	**Spread**
Mid-25% point	______	______
33% difference	______	______
50 %-ile	______	______

3. (a) One of the Altos is 72 inches tall. Change her value to a missing value. (Open Alto and insert a "*" before the 72.) Recompute the statistics for Altos. Report these numbers:

Mean __________ Median __________ St. dev.___________ Interquartile range_________

(b) Considering the effect that removing the 72 inch Alto has on the mean and median values, which statistic would you use to describe the typical height of Altos? Explain your choice.

(Don't forget to return the 72 inch measurement to its non-missing status.)

4. Using the CU Cars dataset compute the mean, median, midrange, and mid-25 percent point of the Displacement variable. Write them below.

Mean ________ Median ________ Midrange ________ Mid-25% point ________

All of these statistics are measures of Center. Why do they differ so much?
(Hint: *Make a histogram and mark on it the median, 25, and 75 percent points.*)

SIMPLE SUMMARIES

NAME ______________________ DATE ______________________

CLASS ______________________ INSTRUCTOR ______________________

5. The data in the Cars dataset are recorded in U.S. units. Thus, for example, Weight is in thousands of pounds, Displacement is in cubic inches, and MPG is in miles/gallon. Suppose we wish to convert some of these variables to other units. First, consider the summaries:

	Unit	Mean	Standard deviation	Interquartile range	Maximum
Weight	1000 lbs	______	______	______	______
Displacement	in^3	______	______	______	______
Horsepower	hp	______	______	______	______
MPG	mi/gal	______	______	______	______
Cylinders	#cylinders	______	______	______	______

6) Fill in the corresponding values for the indicated units. You should be able to find these values with no more than a calculator and the information from Exercise 5. You can also use the calculator ability of Data Desk's ScratchPads to make the calculations easy.

	New Unit	Conversion Factor	Mean	Standard deviation	Interquartile range	Maximum
Horsepower	watts	1watt=746hp	______	______	______	______
Displacement	cubic cm	1c =1in^3/16.4	______	______	______	______
MPG	furlongs/day	1fg/dy =mpg/52	______	______	______	______
Weight	long ton	1ton = 2240lb	______	______	______	______
Cylinders	excess of 4	#Cyl – 4	______	______	______	______

What principles did you apply in arriving at your answers?

7.. On the histograms below, mark the approximate locations of the mean, median, and mode:

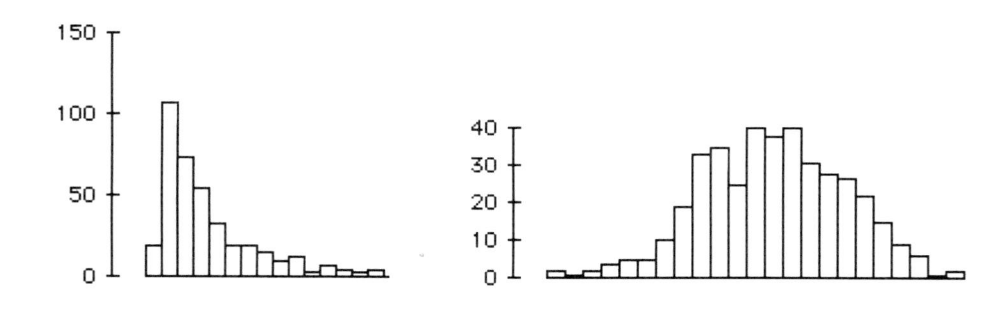

8. Which of these histograms shows a skewed distribution? Would you expect the coefficient of skewness to be positive or negative?

9. Suppose the data in the histograms of exercise 7 were shifted by having 100 added to each value, and were rescaled by multiplying each value by 10. What would the histograms of these new variables look like (compared, for example, with the histograms above)?

__________ __________ __________

DISPLAYING DATA

NAME ______________________ DATE ______________________

CLASS ______________________ INSTRUCTOR ______________________

1. For each of the listed characteristics, draw a sketch of a distribution that is

(a) bimodal

(b) skewed to the left

(c) skewed to the right

(d) multimodal

2. Make a histogram of the Altos variable in the Singers dataset. Which of the descriptions of (1) apply to this histogram?

3. Make a histogram of Soprano heights, keeping the window containing the histogram of Alto heights open. Now, resize both windows so that they can be viewed side-by-side. Describe how they compare with respect to:

• modes

• skewness

4. Make histograms of the heights of Tenor and Bass singers. How do these groups behave with respect to modality and skewness?

5. Close the histogram windows and create a boxplot comparing all four parts.

(a) How do the centers compare among the parts? How can you tell that from the boxplots? Which measures of center are the boxplots informative about? Use a sketch to indicate how to compare centers with boxplots.

(b) How does skewness compare among the parts? How can you tell that from the boxplots? (Sketch boxplots that would indicate symmetry or skewness.)

6. Open the Michelson dataset. It contains measurements of the speed of light recorded in several experimental trials. Display boxplots of the first 5 trials. The boxplots suggest that the first and third trials may be skewed. Construct histograms of the first and third trials and examine the tails of the distributions. Try various scalings for the histograms. Pick scalings that accentuate the skewness in the distributions. Sketch the distributions as they appear on your screen.

DISPLAYING DATA

NAME ______________________ DATE ______________________

CLASS ______________________ INSTRUCTOR ______________________

7. The values reported in the Michelson data are measurements of the speed of light in air (adjusted to make them more convenient for computation). The results are reported in sets of 20 trials. At the time of Michelson's work, measuring the speed of light was near the limits of scientific ability. What explanations can you suggest for the patterns you see here? Sketch any plots that may be helpful to your explanation.

8. (a) Open the Clouds dataset. It contains results from an experiment to see if seeding clouds increases rainfall. In this experiment, clouds were randomly assigned to be seeded or not and the amount of rain they generated was measured. Make side-by-side boxplots of unseeded and seeded clouds. Sketch the boxplots below.

The y-axis of the boxplot is inches of rainfall. Do you see evidence that seeding increases rainfall? Explain your answer.

(b) We could make a scatterplot of seeded versus unseeded clouds. (Feel free to do so.) Explain why such a scatterplot tells us very little about the data. In what ways is it actually misleading to graph this data in this way? (Hint: *Do you think these numbers were recorded in the order given here?*)

9. (a) Open the Cars dataset. Draw a scatterplot of MPG (y-axis) versus Weight (x-axis). Sketch it below.

(b) Describe the relationship between MPG and Weight using words.

(c) Try "stretching" the plot by dragging the size box to the right. Does it still give the same impression?

10. Open the SMSAs dataset and make Normal Probability plots of the Populations and July temperatures of the 60 cities represented there. Sketch the plots here and discuss what shape you expect the distributions to show given what the probability plots show. If you wish, make the corresponding histograms and compare what they show with what the probability plots show.

11. For comparison, open the Random Normals dataset. This dataset contains five samples of 250 numbers from a standard Normal distribution. Make Normal Probability plots of at least two of the samples. In what ways do these plots seem to differ from the ideal Normal probability plot of data drawn from a Normal distribution? Sketch the patterns you are discussing.

DERIVED VARIABLES

NAME ______________________ DATE ______________________

CLASS ______________________ INSTRUCTOR ______________________

1. Which of the following expressions are legal derived variable expressions? For those that are not legal, state why. Assume that the following variables have already been defined: "Wages", "tips", "tax rate", "bonus".

(a) 'Wages' + 'tips'

(b) 'Income' = 'Wages' + 'tips' +' bonus'

(c) Log('Wages' + 'tips'))

(d) ('Wages' + 'tips') * 'tax rate'

(e) 'Wages' + 'tips'
- ('Wages' + 'tips') * 'tax rate'

(f) Sqrt(SSQ('Wages' – mean('Wages'))/
(numNumeric('Wages') – 1))

(g) 'tax rate' ** –1.5

(h) 'tips' * 'tax
rate' + 'bonus'

(i) (((3)))

(j) Sin('Wages'/2π)

DERIVED VARIABLES

2. Several of the functions computed by Data Desk can be found in terms of other functions. For each of the following functions, write out an equivalent derived variable expressing using other functions. Try them out to confirm your answers. Recall that you can type expressions into a ScratchPad and evaluate them there. (Assume that the variable "y" has been defined).

(a) Mean(y)

(b) Variance(y)

(c) StDev(y)

(d) NumNonNumeric(y)

(e) Abs(y) [Hint: use Sign( )

(f) SSQ(y)

DERIVED VARIABLES

NAME ______________________ DATE ______________________

CLASS ______________________ INSTRUCTOR ______________________

3. In August of 1961, the applications of 413 marriage licenses in the city of Seattle with both bride and groom resident of Seattle were classified by the distance between the groom's residence and the bride's residence. The distribution of the distance between residences was as follows:

Distance (miles)	Mid point	Number of # licenses	Fraction of licenses
0.00 - 0.99	0.5	115	.278
1.00 - 1.99	1.5	62	.150
2.00 - 2.99	2.5	48	.116
3.00 - 3.99	3.5	42	.102
4.00 - 4.99	4.5	30	.073
5.00 - 9.99	7.5	76	.184
10.00 - 19.99	15.0	40	.097

(Source: William R. Catton, R. and R.J. Smircich, "A Comparison of Mathematical Models for the Effect of Residential Propinquity on Mate Selection," American Sociological Review, 1964, 29:4, 522-529.)

(a) Construct a derived variable to compute the mean of this sample. (Hint: *The formula for the group mean is* Σ*(freq * value)/*Σ*(freq).*) Write your derived variable expression below. Evaluate the derived variable and report the mean here:

(b) Construct a derived variable to compute the standard deviation of this sample. (Hint: *The formula for the grouped standard deviation is*

$$\sqrt{\frac{\sum_{i=1}^{k} N_i (x_i - \bar{x})^2}{N - 1}}$$

where N_i *is the* i^{th} *group frequency.*)

Write your derived variable expression below. Evaluate the derived variable and report the standard deviation here:

DERIVED VARIABLES

4. Create a variable called "*x*" containing the integers from 1 to 30. (Type in the numbers or use {Manip▷} **Generate Patterned Data....**) Create a derived variable named "*f*(*x*)", type the expression given in (a) below, and make a scatterplot of "*f*(*x*)" *vs* "*x*". (Use the derived variable icon; do not evaluate it first.) Make a sketch of the scatterplot showing its general shape.

(a) 3 + (4 * '*x*')

Now, for each of the following expressions, edit the expression in "f(x)" to correspond to the expression in the exercise and update the scatterplot by using the HyperView popup menu found by clicking on the ❢ in the lower left corner of the plot. (The ❢ will appear only after you edit the derived variable and will clear after the plot is redrawn.) You can leave the derived variable window open during all of this work. Sketch the resulting scatterplots to show their general shape, or print them and paste them in the space provided.

(b) 3 – (4 * '*x*')

(c) ('*x*' – 15) ** 2

(d) '*x*'**2

(e) abs('*x*' - mean('*x*'))

(f) 2 * ('*x*' – mean('*x*'))

DERIVED VARIABLES

NAME ______________________ DATE ______________________

CLASS ______________________ INSTRUCTOR ______________________

5. Repeat Exercise (4) with the following functions:

(a) '*x*'**2

(b) '*x*'

(c) Sqrt('*x*')

(d) Log('*x*')

(e) –1/Sqrt('*x*')

(f) –1/'*x*'

These transformation functions are in order according to the power to which x has been raised (with the logarithm falling at the zero power). What ordering can you detect in the shapes of the plots?

DERIVED VARIABLES

6. Use the **Generate Random Numbers** command in the **Manip** menu to simulate 1 sample of 50 numbers from a Normal distribution with $\mu = 10$ and $s = 3$. (See chapter 8 for more on simulation.) The sample will be placed in a variable named "Norm:1". Rename the variable "*x*". Create a derived variable named "*f*(*x*)", type the expression *x*, and make a histogram of "*f*(*x*)". Use the derived variable icon; do not evaluate it first. Make a sketch of the histogram showing its general shape, or print it out and paste it in the space provided.

Edit the expression in "f(*x*)" to correspond to each of the following expressions, and update the histogram with the HyperView popup menu under the ❢ in the lower left corner of the plot. You can leave the derived variable window open and generate each new plot from the original with the **Redo in New Window** command. Sketch the resulting scatterplots to show their general shape, or print them out and paste them into the space provided.

a) '*x*'**2

(b) Sqrt('*x*')

(c) Log('*x*')

(d) −1/Sqrt('*x*')

(e) −1/'*x*'

What patterns do you see in the relative shapes of the histograms?

DERIVED VARIABLES

NAME ______________________________ DATE ______________________________

CLASS ______________________________ INSTRUCTOR ______________________________

7. To help assess the extent to which a distribution approximates a Normal population, we can examine a called a ______________________________ plot. To construct this scatterplot, we plot the sample values (y-axis) against what function of these values? ____________________.

8. The NScore of the smallest value in a sample of 20 is the smallest value we would expect to observe if we were to draw a sample of _______ values from a ______________ distribution.

9. Use the **Generate Random Numbers** command in the **Manip** menu to simulate 100 observations from a Normal distribution with $\mu = 0$ and $\sigma = 3$. (Refer to chapter 8 for more in-depth treatment of simulation.)

(a) Make a normal probability plot for this variable. Observe the extent to which the plot is straight. Sketch it below.

(b) Repeat this experiment 4 times. Sketch each below.

(c) Select one of the simulated samples that appears to be very nearly Normal and transform it by squaring its values. Make another normal probability plot. How does this compare to the plot associated with the untransformed sample values? Make the histograms corresponding to each plot for additional information.

DERIVED VARIABLES

10. Construct a normal probability plot for the *Displacement* variable in the CU Cars dataset and sketch it below. Do you think displacement is Normally distributed? If not, in which ways does it fail to be Normal? (You may want to make a Histogram of *Displacement* to compare with the normal probability plot.)

11. Simulate 100 observations from a Uniform distribution.

(a) Make a normal probability plot for this sample. Sketch it below.

(b) Transform *Unif1* with the derived variable expression $\sqrt{(Unif1 + 0.5)}$ and make another normal probability plot. Sketch it below.

Discuss what you find.

TABLES

NAME ______________________________ DATE ______________________________

CLASS ______________________________ INSTRUCTOR ______________________________

1. Contingency tables are tables of ______________ used primarily to investigate the dependence of two categorical factors on each other. Each ______________ of the table represents a combination of a category on the row factor and a category on the column factor, so each case falls in one cell. Contingency tables count the number of ______________ falling in each cell and report the counts and related statistics.

2. The variables used to construct contingency tables must be ____________________. They may be numeric, but they are also often identified with text labels.

3. The margins of a contingency table report row and column ______________.

4. If we are concerned with investigating the dependence between two categorical variables, the natural null hypothesis to test is the hypothesis that they are statistically ____________________. Formally, this means that the probability that a specified case falls in a particular cell depends *only* on __________________________ and ____________________________________. Mathematically, this can be written:
Pr (case falls in row_i *and* column_j) = Pr (__________) x Pr (__________).

5. The null hypothesis of independence is tested with the __________________ statistic. Write the algebraic definition of this statistic below, defining your notation. (Hint: refer to Section 13-3).

6. The χ^2 statistic for independence is used in a _______-tailed test. We reject the null hypothesis of independence when the value of the statistic is relatively ______________. Critical values of the statistic depend upon ____.

7. If a contingency table has r rows and c columns, then the associated chi-square statistic has ________________ degrees of freedom. Write the algebraic expression for degrees of freedom:

8. In the calculation of the chi-square statistic, the *observed* counts are compared to the counts you would *expect* if the null hypothesis of independence was true. If the observed counts and expected counts generally differ, then it is likely that the factors are dependent. For a cell that falls in row_i and column_j, write its expected count if the null hypothesis is true. Define your notation.

9. If the null hypothesis of independence is rejected, does the chi-square statistic tell you how the two factors are related? Explain.

10. Although the **Data Desk** does not enforce this restriction, it is recommended that the chi-square statistic be computed only when the expected count in each cell is at least _________.

11. Write the algebraic expression for a cell's standardized residual. Define your notation.

12. The standardized residuals are the individual contributions to the chi-square statistic. For a particular table, the ______________________________ of the standardized residuals across all cells of the table is equal to the χ^2 value.

13. We can simulate a contingency table for which the categorical variables are statistically independent in the following way. Use the **Generate Random Numbers...** command to construct 2 random variables drawn from a binomial distribution. Generate one binomial variable of 100 cases with 3 Trials/Experiment and a success probability of 0.6. Generate a second binomial variable of 100 cases with 2 Trials/Experiment and a success probability of 0.5.

(a) Set Table Options for count, expected count, standardized residual, and χ^2 and make a Contingency Table of the two generated variables. Copy or paste the table (or at least the counts) in the space provided below. Does the χ^2 test indicate significant lack of independence? Should it?

(b) Write out the calculation of the degrees of freedom. Square each of the standardized residuals and check that the sum of squares equals the χ^2 value.

RANDOM NUMBERS AND SIMULATION

NAME______________________________ DATE ______________________________

CLASS______________________________ INSTRUCTOR ______________________________

1. Define the term *simple random sample.*

2.. Suppose we draw a simple random sample from a Normal population. If we use the sample's numbers to construct a summary value (i.e., mean, variance, range, etc.), this summary value is called a *statistic*. A statistic is a sample value. By contrast, a value that describes some aspect of the underlying population is called a

______________________________ .

3. A Normal population is completely described by its ______________________________ and ______________________________. In a standard Normal distribution, these parameters equal ______________ and ______________, respectively.

4. A Normal population has a shape that is (circle all that apply):

(a) multimodal and skewed

(b) bimodal and flat

(c) rectangular

(d) unimodal and symmetric

(e) dependent on the population's mean

5. Suppose that the coin used to decide who should kick off in the football games of your favorite team is biased, landing heads 56% of the time in the long run. You collect data over many games, accumulating a sequence of 30 coin flips. Using the **Generate Random Numbers...** command from the Manip menu, simulate 30 trials from a Bernoulli distribution with Prob(success)=0.56. Repeat the experiment 4 more times. For each experiment, count the number of heads (successes) and write them below:

How strong is the evidence for a biased coin?

What might you do to strengthen the evidence of bias?

RANDOM NUMBERS AND SIMULATION

6. You begin to suspect a nation-wide conspiracy of coin-flipping referees, and enlist 50 friends. Each of them observes 30 game-starting coin flips and reports to you the number of heads. Simulate the 50 numbers they report (still assuming that the probability of heads is 56%). (Hint: *Generate 1 sample of 50 from the Binomial distribution with 30 Bernoulli trials/experiment*). Construct a histogram of the result and sketch it here.

How strong is the evidence from *this* data for biased coin flips?

7. Simulate samples of size 10, 50, 100, and 150 from the uniform distribution and make histograms of them. How do the histograms differ? Does each histogram look like it came from a uniform distribution? What differences do you see as the sample size grows?

9. Generate samples of 40 Poisson observations with $\lambda = 2, 5, 10$, and 30 and make histograms of the four samples. How do they differ? Sketch the histograms below. Can you rescale the histograms to look more alike?

RANDOM NUMBERS AND SIMULATION

NAME ______________________________ DATE ______________________________

CLASS ______________________________ INSTRUCTOR ______________________________

9. Simulate samples of size 10, 50, and 100 from the Normal distribution with $\mu = 0$ and $\sigma = 1$, and make histograms of them. Sketch the histograms below.

Can you rescale the histograms to look more alike?

What differences do you observe as the sample size grows?

10. Generate 2 samples of size 50 from Normal distributions. For one, let $\mu = 0$ and $\sigma = 1$; for the other, let $\mu = -40$ and $\sigma = 40$. Make histograms of the 2 samples. Sketch them below.

How can you tell that the distribution parameters were different for the two samples?

11. Reproduce the Law of Large Numbers experiment from Section 15.10. Repeat the experiment with a sample of 1000 random Uniform values. Do you notice any difference?

RANDOM NUMBERS AND SIMULATION

12. If we were to draw infinitely many samples of a fixed size from the same ________________, compute the value of a ________________, and make a histogram of the collected values, the distribution of values shown by the histogram would be called the ____.

13. Make a histogram to display the sampling distribution of the median for samples of size 10 from the standard Normal distribution. Follow the example in the chapter, but generate 25 samples and use **Summary Statistics Options** to select the Median to use in the **Summary Statistics as Variables** command. You may want to rescale the histogram to find a scale that best depicts the distribution. Repeat the experiment a second time and compare the two trials. Sketch both histograms below. (It is a good idea to discard the random samples generated the first time and empty the Trash before repeating the experiment.)

14. Repeat Exercise 12 using samples of size 5. What differences do you see compared to Exercise 13?

15. The Central Limit Theorem states that the sampling distribution of the sample mean of n independent observations, drawn from a population with mean μ and standard deviation, σ, is approximately Normal with mean _____________and standard deviation ________________, and that the approximation improves as n grows. According to the theorem, do the independent observations upon which the sample means are based need to be drawn from a Normal distribution? ___________

RANDOM NUMBERS AND SIMULATION

NAME ______________________ DATE ______________________

CLASS ______________________ INSTRUCTOR ______________________

16. (a) Simulate the sampling distribution of the mean of 16 independent observations drawn from a Normal distribution with $\mu = -4$ and $\sigma = 8$. Generate 20 samples. Compare the mean and standard deviation of the sample means to what you would theoretically expect given the Central Limit Theorem.

observed mean: __________
observed standard deviation: __________

theoretical mean: __________
theoretical standard deviation: __________

(b) How might you reduce the difference between what you observe in your simulation and what is expected theoretically?

(c) Construct a normal probability plot of the empirical sampling distribution. Sketch it here. Does it appear to be approximately Normal? ______________

(d) What might improve the approximation to Normality?

17. (a) Simulate the sampling distribution of the mean of 16 independent observations drawn from a Uniform distribution. Generate 20 samples. Compare the mean and standard deviation of the sample means to what you would theoretically expect given the Central Limit Theorem. (The mean of a Uniform distribution is 0.5 and its standard deviation is $1/\sqrt{12}$.)

observed mean: __________
observed standard deviation: __________

theoretical mean: __________
theoretical standard deviation: __________

(b) How might you reduce the difference between what you observe in your simulation and what is expected theoretically?

(c) Construct a normal probability plot of the empirical sampling distribution. Sketch it here. Does it appear to be approximately Normal?

(d) What might improve the approximation to Normality?

(e) How do these results compare to the experiment in Exercise 16?

SIMPLE INFERENCE

NAME______________________________ DATE ______________________________

CLASS______________________________ INSTRUCTOR ______________________________

1. Work through the example in Appendix 16A, generating your own set of 20 confidence intervals. How many of the confidence intervals you constructed contained the true population mean μ? ____________ How many would you have expected? __________ Write the intervals you obtained (or print them and attach the output here) and circle the ones that do not contain the population parameter.

2. Suppose tenors' heights are Normally distributed with $\mu = 69$ and $\sigma = 3.2$. Repeat the experiment in exercise (1) for tenors. (Save the random samples you generate for problem (3).)

(a) How many of the intervals contain μ? __________

(b) Are all the intervals of the same width?__________ Should they be? Explain

(c) Suppose we specified a 97% confidence level for the intervals we constructed instead of a 90% level. Would you expect the intervals narrower or wider than before? Explain.

(d) Suppose we drew samples of size $n = 49$ instead of 9. Would you expect the intervals we compute to be narrower or wider than for $n = 9$? __________ By how much? _____

(e) Suppose we thought that σ was 2 (rather than the true value of 3.2) and calculated z-intervals accordingly. Would you expect more or fewer intervals to fail to cover μ? (Try it if you wish.)___________.
Why?

3. Repeat the experiment in Exercise 2 *using the same random samples* but computing t-intervals instead of z-intervals.

(a) How many of the intervals contain μ?

(b) Are all the intervals of the same width?__________ Should they be? Explain

(c) Suppose we specified a 97% confidence level for the intervals we constructed instead of a 90% level. Would you expect the intervals narrower or wider than before? Explain.

(d) Suppose we drew samples of size $n = 49$ instead of 9. Would you expect the intervals we compute to be narrower or wider than for $n = 9$? Why?

(e) Compare individual intervals between exercise 2 and exercise 3. How do the critical values from the tables compare? How do the estimated standard deviations compare to the specified standard deviation in exercise 2?

4. When we construct a confidence interval for some population parameter we first select a *confidence level,* say 95%. What is the correct interpretation of this number? (Circle the letters of all that apply.)

(a) the population value has a 95% chance of falling in the constructed interval
(b) if a very large number of confidence intervals were constructed, roughly 95% of them would contain the population value μ
(c) the population value is equal to 95
(d) if a very large number of confidence intervals were constructed, roughly 5% of them would not contain the population value μ.

SIMPLE INFERENCE

NAME ____________________ DATE ____________________

CLASS ____________________ INSTRUCTOR ____________________

5. Generate a random sample of 25 observations from a Normal distribution with $\mu = 71$ and $\sigma = 2.5$. Construct a confidence interval for μ from the data, specifying a confidence level equal to 93%. Assume that the parent Normal distribution is the true population that describes the heights of bass singers. Write the interval you obtained below and interpret its meaning in *words* and numbers.

6. To construct a confidence interval for μ when σ is unknown, we refer to the _______ distribution. This population is symmetric about its mean, but the peakedness varies according to the __________.

7. When constructing a hypothesis test, we specify the α-level. It is a number that ranges between __________ and __________. Give the definition of the α-level.

8. The general problem of performing several inferential procedures together is called the problem of multiple comparisons. We can apply a Bonferroni adjustment to guarantee that a set of independently constructed inferential statements (*i.e.*, confidence intervals) collectively yield a specified "total confidence level". If, for example, we request that a set of 5 intervals collectively yield a total confidence level of 95%, the Bonferroni adjustment would require that each interval be constructed according to a _____% level.

9. Repeat Exercise 3, but request a Bonferroni adjustment for the confidence intervals.
How many contain the population mean? __________

How many should you have expected? ______________

10. Retrieve the **Singers** dataset. Compute a test of the null hypothesis that the true mean height of bass singers is 70 inches. Specify an alternative hypothesis of your choosing. Assume that the population of bass heights has a standard deviation equal to 3. Perform the test at a 5%-level.

(a) For this problem, should you perform a z-test or t-test? Explain your answer.

(b) Write the results of the test, as presented by Data Desk in the space below.

(c) Using *words* and numbers, interpret what the test tells us about the heights of bass singers.

11. Test the same hypothesis as described in Exercise 10, but assume the true standard deviation of bass heights is unknown. In this case, you should perform a _______ test. Write the results of the test, as presented by Data Desk, in the space below.

12. Test the null hypothesis that the true population variance of soprano heights equals 4. Assume that soprano heights are Normally distributed. For this problem, we refer to the __________ distribution. Perform the test and write the results below.

COMPARING TWO SAMPLES

NAME ______________________________ DATE ______________________________

CLASS ______________________________ INSTRUCTOR ______________________________

1. Pooled-t procedures are appropriate when we can assume that the underlying population ______________________________ are equal. A t-statistic that is based on a pooled estimate of variance has more ______________________________ than a t-statistic that is based on separate estimates of variances.

2. Retrieve the **Clouds** dataset. Recall that the data represent rainfall measured for each of 52 clouds, 26 of which were chosen at random and seeded with silver oxide. Test the null hypothesis that the mean rainfall obtained from seeded clouds equals the mean rainfall obtained from unseeded clouds. Assume that the underlying population variances are unequal.

(a) Choose $\alpha = 0.05$ and test the null hypothesis against a two-sided alternative. Write the results of the test from the Data Desk output table below.

(b) The scientist who conducted this experiment believed that seeding clouds increases rainfall and could not reduce it. For $\alpha = 0.05$, test the null hypothesis against this alternative. Write the results below.

(c) Discuss any differences you observe between the results obtained in (a) and (b).

3. (a) Repeat the test of Exercise 2(a) but, choose $\alpha = 0.10$. Write the results below.

(b) Explain how and why these results differ from those obtained in Exercise 2(a).

4. (a) Construct a confidence interval for the true mean difference between unseeded and seeded clouds. Request a 95% confidence level; assume that population variances are unequal. Write the confidence interval below.

(b) Explain in *words* what this interval tells you .

(c) How does the confidence interval relate to the hypothesis test performed in Exercise 2(a)?

(d) If you assumed that the underlying population variances were equal, how many degrees of freedom would the resulting t-statistic have?

5. Using the **Singers** dataset:

(a) Compute a 95% confidence interval for the difference between the true mean heights of tenor and bass singers. Write your results below.

Explain the meaning of the interval in *words.*

COMPARING TWO SAMPLES

NAME______________________________ DATE ______________________________

CLASS______________________________ INSTRUCTOR ______________________________

6. Using both a pooled-t hypothesis test and a pooled-t confidence interval, compare the heights of the sopranos and altos. Write your findings below.

Does the interval tell you anything more than the hypothesis test? Why or Why not?

7. (a) Assume that the true height of sopranos is 65 inches. Define a derived variable that computes the t-statistic value for the sopranos in the Singers dataset. Write the derived variable expression below.

(b) Compute the value of the derived variable and compare it to the value produced by the **Test...** command. Write these results below.

(c) Assume that the underlying populations of tenors and sopranos are Normally distributed. Test the hypothesis that the underlying variances are equal. Write your findings below.

8. Using the **Labor Force** data:

(a) What are the individual cases in these data? What population(s) might they have been drawn from?

(b) Construct a 95% confidence interval for the difference between the Labor Force Participation rates in 1968 and in 1972 both *without* pooling variance and *with* pooled variance. Write the intervals below and discuss how and why they differ.

(c) Does there appear to be a difference in the LFPR between these two years? How does this interval differ from the pooled-variance interval? Explain your answers.

(d) Define a derived variable that gives, for each city, the difference between the LFPR in 1968 and the LFPR in 1972. Construct a 90% confidence interval for its mean and compare it to a paired-t confidence interval computed for the same data.

COMPARING TWO SAMPLES

NAME ______________________ DATE ______________________

CLASS ______________________ INSTRUCTOR ______________________

9. Using the **Hearing** data:

(a) Test the null hypothesis that the first and fourth list are equally difficult to understand in the presence of noise. Write your results below.

(b) Compute a two-sample t-interval for the mean difference in hearing perception between List 1 and List 4 . Write the interval below.

(c) Use a paired-t procedure to compare hearing perception for List 1 and List 4. Write the results below.

(d) Which of the three ways of comparing means is most appropriate for these data? Why?

10. Generate two samples of 35 cases from a Normal distribution with $\mu = 10$ and $\sigma = 2$. Suppose we want to compare these samples to test whether their underlying populations have the same mean.

(a) What is the appropriate null hypothesis? Is it true for these data? Explain your answer.

(b) If we perform a pooled-t test at $\alpha = 0.10$, what is the probability that we will reject the null hypothesis for these samples?

(c) Perform the test and report your result.

COMPARING SEVERAL MEANS WITH ANOVA

NAME ______________________________ DATE ______________________________

CLASS ______________________________ INSTRUCTOR ______________________________

1. The analysis of variance (ANOVA) procedure attempts to answer the question, "Do all of the groups have the same ____________?". To perform an ANOVA we must have random samples drawn from ____________ distributed populations. We must also assume that the groups being compared have the same underlying ____________, even if they have different means. The ANOVA procedures examines the ____________ among groups means to determine if this value is large relative to the variability inherent in the observations themselves.

(a) Write the null hypothesis appropriate for an ANOVA.

(b) Write the alternative hypothesis. (Words are sufficient).

(c) In an ANOVA, the test statistic follows an ______ distribution. When the null hypothesis is true, the test statistic has an expected value equal to ______.

(d) For the category variable, the ________________ is the sum of squared differences among the group means from the overall mean of all the measurements.

(e) In the ANOVA table, the MS column is obtained by dividing the ______________ values by the corresponding ______________.

(f) The ____________ for error behaves like the pooled estimate of variance using all the observations, regardless of their level in the factor.

2. Retrieve the **Michelson** data and omit the first run. Perform an ANOVA.

(a) State the null hypothesis.

(b) What can you conclude from the test?

(over)

2(c) Can you suggest any reason for what you found?

(d) The currently accepted value for the true speed of light in air is 299,792.5 km/sec. When this value is converted to one that corresponds to those in the dataset it is 734.5. Perform a test of the hypothesis $\mu = 734.5$ for these data with and without the first trial. Write the results from both tests below and discuss what you find.

3. Using the **Graduation** dataset:

(a) Investigate whether there is a difference in the rates of on-time graduation among the various colleges (in the variable "school") at this University. First, state the null and alternative hypotheses.

(b) Perform the test and write the results below.

(c) Explain *in words* what the test tells you.

(d) Investigate whether there has been a change in on-time graduation rates from year to year. How does the null hypothesis differ from the one used in (a) above?

(e) Perform the test and write the results below.

COMPARING SEVERAL MEANS WITH ANOVA

NAME ____________________ DATE ____________________

CLASS ____________________ INSTRUCTOR ____________________

4. The **Hearing** dataset contains data from a study comparing the hearing perception among matched lists of words. These lists are used to test hearing aids and are matched to be equally difficult to understand when played on a tape player at low volume. The researcher played the tapes for subjects with normal hearing but added background noise. The question of interest is whether the lists retain the property of being equally hard to understand in the presence of background noise. The data reported are percent of words understood by each of 24 subjects. State and test the appropriate hypothesis.

(a) State the null and alternative hypotheses.

(b) Perform the test and write the results below.

(c) Explain in *words* what you have learned about hearing perception among the matched lists in the presence of background noise.

5. Generate a sample of 50 observations from a Normal distribution with $\mu = 10$ and $\sigma = 2$. Use **Generate Patterned Data...** to generate a variable that contains the numbers 1, 2, 3, 4, 5, 1, 2, 3, 4, 5,... with the 5-number sequence repeated 10 times. Now pretend that the random numbers were data on 5 groups identified by the patterned variable, and investigate whether the 5 groups have different means.

(a) What *should* you find if you perform an ANOVA to compare the 5 groups?

(b) What is the probability that the F-value will be significant at the 5% α-level?__________

(c) What 3 basic assumptions of the ANOVA procedure have been assured in this experiment?

(d) Perform the ANOVA and report what you find.

MULTI-WAY ANOVA

NAME ______________________________ DATE ______________________________

CLASS ______________________________ INSTRUCTOR ______________________________

1. The one-way analysis of variance (ANOVA) studied in Chapter 11 compares the means of several groups, represented by a single factor. The two-way ANOVA introduces a second factor. Typically, the two factors might affect the response variable both individually and jointly through some interaction. In a ________________ ANOVA there are an equal number of measurements for every combination of levels of the two factors.

2. In a two-way ANOVA, the ________________ mean square can be used to test whether the difference among the means of the response variable due to one of the factors changes at different levels of the other factor. This behavior can only be tested when there are at least ______ measurements recorded for each combination of factor levels.

3. Suppose you have a balanced two-way ANOVA design. One factor (Treatment A) contains 5 levels and the second factor (Treatment B) contains 4 levels. Suppose that for each combination of factor levels, there are 4 measurements recorded.

How many degrees of freedom are available for the following terms when interaction is included and when interaction is not included:

Interaction included		**Interaction not included**	
Treatment A	__________	Treatment A	__________
Treatment B	__________	Treatment B	__________
Interaction	__________		
Error	__________	Error	__________
Total	__________	Total	__________

4. Simulate the design of Exercise 3.

(a) Generate four random samples of size 20, each drawn from Normal populations with different means and $\sigma = 1$. Generate side-by-side boxplots of the samples to check that you simulated the data correctly. Sketch the boxplots below.

Use the **Append & Make Group Variable** command to append all four samples into a single response variable. (For convenience, name the appended data "Response" and the group variable "Treatment B"). Create the levels of Treatment A in the following way. First, use the **Generate Patterned Data...** command to create a variable consisting of "1, 1, 1, 1, 2, 2, 2, 2,...,5, 5, 5, 5" repeated four times. Rename the variable "Treatment A". To convince yourself that you have created a balanced two-way ANOVA, open all 3 variable: Response, Treatment A, and Treatment B. Notice that for every combination of Treatment A and Treatment B, there are four measurements. You might also make a Contingency Table of Treatment A by Treatment B to see that each cell count is 4.

(b) These data conform to the 3 basic assumptions of ANOVA. What are these assumptions?

(c) If we compute an ANOVA for the data created in (a), do you expect to observe a treatment "effect"? If so, for which treatment? Why?

MULTI-WAY ANOVA

NAME ______________________ DATE ______________________

CLASS ______________________ INSTRUCTOR ______________________

(d) **Select ANOVA▸ANOVA** (without interactions) for these data and write or paste the resulting table below.

(e) Check the degrees of freedom with the answers you gave in (a) above. Explain the results of the test.

5. Select ANOVA▸ANOVA with Interactions for these data. Check the degrees of freedom with the answers you gave in Exercise 3, and write the resulting ANOVA table below.

6. For these simulated data, interaction would be present if measurements on levels of Treatment A depended upon Treatment B.

(a) Suppose the first level of Treatment A and the first level of Treatment B behave differently from all other combinations. To demonstrate the effect on the ANOVA interaction term, open the response variable and add 10 each response where Treatment A = 1 and Treatment B = 1. Recompute the ANOVA using the HyperView menu **Redo in New Window** command on the ANOVA table, and write or paste the resulting table below. (Alternatively, select the variables as before and issue a new **ANOVA with Interactions** command.)

(b) Compare this table to the others computed in this exercise. Explain what you have seen.

MULTI-WAY ANOVA

NAME ______________________ DATE ______________________

CLASS ______________________ INSTRUCTOR ______________________

7. In the **Graduation** dataset, perform the two-way ANOVA of on-time graduation percentage with College and Year as the two factors.

(a) Write or paste the results below.

(b) Is there evidence of differences among colleges? __________ Explain your answer.

(c) Is there evidence of changes across years? __________ Explain your answer.

(d) State the null and alternative hypotheses carefully.

8. We might be tempted to perform a two-way ANOVA on the **Michelson** data using trial number and experiment number as the two factors. Explain why this would be inappropriate.

9. In chapter 19's exercises, we performed the one-way ANOVA comparing the four word lists in the **Hearing** data. Repeat this procedure and compare it to the two-way analysis performed in this chapter's example.

(a) How are the results different?

(b) Why are the results different?

(c) Which analysis would you prefer for these data?

(d) Why?

SIMPLE REGRESSION

NAME________________________________ DATE________________________________

CLASS________________________________ INSTRUCTOR________________________________

1. In the regression equation $\hat{y} = a + bx$ the coefficient a describes the regression line's ________________________. The coefficient b describes the line's ________________________.

2. Create a variable called "x" containing the integers in order from –10 to 20. (Use the **Generate Patterned Data...** command in the **Manipulate** menu). Create derived variables using the following expressions and plot them against "x". Resize the scatterplots so that the y-axes are comparable. Sketch the plots below.

(a) x

(b) $-x$

(c) $3 + x$

(d) $3 + 2 * x$

(e) $3 - 3 * x$

SIMPLE REGRESSION

3. The most common statistical criterion for fitting a regression line is called the *least squares criterion*. Give its definition.

4. In 1975, there were an estimated 60,000 deaths in the United States attributable to lung cancer. One 1985 study estimated an annual increase of 5,000 lung cancer deaths during the decade.

(a) Write the equation of the line that describes this trend using *words* and *numbers*.

(b) Using this prediction equation, what were the expected number of deaths in 1980?

(c) If, in fact, there were 90,000 lung cancer deaths in 1980, find the residual value for the year 1980.

(d) Suppose smoking had become decidedly unpopular during the decade so that the prediction changed to an annual *decrease* of 5,000 deaths from lung cancer. Write the new regression equation below using words and numbers.

(e) Using the new equation, what would be the expected number of deaths

In 1980? ______________________ In 1995? ______________________

(f) Why is the prediction for 1995 in (d) meaningless? What does this tell you about the use of regression equations for predicting real events?

SIMPLE REGRESSION

NAME____________________ DATE____________________

CLASS____________________ INSTRUCTOR____________________

5. Using the **Nuclear Plants** dataset, define a new derived variable named *Date67* with the expression *'Date' - 67*. This variable is measured in years since 1967. Plot *Cost* versus *Date67*.

(a) Print the plot and draw a line on the plot *by hand* to describes the relationship. Paste or sketch the plot:

(b) Compute the slope and intercept of your line. Write them below.

(c) Define a derived variable that would compute the residuals from your line. Write its equation here:

(d) Modify the derived variable in (c) to compute the sum of squared residuals from your line. (Hint: *You may find the SSQ function useful.*) Evaluate the sum of squared residuals with the **Compute Derived Variable** command. Write the equation you are using and the resulting sum:

(e) Compute the regression of *Cost* on *Date.* Write the regression equation below and compare it to your fit by eye.

(f) Find the sum of squared residuals from the regression. (Hint: *You can find it on the regression summary table or compute it from the saved Residuals.*) Compare this sum to the sum of squared residuals for your fit by eye. How to they compare?

(g) Write a brief sentence that describes the meaning of the slope and intercept of the regression equation as you might report it to someone who knew no statistics.

6. As with other statistics, the least squares regression coefficients have known sampling distributions, provided certain conditions hold. These conditions are:

(a) The underlying relationship between the dependent variable, y, and the independent variable, x, is ________________________.

(b) The true residuals are mutually ________________________________. This assumption is sometimes violated if the data is measured over time.

(c) The true residuals have the same _________________________ for all values of the independent variable x. This is equivalent to the statement that the variance of the population from which values of the dependent variable y are sampled is the same for all values of x.

(d) The true residuals follow a __________________ distribution with mean _____ and variance σ^2.

7. If all the assumptions in (6) are satisfied, then the ratio of the regression coefficient to its standard deviation has a _______ distribution.

8. (a) The R^2 statistic is an overall measure of the success of the regression in predicting y from x. R^2 measures the ______________ of the variability of y accounted for by ________________________________ ____________________________________.

(b) R^2 is always between zero and 1. A value of zero indicates that ____________________. A value of 1 indicates that ________________________________

(c) The square root of R^2 is the __________________________ coefficient.

9. Using the **Olympic Gold** dataset answer the following: The modern Olympic series actually started in 1896. In that year the gold medal in the long jump was awarded for a jump of 249.75 inches.

(a) Compute the residual corresponding to this jump using the regression model of the example in this Chapter either by hand or with a derived variable.

(b) Discuss what you find: Is the residual large or small? Was the jump particularly short, long, or about as expected? Can you offer any explanation of why this might have happened?

SIMPLE REGRESSION

NAME______________________ DATE______________________

CLASS______________________ INSTRUCTOR______________________

10. Use the regression equation in the text to predict how long a long jump will be necessary to win the gold medal in the Olympics of the year 2000. Show your work or explain how you used derived variables.

11. Omit Beamon's 1968 performance and predict long jump distance from Olympic year. (To omit a point, open the "year" variable, click just in front of "68", and type any nonnumeric character such as a "*".)

(a) Write the new regression equation below. How does it differ from the equation based on the full dataset?

(b) What prediction would you make from this regression for the gold medal performance in 2000?

(c) Which of the two equations do you think would provide the best prediction? Why?

12. Some people have suggested that one factor contributing to Beamon's remarkable performance in 1968 is that Mexico City is at such a high altitude that the force of gravity may have been slightly less. Do you see any evidence of improved performance in the high jump in 1968? (Hint: *It is probably better to examine the trend in the residuals rather than in the original data. Why?*) Report your findings here, along with the analyses you performed or sketches of plots you made.

13. Using the **CU Cars** dataset, compute the regression of *MPG* versus *Weight*.

(a) Write a sentence describing the relationship, using the coefficients in the regression.

(b) Make a histogram of the residuals and plot them against the predicted values and sketch it here.

(c) Do these residuals seem to satisfy the regression assumptions? Explain your answer. Be specific.

(d) Write a brief sentence describing what the R^2 value means.

SIMPLE REGRESSION

NAME________________________________ DATE________________________________

CLASS________________________________ INSTRUCTOR________________________________

14. Create the variable x containing the integers from 1 to 30. Generate a random sample of 30 values from a Normal population with $\mu = 0$ and $\sigma = 10$ (creating the variable *Norm1*). Define the derived variable y as
3 + 2*x + '*Norm1*'

(a) Plot y versus x. Sketch or paste the plot here:

(b) How do y and x relate to the regression model assumptions?

(c) Perform the regression of y on x. How close do the estimated coefficients come to the true values?

(d) Repeat the experiment four more times using new random samples. How much do the estimated coefficients vary? (Note: The easiest way to do this is to generate four samples from the Normal(0, 10) distribution and then edit the definition of y in the derived variable to refer to *Norm2*, *Norm3*, and so on. Renaming the random samples to *Norm1* will not work because the derived variable identifies its underlying variables at the time that it is defined. Editing the derived variable makes it re-evaluate the variable names.) Discuss what you find.

(over)

(e) Suppose the random numbers had been generated with $\mu = 0$ and $\sigma = 50$. How would you expect each of the following statistics to change compared to what you saw in the first experiment? Should it get larger, smaller, or not change?

R^2	________________	F	________________
SSresidual	________________	s	________________
a	________________	b	________________

You may perform the experiment to check your answers.

(f) Suppose the random numbers had been generated with $\mu = 10$ and $\sigma = 1$. How would you expect each of the following statistics to change compared to what you saw in the first experiment? Should it get larger, smaller, or not change?

R^2	________________	F	________________
SSresidual	________________	s	________________
a	________________	b	________________

You may perform the experiment to check your answers.

CORRELATION

NAME________________________ DATE________________________

CLASS________________________ INSTRUCTOR________________________

1. Define a variable named x containing the integers from –10 to 10. (Use the **Generate Patterned Data...** command in the **Manipulate** menu). Define derived variables for the following functions. For each one, make a scatterplot against x and compute its correlation with x. Sketch each plot below and write the correlation next to the plot.

(a) x^2

(b) abs(x)

(c) sqrt $(100 - x^2)$

(d) sign(x - 5) + sign(x + 5)

(e) cos(π * x/30)

CORRELATION

2. The correlation coefficient measures the degree of ______________ association between two variables, x and y. Its value ranges from ________ to ________. A positive value indicates that y ___________ as x increases. A negative value indicates that y ___________ as x increases. A value equal to0.0 indicates that ____. A value of +1 or –1 indicates that there is a ____________ ____________ relationship between y and x. The correlation of any variable with itself equals ____________.

3. Generate 21 random numbers from the Normal (0, 1) distribution, and rename the variable "normals". Using x from Exercise 1 above, construct derived variables for the following functions. For each one, make a scatterplot against x and find its correlation with x. Sketch each plot and report the correlation coefficient.

(a) x + normals

(b) x + 4 * normals

(c) x + 8 * normals

(d) x + 16 * normals

4. The correlation of y and x is also the correlation of the __________________ and __________________ values obtained from the regression of y on x.

CORRELATION

NAME________________________________ DATE________________________________

CLASS________________________________ INSTRUCTOR________________________________

5. Using the **Nuclear Plants** dataset, show that the correlation of Cost and Date is the same as the slope coefficient of the regression of Cost / Std(Cost) on Date / Std(Date). Write the results below.

6. Correlation tables can be dangerous because each of the correlations carries the assumption of _______________________. It is easy to compute many correlations without checking the correctness of this assumption.

7. Spearman's rho is the correlation between the _________________ of two variables. If the value of this statistic is near –1 or +1, we can assume that the underlying variables are _____________________________ . Spearman's rho is less affected by _________________ than the Pearson correlation coefficient.

8. Generate 10 variables of 20 observations each from a Normal distribution. Compute the correlation table for these variables. Write or paste it below.

What is the underlying "true" value of these correlations? Why?

Do the values you observe differ much from what you expected? Why should they differ at all?

9. In the **Olympic Gold** dataset, compute the correlation table for the three gold medal performance variables both with and without Beamon's 1968 long jump. Write the results below.

Explain any differences you observe between the two tables.

10. Using the **CU Cars** dataset, make a scatterplot between MPG and Displacement.
(a) Sketch it below.

(b) Compute the correlation coefficient and Spearman's rho between these 2 variables.

correlation value ________

Spearman's rho value ________

(c) Which statistic is more appropriate for this relationship? Why

MULTIPLE REGRESSION

NAME____________________ DATE____________________

CLASS____________________ INSTRUCTOR____________________

1. Working with the Nuclear Plants dataset compute a regression to predict *Cost* in terms of *Date* and *MWatts.*

(a) Consider the statistics in the first three lines of the regression summary table. For each of these statistics, give its value and write a sentence interpreting it. How to these statistics compare to the corresponding statistics in the regression of *Cost* on *Date* from Chapter 14?

R^2

R^2(Adjusted)

s

(b) Consider the table of regression coefficients in the last four lines of the regression summary table. For each of the statistics listed below, give its value and write a sentence interpreting it.

Intercept coefficient

regression coefficient for MWatt

s.e. of the regression coefficient for MWatts

t-statistics for the regression coefficient for MWatts

2. Generate five random samples of 30 numbers from a Normal distribution (with any mean and standard deviation you wish). Compute the regression to predict one of them from the other four.

(a) What is the "true" value of the underlying coefficients? Why?

(b) Compare R^2 and R^2(adjusted). Which one describes the success of the regression more accurately?

(c) What value did you compute for F? _________.

(d) Examine the residuals to see if they seem to satisfy the regression assumptions. Sketch or paste an appropriate display below. Knowing how the data were generated, would you say that the regression assumptions are satisfied?

(e) If we were to generate 20 random samples of 30 rather than 5 and compute the regression of one on the others, how would you expect the following regression statistics to change?

R^2

R^2(Adjusted)

s

MULTIPLE REGRESSION

NAME________________________________ DATE________________________________

CLASS________________________________ INSTRUCTOR________________________________

3. Using the CU Cars dataset, compute the regression of *MPG* on *Weight* and *Horsepower.* What are the units of these three variables?

Write a few sentences describing what the coefficients mean.
Weight coefficient

Horsepower coefficient:

4. Now compute the regression of *MPG* on *Weight*, *Horsepower*, and *#Cylinders*. Does this appear to be a more or less useful regression? Why?

5. Compute the regression of *MPG* on *Weight* and *Displacement*. Compare the regressions of exercises 3, 4, and 5. Which appears to be the most useful regression? Why? Be specific.

6. (a) Consider the regression of *MPG* on *Weight* and *Displacement* computed in Exercise 5. Report the regression coefficients and any useful associated statistics below:

(b) The coefficient of *Displacement* in this regression is positive and has a large t-statistic. Does this mean that bigger engines (with larger displacement) get better gas mileage? If so, how can you account for that? If not, what does it mean?

(c) Compute the regression of *MPG* on *Displacement*. Discuss the relationship of the two regression models.

(d) Construct a partial regression plot for *Displacement* removing the linear effects of *Weight*. (Use the HyperView on the regression summary table under the *Displacement* coefficient in the appropriate regression.) Compare it to a scatterplot of *MPG* versus *Displacement*. Sketch or paste both plots below. Discuss their differences and what they tell you.

MULTIPLE REGRESSION

NAME________________________________ DATE________________________________

CLASS________________________________ INSTRUCTOR________________________________

7. (a) Examine the residuals from the regression of *MPG* on *Weight* and *Displacement* computed in exercise 5 by plotting residuals against predicted values. (The global HyperView in the Summary table makes this easy.) Sketch or paste the scatterplot below:

(b) The curved pattern indicates that the linearity assumption may have been violated. Define a derived variable named *GPM* as *-100/MPG*. (Multiplying by 100 gives numbers that are of a convenient size, computing "gallons per 100 miles". The minus sign preserves the order so that larger numbers still represent better gas mileage.) Now compute the regression of *GPM* on *Weight* and *Displacement* and sketch the scatterplot of residuals against predicted values below:

(c) What other differences do you notice between the two regressions? (We expect the coefficients to change because of the transformation. What other statistics have changed?)

MULTIPLE REGRESSION

NAME ______________________ DATE ______________________

CLASS ______________________ INSTRUCTOR ______________________

8. (a) Perform a regression of MPG on Weight, Drive Ratio, Horsepower, and Displacement. Make a histogram of the DFFits or Cook's distance values by using the appropriate command in the globally HyperView menu on the Summary table. Sketch or paste the histogram below.

(b) Identify the extraordinary car. Find its name by opening the Car variable, selecting the extraordinary point in the plot, and finding the selected car. Which car is it?

(c) Now omit the extraordinary car and repeat the regression. (Open MPG and edit the value to be missing , for example by typing a * next to its value.) How do each of the following statistics change? Why?

R^2

R^2(Adjusted)

s

coefficient of *Weight*

t-statistics for the coefficient of *Weight*

F